T0145399

Studies in Big Data

Volume 106

Series Editor

Janusz Kacprzyk, Polish Academy of Sciences, Warsaw, Poland

The series "Studies in Big Data" (SBD) publishes new developments and advances in the various areas of Big Data- quickly and with a high quality. The intent is to cover the theory, research, development, and applications of Big Data, as embedded in the fields of engineering, computer science, physics, economics and life sciences. The books of the series refer to the analysis and understanding of large, complex, and/or distributed data sets generated from recent digital sources coming from sensors or other physical instruments as well as simulations, crowd sourcing, social networks or other internet transactions, such as emails or video click streams and other. The series contains monographs, lecture notes and edited volumes in Big Data spanning the areas of computational intelligence including neural networks, evolutionary computation, soft computing, fuzzy systems, as well as artificial intelligence, data mining, modern statistics and Operations research, as well as self-organizing systems. Of particular value to both the contributors and the readership are the short publication timeframe and the world-wide distribution, which enable both wide and rapid dissemination of research output.

The books of this series are reviewed in a single blind peer review process.

Indexed by SCOPUS, EI Compendex, SCIMAGO and zbMATH.

All books published in the series are submitted for consideration in Web of Science.

More information about this series at https://link.springer.com/bookseries/11970

Witold Pedrycz · Shyi-Ming Chen
Editors

Recent Advancements in Multi-View Data Analytics

Editors
Witold Pedrycz
Department of Electrical and Computer Engineering
University of Alberta
Edmonton, AB, Canada

Shyi-Ming Chen
Department of Computer Science and Information Engineering
National Taiwan University of Science and Technology
Taipei, Taiwan

ISSN 2197-6503 ISSN 2197-6511 (electronic)
Studies in Big Data
ISBN 978-3-030-95241-9 ISBN 978-3-030-95239-6 (eBook)
https://doi.org/10.1007/978-3-030-95239-6

This Springer imprint is published by the registered company Springer Nature Switzerland AG
The registered company address is: Gewerbestrasse 11, 6330 Cham, Switzerland

Preface

In real-world problems, there is a visible shift in the realm of data analytics: more often we witness data originating from a number of sources, on a basis of which models are to be constructed. Data are generated by numerous locally available sensors distributed across some geographically distant areas. Data come from numerous local databases. All of them provide a valuable multi-view perspective at real-world phenomena. Each of these views provides an essential contribution to the overall understanding of the entire system under analysis. Quite commonly data might not be shared because of some existing technical or legal requirements. Because of the existing constraints, there arises a timely need for establishing innovative ways of data processing that can be jointly referred to as a multi-view data analytics. A multi-view nature of problems is manifested in a variety of scenarios including clustering, consensus building in decision-processes, computer vision, knowledge representation, big data, data streaming, among others. Given the existing constraints, the aim is to analyze and design processes and algorithms of data analytics addressing the specificity of this class of problems and the inherent structure of the data. Given the diversity of perspectives encountered in the problem at hand, it becomes imperative to develop efficient and interpretable ways of assessing the performance of results produced by multi-view analytics.

The objective of this volume is to provide the reader with a comprehensive and up-to-date treatise of the area of multi-view data analytics by focusing a spectrum of methodological and algorithmic issues, discussing implementations and case studies, and identifying the best design practices as well as assessing their performance.

Given the diversity of the topics being covered by the chapters, there are several groups focusing on some focal points of the research area. The first two, on psychology of conflictive uncertainty and uncertainty processing bring an interesting and general perspective at the notion of the multi-view way of coping with real-world phenomena and data. A group of papers composed of seven papers addresses a timely issue of clustering regarded as a paradigm supporting a spectrum of activities in data analytics; these papers bring about concepts, new views and algorithms of collaborative clustering. In the sequel, three chapters show innovative ways of applications to EMG signals, performance profiling, and pedestrian detection.

We are truly indebted to the authors for presenting their timely research findings, innovative thoughts, and compelling experimental studies. We would like to express our thanks to the reviewers whose constructive input and detailed comments were instrumental to the rigorous quality assurance process of the contributions. We hope that this volume will serve as a timely and handy addition to the timely and rapidly growing body of knowledge in the domain of multi-view of intelligent systems.

Edmonton, Canada
Taipei, Taiwan

Witold Pedrycz
Shyi-Ming Chen

Contents

The Psychology of Conflictive Uncertainty

Michael Smithson

Abstract In the literature on multi-view learning techniques, "view disagreement" is listed as among the major challenges for multi-view representation learning. View disagreement is distinct from nonshared features in alternative views, because it gives rise to a type of uncertainty that humans find especially aversive, i.e., "conflictive uncertainty". This chapter presents an overview of the psychological effects of conflictive uncertainty, and then provides some guidance for resolving and communicating about conflictive uncertainty that may avoid its problematic impacts on decision making and source credibility. Implications are discussed for developing explainable multi-view methods in the face of view disagreement.

Keywords View disagreement · Conflictive uncertainty · Nonshared information · Ambiguity · Trust · Credibility

1 Introduction

Data from multiple sources may generate conflicting (disagreeing) views, estimates, or predictions, thereby requiring methods for resolving such disagreements. Recent reviews of multi-view learning techniques [1, 2] list " view disagreement" as one of the challenges facing multi-view representation learning. However, these surveys do not treat this problem as distinct from nonshared information across views, and instead discusses techniques as though alternative views always provide "complementary" information that can be safely resolved via techniques for data fusion and modeling associations among variables, or "nonshared" information that may be discarded in the process of resolving divergent views. Such approaches may suffice when disagreement is a matter of degree (e.g., drug X is 60% effective according to one study but 40% effective according to another) but not when two or more alternatives cannot simultaneously be true (e.g., drug X is effective according to one study but ineffective and harmful according to another).

M. Smithson (✉)
Research School of Psychology, The Australian National University, Canberra ACT 0200, Australia
e-mail: Michael.Smithson@anu.edu.au

W. Pedrycz and S. Chen (eds.), *Recent Advancements in Multi-View Data Analytics*, Studies in Big Data 106, https://doi.org/10.1007/978-3-030-95239-6_1

The position taken in this chapter is that view disagreement is distinct from nonshared features in alternative views, because it gives rise to a type of uncertainty that humans find especially aversive. In this chapter uncertainty arising from disagreeing or conflicting information will be called "conflictive uncertainty", or CU. While developing normatively adequate methods for resolving CU is undoubtedly important, it also is important to ensure that these methods are explainable and well-suited to the decision-makers who will use them. For instance, in a meta-analysis of factors influencing trust in human–robot interaction, Hancock et al. [3] recommend "transparency" in system designs and algorithms that are accessible and clear to human users, and currently explainable artificial intelligence is a fast-growing area of research and development.

A key influence on decision-makers' attitudes towards alternative methods of resolving CU is their attitude toward CU itself. Understanding how decision makers think about and respond to uncertainty arising from conflicting information may provide guidance for designing and implementing such methods. It turns out that people treat CU differently from other kinds of uncertainty. Psychological research on this topic over two decades has consistently found that people view CU as more aversive than uncertainty arising from ambiguity, vagueness, or probability. CU also has been shown to have potentially deleterious consequences for decision making. Lastly, there are communications dilemmas for sources that provide inputs potentially leading to CU because communications resulting in CU can decrease trust in those sources.

These findings have implications for best design-practices in resolving CU and communicating about it to non-specialists. This chapter presents an overview of the relevant literature on the psychological effects of CU, and then provides some guidance for resolving and communicating about CU that may avoid or reduce effects that are problematic for decision making and source credibility.

2 Conflictive Uncertainty

Conflictive uncertainty refers to uncertainty arising from disagreement about states of reality that the cognizer believes cannot be true simultaneously. If one source tells us that today is Tuesday and another tells us that today is Thursday, then we will regard these statements as conflicting if we believe that Tuesday and Thursday cannot occur simultaneously. If we do not know what day it is, then the two statements will arouse CU for us. Thus, CU occurs when two or more hypothetical states that cannot simultaneously be true are stated as true by separate sources and/or the same source on separate occasions, and the recipient of these statements does not know which (if either) to believe.

Conflict is related to ambiguity and vagueness, and the distinction between them is somewhat blurry. An early definition of ambiguity defined it as a condition in which a referent has several distinct possible interpretations, but these may be able to be true simultaneously [4]. Smithson [5] gives an example of ambiguity in the statement "this food is hot", potentially referring to the food having high temperature, or being spicy, or sexy, or fashionable, or having been stolen. All of these states could hold about the food at the same time, so the statement is ambiguous but not conflictive. Thus, conflicting states seem to be a special case of ambiguous states.

Some formal perspectives allow this kind of distinction, but not all of them do. For instance, Shafer's belief function framework distinguishes between conflictive and non-conflictive forms of uncertainty [6]. In that framework, conflict occurs when nonzero weights of evidence are assigned to two or more disjoint subsets of a universal set. However, there also are generalized probability frameworks that deal in sets of probabilities, where the distinction between ambiguity and conflict appears unnecessary or irrelevant (an accessible survey of such frameworks is provided in [7].

The connection between conflict and ambiguity regarding psychological states of mind was first suggested in Einhorn and Hogarth's pioneering study when they claimed that conflict can be experienced as ambiguity [8]. However, it was not clear at the time whether the uncertainty aroused by conflicting information would be equivalent to that evoked by ambiguity. More recently, a popularization of this type of uncertainty [9] refers to it as "noise" without distinguishing it from ambiguity or other kinds of uncertainty. Nevertheless, we shall see that people treat the distinction between them as real. There are clues about this possibility in the psychological literature as far back as Festinger's [10] discussion about people's aversion to inconsistency, and attribution theory's inclusion of consensus among the three major determinants of causal attribution [11].

Whether alternative states are *perceived* as conflicting crucially depends on beliefs about when or whether different states can hold at the same time and attributions regarding why they differ. A vaccine cannot be effective and ineffective at the same time for the same person, but it can be effective for one person and ineffective for another. If the vaccine is effective for 50% of the persons tested in clinical trial A and also for 50% tested in trial B then this usually would not be regarded as conflicting information, but if it was 100% effective in trial A and 0% ineffective in trial B then this would be regarded as conflictive.

Likewise, agreed-upon ambiguity or vagueness from multiple sources is not perceived as conflicting information. If experts C and D both claim that a vaccine may be effective for either 30% or 60% of a population then laypeople would be unlikely to perceive conflict there because the experts are in agreement. However, they would perceive conflict if C says the vaccine is 30% effective while D says it is 60% effective. The primary feature distinguishing conflict from ambiguity is disagreement among sources or inconsistency by a source over time, and we shall see that it is this distinction that seems to be most important psychologically.

3 Conflictive Uncertainty Aversion

The first type of non-probabilistic uncertainty found to influence decision makers was ambiguity of a particular kind [12]. The Ellsberg-type demonstration that ambiguity is psychologically reactive involves asking people to choose between betting on drawing, say, a red ball from a "risky" urn that has 50 red and 50 white balls therein, versus betting on drawing a red ball from an "ambiguous" urn with 100 red and white balls of unknown composition. A uniform (or even a symmetric) prior over the probability of drawing a red ball from the ambiguous urn should result in a rational agent being indifferent between the unambiguous 50–50 urn and the ambiguous urn. However, numerous experiments of this kind have shown that most people prefer the unambiguous risky urn (see the review in [13]). This phenomenon often is referred to as "ambiguity aversion".

Reactions to uncertainty arising from ambiguity have been much more widely studied than reactions to CU, partly because the latter was not systematically investigated until the late 1990s. Some of the findings in the ambiguity literature apply to CU, although others do not, and their comparison is instructive. We therefore will briefly survey the main findings regarding responses to ambiguity and discussions about whether or when such responses are irrational.

There has indeed been considerable debate over whether or when ambiguity aversion is irrational. Briefly, most of the arguments for its irrationality focus on just the first moment of the distributions of outcomes for repeated gambles under ambiguity versus risk, i.e., their respective expected values. In the classic Ellsberg setup the expected values of both urns are identical, so by this line of reasoning concern over ambiguity is irrelevant. Some arguments against irrationality draw attention to the second distribution moment, observing that draws from an urn whose composition randomly (and symmetrically) varies around 50–50 will be greater than draws from a constant 50–50 urn, and if an agent assigns less utility to greater variance then they are not irrational in preferring the 50–50 risky urn. It is not difficult to imagine scenarios where concern about variability in outcomes would make sense [14]. For instance, humans would be well-advised to strongly prefer dwelling in a room whose temperature is constantly, say, 22 °C over a room whose average temperature also is 22° but assigns its daily temperature via random draws from a Gaussian distribution with a mean of 22 and a standard deviation of 40.

There also is a long-running debate over whether rational agents must have precise credences (subjective probabilities), or whether they may still be rational if they have imprecise (ambiguous) credences. The default model for a Bayesian decision making agent assumes that the agent has precise probabilities [15], but there has been increasing allowance for rational agent models in which the agent deals in sets of probability functions [16, 17]. Some scholars in this debate even have argued that rationality actually requires credences to be indeterminate and therefore imprecise [18].

Returning now to ambiguity aversion, [12] speculated and subsequent research has confirmed that it is neither universal nor ubiquitous. It holds generally for prospects of gains with moderate to high likelihoods, but even there many people appear to be indifferent to ambiguity versus risk. Moreover, experimental evidence on ambiguity attitudes where the prospective outcome is a loss indicates ambiguity aversion mainly for low likelihood losses, but ambiguity seeking for moderate to high likelihood losses [19, 20]. Likewise, ambiguity seeking often is found for low likelihood gains, especially when the stakes are high [13]. Using an elaborate experimental design, [21] report widespread ambiguity-neutrality under all conditions deviating from a gain prospect with moderate likelihood, and some evidence of ambiguity-seeking under low-likelihood prospects.

To what extent do CU reactions parallel reactions to ambiguity? And are they at similar levels, i.e., if people are given a choice between informationally equivalent agreeing- ambiguous sources versus disagreeing-unambiguous sources, do they exhibit a preference? Two early, independent, investigations produced evidence that people prefer agreeing-ambiguous information over disagreeing-unambiguous information. These investigations were the first to suggest that people distinguish between uncertainty arising from ambiguity and uncertainty arising from conflicting information.

Viscusi [22] employed an experimental setup in which respondents considered the choice of moving to one of two locations, each posing a cancer risk from air pollution. One area had full information regarding the risk, whereas the risk information for the other area came from two sources. Viscusi set up these conditions so that a Bayesian learner would be indifferent between the two areas' risks, with the expected utility functions equating an imprecisely assessed probability to one for which there is expert consensus. The results showed a consistent preference for the full-information area. Viscusi also reports that participants in the experiment devoted "excessive" attention to worst-case scenarios. He concludes that disagreeing risk information from experts results in greater risk overestimation by laypersons than risk information on which experts agree.

Smithson [5] experimentally investigated ambiguity vs conflict preferences in several hypothetical scenarios. His first experiment adapted a scenario from [8], offering participants a choice between two situations as jury members in a trial for armed robbery with testimony from two eyewitnesses: One witness saying that the getaway vehicle was green but the other saying it was blue, vs both witnesses saying that the vehicle was either green or blue. His second scenario involved precise but conflicting computer forecasts of the path of a cyclone versus agreeing but ambiguous forecasts, both of which were informationally equivalent about the cyclone's possible paths. Smithson reports strong preferences for ambiguity over conflict in both scenarios, thereby also demonstrating that this effect holds for nonhuman as well as human information sources. A second round of experiments shows this preference holding to a similar degree regardless of whether the decisions had consequences for the decision maker, another person, or the environment. Smithson [5] called this preference for ambiguity over conflict "conflict aversion".

The Smithson and Viscusi papers stimulated two streams of research: Further tests and extensions of the conflict aversion hypothesis and possible explanations for it, and investigations into the consequences of communicating CU for the credibility and trustworthiness of its sources. The latter stemmed from [5] reporting a strong tendency for subjects to see ambiguous but agreeing experts as more knowledgeable than precise but disagreeing ones. The remainder of this section surveys the first line of research, and the second line of research is reviewed in the next section.

The conflict aversion hypothesis has been verified in numerous studies, including several in realistic settings. Cabantous [23] obtained data from professional actuaries (from the French Institute of Actuaries), demonstrating that they assigned higher premiums for insurance against risks with ambiguous loss probabilities than precise probabilities, and still higher premiums if the indeterminacy in the probabilities stemmed from disagreeing estimates. Cabantous, et al. [24] followed this initial study with data from American insurers, finding that they also would charge higher premiums under ambiguity than under precisely estimate risk. While they also charged more under conflict than ambiguity for flood and hurricane hazards, this did not hold for fire hazards. Cabantous, et al. report that under ambiguity insurers were more likely to attribute the uncertainty to the difficulty of the risk assessment problem, whereas they tended to attribute conflicting estimates to incompetence or unreliability in the assessors.

Han et al. [25] investigated the impact of conflict aversion on uptake of medical tests. They presented people with one of two vignettes describing a hypothetical new screening test for colon cancer: A "missing information" vignette in which they were told that the new test was potentially better than existing tests but only a few small studies had so far been conducted; and a CU vignette in which they were told that studies of the screening test produced differing results and experts disagreed about recommending it. They report that respondents in the CU vignette were less willing to undergo the test than those in the missing-information vignette.

Smithson [5] investigated framing effects on conflict aversion along lines suggested by prospect theory [26]. Prospect theory asserts that people are risk-averse under prospects of gain and risk-seeking under prospects of loss, and it has received substantial empirical support. Smithson reports a reduced degree of conflict aversion under prospect of loss, including a modest tendency to prefer conflict over ambiguity when there is a high likelihood of a negative outcome, but otherwise finds that conflict aversion prevails [5].

Smithson's findings echoed prospect theory's predictions to some extent. In an unpublished experiment, [27] presented participants with hypothetical medical scenarios in which the prospective gain was curing victims of an illness and the loss was the victims remaining ill. Participants were randomly assigned to choosing between one of the three possible pairs (risky vs ambiguous, risky vs conflicting, and ambiguous vs conflicting) of estimates of the probability of either the gain or the loss. The results exhibited both ambiguity and conflict aversion, along with a framing effect that was similar for all pairs, namely a tendency to weaken the preference for precisely specified risk under a prospect of loss.

Lohre, et al. [28] identified another framing effect, involving the use of directional terms (e.g., "over 50%" vs "under 50%") for imprecise probabilities. They find that disagreement is perceived as greater when sources use opposite directional terms than when they use terms in the same direction. For instance, an estimate that P(E) is "over 40%" is perceived as disagreeing more with an estimate that P(not E) is "under 30%" than with a logically identical estimate that P(E) is "over 70%". However, it is not clear whether this effect arises from confusion about comparing the probability of an event with the probability of its complement. Smithson, et al. [29] identify a tendency for laypeople to be less consistent and to have less of a consensus in their numerical translations of verbal probability expressions when these expressions are negative (e.g., "unlikely") than when they are positive (e.g., "likely").

Conflict aversion occurs for indeterminate outcomes as well as probabilities of outcomes. Smithson et al. [30] report two studies where judges encounter ambiguity or CU in the sampled outcomes. Examples of an ambiguous outcome are an inconclusive blood test, or an inconclusive expert assessment of the provenance of an artwork. They find that ambiguity aversion is not less than when people are given a range of probabilities of the outcomes without reference to ambiguous outcomes. They also find that conflict also does not decrease when the uncertainty is in the outcomes rather than in the probabilities.

What are possible explanations for conflict aversion? Again, we may borrow some ideas from the more extensive literature on ambiguity attitudes. The most popular explanations already have been described. The first of these is sensitivity to variability in outcomes and/or outcome probabilities. Rode, et al. [14] present evidence from the literature on foraging and their own experiments that people avoid alternatives with high outcome variability even when probabilities are not explicitly stated, except when their level of need is greater than the expected value of the outcome.

A related explanation that can be applied to CU is that it violates expectations that sources will agree, or at least that any differences between them will be within a tolerated range. Kahneman, et al. [9] observe that there is substantially more disagreement among professional and expert judgments that is either expected or tolerated in fields ranging from jurisprudence to medical diagnosis to actuarial assessments. In one of their surveys they asked executives and underwriters from an insurance company about the amount of variation they would expect between two underwriters' assignments of a premium to the same case, the most popular estimate was 10% of the premium. When they put this to a test, they found that the median difference was 55%.

Although there is, to my awareness, no systematic empirical evidence for this in the general sense, it seems plausible that when expectations for agreement among judges are violated, people will find this violation more aversive when the judgments involve evaluations and/or consequential decisions than when they are only estimates or predictions. For example, mounting evidence of considerable differences among American judges in the sentences they would deliver for identical crime cases resulted in attempts to standardize sentencing by the Federal government during the 1980's. The tone of the 1983 Senate Report [31] in its leadup to recommendations conveys a level of outrage:

> ... every day Federal judges mete out an unjustifiably wide range of sentences to offenders with similar histories, convicted of similar crimes, committed under similar circumstances. One offender may receive a sentence of probation, while another-convicted of the very same

crime and possessing a comparable criminal history-may be sentenced to a lengthy term of imprisonment. (pg. 38)

The recommendations thereafter highlight a reason for why inconsistencies in consequential evaluations may be especially aversive: Violations of fairness or justice. The Senate Report makes this concern explicit in their criteria for sentencing law reformation: "… it should assure that sentences are fair both to the offender and to society, and that such fairness is reflected both in the individual case and in the pattern of sentences in all Federal criminal cases." (pg. 39).

A second explanation for conflict aversion is that CU invokes pessimistic probability estimates, as initially hypothesized by [8]. Smithson, et al. [30] find that ambiguity and conflict aversion are partly (but not entirely) explained by more pessimistic outcome forecasts by participants in their experiments. This holds regardless of whether the conflictive uncertainty is presented in the form of indeterminate probabilities or indeterminate outcomes. They further report that pessimism may be due to uncertainty about how the chance of a desirable outcome in an ambiguous or conflictive setting compares with an equivalent alternative with precise probabilities.

Third, attributions regarding the causes of ambiguity also have been studied as possible influences on ambiguity attitudes. Stuart et al. [32] report experimental evidence that when ambiguity (or even possibly CU, which they do not distinguish from ambiguity) is believed to be due to something that the decision maker can control or that they are optimistic about, then the decision maker will exhibit ambiguity-seeking. Du et al. [33] proposed and tested a hypothesis that investors prefer ambiguous (vague) earnings forecasts over precise forecasts for an investment when their prior belief is that little is known about the performance of the investment.

The most common CU-specific explanation for conflict aversion is attributions of incompetence or other detrimental inferences about the sources and/or information, resulting in a decline in their credibility or trustworthiness [5, 24, 34], and thereby a discounting of their judgments or predictions. Another is that CU can require decision makers to "take sides" (e.g., choosing one estimate and discarding all others), especially if a compromise or middle-ground resolution is not available [5]. It seems plausible that most of the influences on CU attitudes that are not shared by ambiguity attitudes will involve social factors regarding perceptions of the sources and the perceiver's relationships with the sources.

Summing up, conflictive uncertainty appears to be more aversive to many people than either risk (in the sense of probabilities) or ambiguity. Is conflict aversion irrational, or does it lead to irrational behavior? There has been relatively little discussion about this, with several authors taking the position that it results in irrational decision-making, although others seem more agnostic on the topic.

People tend to be more pessimistic under CU than under ambiguity or risk, and they put greater weight on pessimistic forecasts. This tendency can be irrational, but under some conditions it can be prudent. Several researchers observe that CU can result in irrationally "alarmist" responses, such as placing disproportionate weight on high-risk estimates when given multiple disagreeing risk estimates [5, 22, 35]. Baillon et al. [35] demonstrated that this effect does not occur under ambiguous uncertainty. These sets of findings underscore a tendency for people to be more risk-averse under CU than under other kinds of uncertainty, whether greater risk-aversion is rational or not. Nevertheless, it is not difficult to find justifications for conflict aversion, particularly when the sources are experts. Laypeople are not unreasonable in expecting experts to agree on matters within their domain of expertise. As Viscusi and Chesson [36] point out, agreement among experts suggests that we should have more confidence in their assessments than if they disagree.

Various computational models have been proposed and tested to account for ambiguity attitudes, but most of these specialize in ambiguity about probability estimates. Several models attempt to jointly model ambiguity and CU attitudes in more general settings, and these are briefly surveyed here. Smithson [37] proposed two types of models, variance and distance based. The simplest versions of these assume that there are two judges (or sources), each providing ambiguous quantitative estimates in the form of intervals, where p can be any quantity (i.e., not limited to probabilities), and $k = \{1,2\}$ and indexes the judges. Following [14], Smithson defined ambiguity for each judge as the variance of their interval limits:

$$A_k = \sum_{j=1}^{2} \left(p_{kj} - \overline{p}_{k.}\right)^2 / 2 \tag{1}$$

where $\overline{p}_{k.}$ denotes the arithmetic mean taken over j, so that the total ambiguity is the sum of the A_k. One measure of conflict, then, is the between-judges variance,

$$C_1 = \sum_{k=1}^{2} \left(\overline{p}_{k.} - \overline{p}_{..}\right)^2 / 2. \tag{2}$$

However, another measure is the variance among the order-statistics of the same rank (recalling that $p_{k1} \leq p_{k2}$):

$$C_2 = \sum_{k=1}^{2} \sum_{j=1}^{2} \left(p_{kj} - \overline{p}_{.j}\right)^2 / 4 \tag{3}$$

The first model predicts that a pair of interval estimates with the same midpoints will not be perceived as conflictive, regardless of differences in the interval widths, whereas the second model predicts that they will be conflictive. Smithson's experimental evidence indicated that the conflict measure in Eq. (3) predicts conflict attitudes better than the one in Eq. (2), suggesting that people are sensitive to disagreements about the uncertainty of an estimate.

Smithson's [37] distance-based models evaluate ambiguity and conflict in terms of distances between order statistics. An index of ambiguity using Euclidean distance is

$$A_k = \sum_{j_1=1}^{2} \sum_{j_2=1}^{2} \left(p_{kj_1} - p_{kj_2}\right)^2/4 \tag{4}$$

As before, conflict can be evaluated in two ways. First is the absolute value of the sum of the differences over ranks:

$$C_1 = \sum_{k_1=1}^{2} \sum_{k_2=k_1+1}^{2} \left| \sum_{j=1}^{2} \left(p_{k_1 j} - p_{k_2 j}\right) \right| \tag{5}$$

Second is the sum of the absolute differences between pairs of order-statistics of the same rank:

$$C_2 = \sum_{k_1=1}^{2} \sum_{k_2=k_1+1}^{2} \sum_{j=1}^{2} \left| p_{k_1 j} - p_{k_2 j} \right| \tag{6}$$

Similarly to the two variance-based conflict indexes, C_1 predicts that a pair of interval estimates with identical midpoints will not be perceived as conflictive, whereas C_2 predicts that they will be, and again C_2 performed better on empirical data.

Gajdos and Vergnaud [38] also defined a model of decision making under ambiguity and conflict based on the maxmin framework. It was originally limited to dealing with probability estimates, and more importantly, was intended as a model of a rational agent taking account of both ambiguity and conflict aversion. Smithson [37] generalizes the two-state, two-judge special case of their model to judgments of magnitudes and tested it empirically along with the models described above. In the variance and distance models above, two weight parameters, α and θ, are used to estimate both attitude and sensitivity toward the ambiguity and conflict indexes. In the [38] model, these modify the order statistics of each judge. The θ parameter shrinks the width of the $[p_{k1},\ p_{k2}]$ interval by $1 - \theta$, so that the lower and upper bounds become

$$\begin{aligned} \pi_{k1} &= p_{k1}(1+\theta)/2 + p_{k2}(1-\theta)/2 \\ \pi_{k2} &= p_{k1}(1-\theta)/2 + p_{k2}(1+\theta)/2 \end{aligned} \tag{7}$$

Gajdos and Vergnaud do not define an ambiguity measure but Smithson constructs one by summing the differences $\pi_{k2} - \pi_{k1}$. Likewise, their model treats α as contracting the pairs of interval endpoints p_{kj} and p_{mj} around their mean at the rate $1 - \alpha$. The order statistics are modified as follows:

$$\begin{aligned} \gamma_{kj} &= p_{kj}(1+\alpha)/2 + p_{mj}(1-\alpha)/2, \\ \gamma_{mj} &= p_{mj}(1+\alpha)/2 + p_{kj}(1-\alpha)/2. \end{aligned} \tag{8}$$

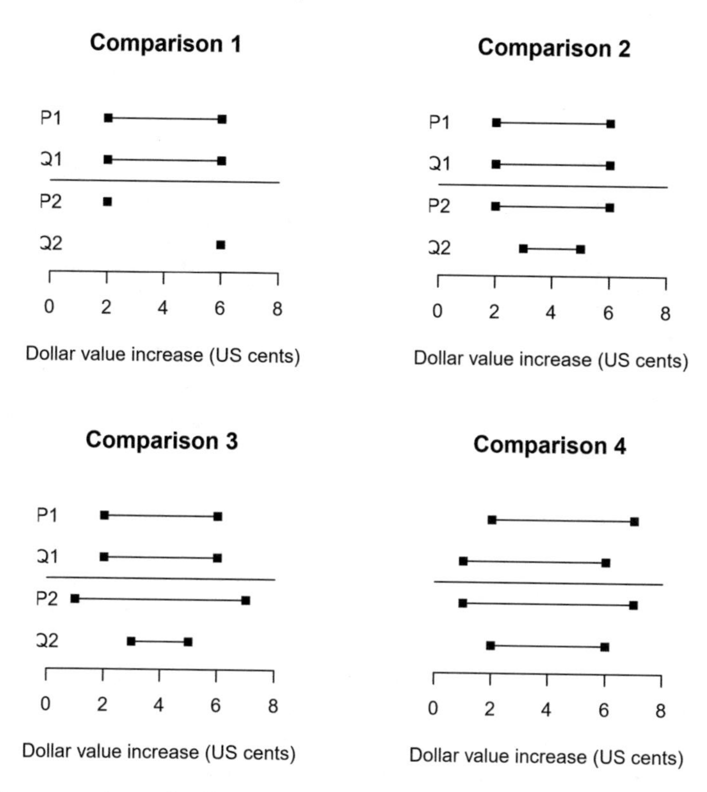

Fig. 1 Four comparisons of ambiguous-conflicting estimates

A conflict measure can be constructed by summing the absolute values of the $\gamma_{kj} - \gamma_{mj}$ differences, which yields an index similar to the one in Eq. (6).

Smithson [37] tested these models by presenting participants with choices between two pairs of interval estimates, with four sets of these as shown in Fig. 1. The second kind of variance and distance models and the [38] model outperformed the first kind of variance and distance models, and the relative effects of the ambiguity and conflict indexes on the preference-rates exhibited by experimental participants suggested that Conflict aversion and ambiguity aversion operate relatively independently of one another.

The best-performing models correctly accounted for the tendency to prefer pairs of estimates with identical interval widths over those whose widths differed (Comparisons 2 and 3). However, none of the models succeeded in accounting for people's tendency to prefer the top pair of intervals with mismatching midpoints over the bottom pair with identical midpoints in Comparison 4. This finding indicates that people may sometimes be more sensitive to mismatching uncertainties than to mismatching estimates.

4 Consequences of Conflictive Uncertainty for Risk Communication

As mentioned earlier, the second line of research inspired by [22] and [5] has focused on the consequences of communicating CU for its recipients' appraisals of the information and its sources. Smithson reported tendencies for people to downrate the knowledgeability of disagreeing sources [5]. If generally true, then experts' communications resulting in CU (or having it attributed to them) could suffer losses of credibility and trust from the public.

Science communicators are not unaware of this possibility, and some of them realize that they face a "Chicken-game" dilemma when considering how to frame statements of their views on issues where there is scientific controversy. Forthright statements from them and those on the other side of a controversy run the risk of yielding discreditation of all sources involved in the debate. On the other hand, softening one's position by appearing to agree on some points with an opponent runs the risk of exploitation by the opponent. Indeed, [39] identified evidence that at the height of controversy over anthropogenic climate change, scientists over-emphasized scientific uncertainty, and that even when refuting contrarian claims they often do so in a manner that reinforces those same claims.

The question of whether scientific experts should or should not communicate uncertainty about theories or research to the public has been the subject of considerable research. Nonetheless, in a review of 48 articles on the effects on science communication generally, Gustafson and Rice [40] observe that the literature on science communication is divided on whether communicating scientific uncertainty generally will have positive or negative effects (e.g., on trust or credibility accorded to scientists or to science generally). As a result, advice from this literature for science communicators is confusing at best.

Gustafson and Rice [40] examine what they identify as four kinds of uncertainty: Deficient, technical, scientific, and consensus (the latter is CU). Deficient uncertainty refers to gaps in knowledge, technical uncertainty mainly to imprecision in estimates or measurement, scientific uncertainty to the notion that all scientific knowledge is tentative, and consensus uncertainty to disagreement or controversy. Are the consequences of CU different from other kinds of uncertainty for risk communication?

We already have reviewed evidence that people prefer agreeing but ambiguous or vague sources over precise but disagreeing sources, even though their collective estimates or accounts are informationally identical. Evidence that people also regard disagreeing sources as less credible or trustworthy often has been borne out in studies of the effects of disagreements among scientists on public attitudes toward scientists and, indeed, science. For instance, [41] observed that in the context of the Swedish acrylamide scare, public disagreements between epidemiologists and toxicologists over the link between acrylamide and cancer led to public distrust in scientists. Similarly, Regan et al. [42] Investigated the effects of third-party communication on trust in risk messages. New information emphasizing the benefits of red meat and contradicting a previous risk message led to judgments that the original risk message was less credible. Evaluation of the new message was not affected by any apparent conflict with the original risk message. Instead, the trustworthiness of the third-party communicating the new message influenced its credibility. In an experimental study, Gustafson and Rice [43] tested the effects of their four uncertainty types in three scientific topics, and found that CU was the only kind that had consistent significant effects on beliefs, risk perceptions, or behavioral intentions, and all of these effects were negative.

To some extent, the impact of CU created through disagreement among scientists may depend on contextual factors, such as the topic. For example, Jensen and Hurley [44] observe that uncertainty about the health effects of dioxin (a possible carcinogen) increased the credibility and trustworthiness of scientists, whereas they report a deleterious effect for conflicting claims about risks regarding wolf reintroduction in the USA. Moreover, Gustafson and Rice's [43] negative findings about CU pertained mainly to just one of their three topics, climate change, but not to either GMO food labeling or machinery risks. Likewise, [45] report no detrimental effects from CU on perceived trustworthiness of experts' messages about genetically modified food risks, although they did find that consensus reduced the perceived risks themselves. Unfortunately, the current understanding regarding when and why people find CU aversive and its sources untrustworthy is in an inconclusive and confused state, with no convincing overview or synthesis yet.

Nevertheless, the [40] survey of the literature identified CU as the only type of uncertainty that had consistently negative effects on public perceptions of scientists and/or science. Communications of technical uncertainty showed no negative effects, and often enhanced public trust and esteem in scientists instead. Deficient and scientific uncertainties exhibited a mix of effects, some of which were moderated by the beliefs and prior knowledge of those receiving the communication. However, the negative impact of CU has been shown in some instances to spread discreditation beyond its immediate sources. For instance, Nagler's [46] survey revealed that people reporting greater exposure to CU regarding the effects of consumption of various kinds of food were more likely to discount nutrition research altogether.

Taking all this into account, CU is arguably the most corrosive kind of uncertainty whose psychological effects have been investigated systematically. Public exposure to CU has increased in recent times. For the past several decades, the public in a variety of countries has increasingly been exposed to divergent risk messages in various salient domains, such as financial investment, health risks, terrorism, and climate change [47]. A RAND report highlights increasing disagreements about facts and analytical interpretations of data, a blurring of the line between opinion and fact, an increasing influence of opinion over fact, and a declining trust in formerly respected sources of fact [48].

These trends can be partly attributed to the so-called "democratization" of journalism, but even traditional journalism prior to the advent of the internet tended to aggrandize disagreements. Normal journalistic coverage of disagreements or controversies tends to give equal exposure to all sides, regardless of their expertise or evidentiary basis, thereby often amplifying laypeople's concerns and uncertainties. Viscusi

[22] observes that "the media and advocacy groups often highlight the worst case scenarios, which will tend to intensify the kinds of biases observed here." Stocking [49] has a somewhat more balanced account, pointing out that journalists often are under competing pressures to simplify science, which often entails omitting caveats and other expressions of uncertainty, but also to exercise impartiality and thoroughness in presenting alternative viewpoints on issues where conclusions have yet to be reached.

Moreover, the detrimental effects of CU on trust have been exploited, in at least some cases deliberately, by politically-motivated agents in public policy debates and negotiations on issues such as the link between tobacco smoking and lung cancer [50] and, more recently, climate change [51]. Even scientists in such debates have been found to revise their positions in ways they would be unlikely to take in the absence of outspoken contrarian opposition [39]. As mentioned earlier, they face a dilemma between decreased public trust in their expertise by sticking with "hard-line" risk messages versus conceding points to their opponents.

5 Dealing with and Communicating About Conflictive Uncertainty

The evidence from psychological research on conflictive uncertainty generally lends support to the following propositions:

1. People distinguish between CU and other kinds of uncertainty such as ambiguity and probability.
2. They usually find CU more aversive than other kinds of uncertainty and may be willing to trade CU for an alternative kind of uncertainty.
3. Communications from sources or inconsistent communications from a single source resulting in CU tend to erode the credibility and trustworthiness of those sources.

Understanding the psychology behind reactions to CU can contribute to the effectiveness of methods for resolving conflicting information and communicating about resolutions and/or decisions under conflictive uncertainty in the following ways:

- Knowing the aspects of CU that amplify its aversiveness can aid the choice and/or development of methods for dealing with CU to diminish or eliminate those aspects.
- Knowing how people prefer to deal with CU can provide the means for tailoring CU resolution methods to match those preferences where possible.
- Understanding the reasons behind the erosion of trust in CU sources can guide communicators about CU and its resolution in finding ways to prevent or minimize that erosion.

Conflict aversion seems to stem from (and to be exacerbated by) the following contingencies:

1. Violation of expectations of agreement (e.g., among experts),
2. Perceptions that no compromise or middle-ground position is available to resolve the disagreements,
3. Perceptions that no additional information is available that might resolve the disagreements,
4. Perceptions that the conflicting information involves evaluations or consequential decisions, and
5. Personal relevance, especially if any of the conflicting information also disagrees with one's own prior beliefs.

The first three of these aspects are amenable to being mitigated by the ways in which CU situations are framed.

The sense of violated expectations of agreement may be reduced or prevented by framing situations so that people know in advance to expect disagreements. If they perceive differing views and debates as normal and to be expected, then CU may not be as aversive. For instance, [52] present evidence that people respond more positively to uncertainty about research if they see science as a matter of engaging in debates with constant revisions and improvements than if they see science as mainly about arriving at absolute truths.

It may be beneficial to revise expectations regarding CU among the sources of the conflicting judgements or estimates producing CU. Observing that the extent of actual variability in judgements by experts often flies in the face of beliefs about the consensus levels among the experts themselves, [9] recommend conducting what they call "noise audits". These amount to experiments with appropriate designs and controls to assess the variability among experts in their assessments of the same cases or problems. Properly conducted, noise audits can provide experts with realistic perspectives on the extent to which disagreements are likely to occur in their domain, which in turn pave the way to communicating those perspectives to their clients or to the public. If [9] are correct in their assertions that many communities of professionals and experts under-estimate the extent to which their professional judgments are likely to disagree with one another, then the revelations of a noise audit may also motivate these communities to seek ways to reduce unnecessary or avoidable disagreement, thereby reducing the incidence of unwanted CU.

What approaches to resolving CU do ordinary people take? People's preferences for ways of eliminating or reducing CU have yet to be fully systematically studied. However, there is some data on how people go about resolving CU in everyday life. Smithson [27] elicited descriptions of everyday episodes of CU from a sample of 308 adults from the UK and asked how they went about resolving the conflicting information. The most popular responses were seeking more information (21.6%), finding a compromise and/or decide that the alternative positions could be true simultaneously (19.0%), or choosing one of the alternatives and discounting the others (14.0%). This third alternative seemed to be chosen only when participants regarded one source as more credible than the others.

The kinds of disagreement likely to pose the greatest difficulties for people are those presented as mutually exclusive states (i.e., "zero-sum") with no prospect of a compromise or middle-ground position. On the other hand, various kinds of "averaging" have intuitive appeal to people, both in terms of understandability and also fairness. Where possible, framing conflicting positions as having the potential for middle-ground or compromise resolutions is likely to make CU less aversive and such resolutions more acceptable and believable.

In some settings, it may not be feasible to resolve disagreeing judgments by arriving at a precise or single resolution. Instead, it may be necessary to retain a range or set of judgments. Nevertheless, given the evidence that people prefer ambiguity to CU, employing deliberate ambiguity or vagueness to absorb disagreement can aid the construction of a workable consensus in the face of CU. When people believe that full resolution is impossible, they may find an ambiguous resolution more plausible than a precise one. Joslyn and LeCerc [53], for instance, report that quantitative displays of uncertainty in the form of interval estimates result in greater trust in risk messages about climate change than pointwise estimates. The precise estimates may be violating public expectations about the uncertainty involved in such estimates.

Additionally, a focus on processes and procedural fairness instead of solely on outcomes can reduce conflict aversion. If people trust the processes by which assessments have been arrived at, and if they believe that reasoned discussion will continue as part of the resolution process then they are more likely to accept the eventual outcome. Where possible, the methods by which CU is resolved should be explicit and explicable to laypeople. For example, the arithmetic average of two alternative probability estimates is far more likely to be comprehended by laypeople than the [16] linear-vacuous model or even a geometric average. Recalling the words of the U.S. Senate Report [31] recommendations on sentence law reform, a resolution of CU should provide people with reasons for choosing it rather than relevant alternative resolutions.

Turning now to the issue about CU that involves evaluations or consequential decisions, this kind of situation exacerbates CU aversiveness because it raises or intensifies moral considerations regarding both the judgments and the judges. This effect is not unique to CU. Generally, for example, uncertainties regarding reversible or steerable decisions are less detrimental to trust and assurance than uncertainties about irrevocable decisions [54, 55].

The most common moral consideration regarding resolutions of uncertainty of any kind is fairness, and fairness in algorithmic decision making currently is a widely discussed issue. The topic initially arose when deliberately built-in biases in algorithms for assigning prices to consumer goods were detected [56], but attention rapidly shifted to unintentional biases built into automata such as algorithms for assessing recidivism risk, allocating health care, and selecting candidates for recruitment [57]. Inadvertent bias or discrimination can arise in multiple ways, including the nature of the training data, the variables selected for risk assessment, and the criteria for optimization. Worse still, "fairness" turns out to have multiple definitions and criteria (e.g., achieving identical "at risk" assignments across subpopulations but also attaining identical false-positive and false-negative rates across the same subpopulations), and some of these cannot be achieved simultaneously [58].

Fairness is very likely to be a concern with CU and its resolution, and algorithmically implemented resolutions will need to be transparent to users about how fairness is dealt with. Turning to a simple hypothetical example, suppose we have two equally credible sources estimating the probability of event E, and source 1 produces an estimate $p_1(\mathrm{E}) = 0.1$ whereas source 2 produces $p_2(\mathrm{E}) = 0.6$. The familiar "best" resolution of this disparity is their arithmetic mean, $p(\mathrm{E}) = 0.35$. An algorithm using this resolution also will be perceived as being "fair" by many laypeople because it gives equal weight in averaging to the two equally credible sources.

But now suppose that the algorithm is using the linear-vacuous model for lower–upper probabilities [16], which takes the interval width, $p_2(\mathrm{E}) - p_1(\mathrm{E}) = w = 0.5$, to be the probability that both sources actually are ignorant of $p(\mathrm{E})$ and that the real state of knowledge about this probability is the "vacuous" interval [0,1]. The linear-vacuous model then has

$$\begin{aligned} p_1(\mathrm{E}) &= (1 - w)p(\mathrm{E}) \\ p_2(\mathrm{E}) &= (1 - w)p(\mathrm{E}) + w \end{aligned} \tag{9}$$

and its resolution therefore is $p(\mathrm{E}) = p_1(\mathrm{E})/(1 - w) = 0.2$. This resolution will not only be unfamiliar to laypeople, but it also may seem "unfair" because, from their perspective of averaging, it appears to give greater weight to source 1 than to source 2. Moreover, the linear-vacuous resolution will seem pessimistic if event E is desirable or optimistic if E is undesirable.

Turning now to remaining considerations about how best to communicate CU and how to persuade clients or the public to accept and trust a method for resolving it, communicating uncertainty can be thought of as a way of framing science communication [59. 60]. We already have seen several ways in which the aversiveness of CU can be reduced by framing it: As expected and normal, amenable to resolution via a middle-ground or compromise positions, resolvable in a way that is transparent, sensible, understandable, and fair; and both its genesis and resolution framed as products of reasoned, regulated, and fair discussion or debate. Communications about CU and its resolutions will be better received when they employ these frames wherever possible.

Finally, one concept that communicators should keep in mind is that recipients of their messages are likely to engage in what psychologists call "motivated reasoning" as they try to make sense of those messages and also to deal with their own reactions to them. Motivated reasoning [61] refers to the selective use of cognitive strategies and attention to evidence as determined by motivational factors. For instance, Chang [62] suggests that because CU induces discomfort, motivation to reduce that discomfort drives reasoning about the credibility of the sources and evidence involved. A readymade way of reducing this discomfort is to discount the evidence and/or sources as untrustworthy, and Chang's studies show that this is a commonplace response.

On the one hand, findings such as Chang's can be regarded as good news because they indicate that the public is not entirely gullible. On the other hand, discreditation of expert sources and/or carefully martialed evidence often is not a desirable outcome. An effective counter-measure against reasoning dominated by a motive to reduce discomfort from CU is for communications about CU and its resolution to catalyze other motives. Kunda's [61] review highlights research showing that when people are more strongly motivated to find the most accurate view or estimate, they are more likely to engage in deeper and more impartial reasoning. Impartiality in reasoning also is increased when people are motivated to be fair or just in their assessments or decisions. Finally, if the recipient is having to make decisions under CU, it is helpful if framing can remove concerns about blameworthiness and enhance motivation to produce the best outcomes for those affected by the decisions.

6 Conclusion and Suggestions for Multi-View Modeling Practices

We conclude with three recommendations for further developments in multi-view learning. First of all, more attention should be devoted to the problem of view disagreement, and to the question of whether it requires techniques that differ from those employed in its absence. The extent of this problem needs greater acknowledgement than has appeared throughout much of the multi-view literature and it should be treated as separate from nonshared or ambiguous information. While problems of missing and ambiguous data commonly feature in this literature, outright disagreement seldom is squarely faced and oftentimes simply goes unmentioned. Moreover, in some approaches it is essentially swept under the carpet. Consider, for example, one of the assumptions underpinning the so-called "information bottleneck" method of unsupervised learning, namely that each view has the same task-relevant information and therefore a "robust" representation can be generated by "abandoning the information not shared by views, i.e., removing the view-specific nuisances" [2]. This assumption equates disagreement with nonshared information and thence irrelevance (nuisance).

Second, more attention also needs to be devoted to developing explainable multi-view methods, especially in the face of view disagreement and the potential for CU. The importance of explainability is crucial, as [63] demonstrated that users distrust even a high-performing automated system unless they are provided with reasons for why performance errors have occurred. A recent survey concludes with this observation about the state of the art for explainable multi-view learning models: "Although existing deep MVL models have shown superior advantages in various applications, they fail to provide an explanation for the decision of different models." [2]. The currently fashionable deep-learning models pose an even greater difficulty regarding explainability than older techniques based on multivariate statistical approaches such as canonical correlation.

Finally, in both the design and implementation of multi-view techniques, greater use should be made of knowledge about human attitudes toward and responses to automation under uncertainty. This recommendation is a corollary of an admonition voiced by several researchers investigating issues of machine learning trustworthiness, e.g.: "… the fundamental tensions between adversarial robustness and model accuracy, privacy and transparency, and fairness and privacy invite more rigorous and socially grounded reasonings about trustworthy ML." [64]. The main goal in writing this chapter has been to pave the way toward this third recommendation, i.e., incorporating knowledge available from psychology about the nature of CU and human responses to it into the development and implementation of multi-view learning algorithms.

References

1. Li, Y., Yang, M., Zhang, Z.: A survey of multi-view representation learning. IEEE Trans. Knowl. Data Eng. **31**(10), 1863–1883 (2018)
2. Yan, X., Hu, S., Mao, Y., Ye, Y., Yu, H.: Deep multi-view learning methods: a review. Neurocomputing **448**, 106–129 (2021)
3. Hancock, P.A., Billings, D.R., Schaefer, K.E., Chen, J.Y.C., de Visser, E., Parasuraman, R.: A meta-analysis of factors affecting trust in human-robot interaction. Hum. Factors **53**, 517–527 (2011)
4. Black, M.: Vagueness: an exercise in logical analysis. Philos. Sci. **4**, 427–455 (1937)
5. Smithson, M.: Conflict aversion: Preference for ambiguity vs conflict in sources and evidence. Organ. Behav. Human Decis. Proces. **79**, 179–198 (1999)
6. Shafer, G.: A mathematical theory of evidence. Princeton University Press, Princeton (1976)
7. Augustin, T., Coolen, F., de Cooman, G., Troffaes, M. (eds.): An introduction to imprecise probabilities. Wiley, London (2014)
8. Einhorn, H.J., Hogarth, R.M.: Ambiguity and uncertainty in probabilistic inference. Psychol. Rev. **92**, 433–461 (1985)
9. Kahneman, D., Sibony, O., Sunstein, C.R.: Noise: a flaw in human judgment. William Collins, London (2021)
10. Festinger, L.: A theory of cognitive dissonance. Row, Peterson, Evanston, IL (1957)
11. Kelley, H.H.: Attribution theory in social psychology. In: Levine, D. (ed.) Nebraska Symposium on motivation. University of Nebraska Press, Lincoln (1967)
12. Ellsberg, D.: Risk, ambiguity, and the Savage axioms. Quart. J. Econ. **75**, 643–669 (1961)
13. Trautmann, S.T., Van De Kuilen, G.: Ambiguity attitudes. The Wiley Blackwell handbook of judgment and decision making. Wiley, London (2015)
14. Rode, C., Cosmides, L., Hell, W., Tooby, J.: When and why do people avoid unknown probabilities in decisions under uncertainty? Testing some predictions from optimal foraging theory. Cognition **72**, 269–304 (1999)
15. Jeffrey, R.: Probability and the art of judgment. Cambridge University Press, Cambridge (1992)
16. Walley, P.: Statistical reasoning with imprecise probabilities. Chapman Hall, London (1991)
17. Seidenfeld, T., Wasserman, L.: Dilation for sets of probabilities. Ann. Stat. **21**(3), 1139–1154 (1993)
18. Hájek, A., Smithson, M.: Rationality and indeterminate probabilities. Synthese **187**(1), 33–48 (2012)
19. Kahn, B.E., Sarin, R.K.: Modeling ambiguity in decisions under uncertainty. Journal of Consumer Research **15**, 265–272 (1988)
20. Hogarth, R.M., Einhorn, H.J.: Venture theory: A model of decision weights. Manage. Sci. **36**, 780–803 (1990)
21. Kocher, M.G., Lahno, A.M., Trautmann, S.T.: Ambiguity aversion is not universal. Eur. Econ. Rev. **101**, 268–283 (2018)

22. Viscusi, W.K.: Alarmist decisions with divergent risk information. Econ. J. **107**(445), 1657–1670 (1997)
23. Cabantous, L.: Ambiguity aversion in the field of insurance: Insurers' attitude to imprecise and conflicting probability estimates. Theor. Decis. **62**(3), 219–240 (2007)
24. Cabantous, L., Hilton, D., Kunreuther, H., Michel-Kerjan, E.: Is imprecise knowledge better than conflicting expertise? Evidence from insurers' decisions in the United States. J. Risk Uncertain. **42**(3), 211–232 (2011)
25. Han, P.K., Reeve, B.B., Moser, R.P., Klein, W.M.: Aversion to ambiguity regarding medical tests and treatments: measurement, prevalence, and relationship to sociodemographic factors. J. Health Commun. **14**(6), 556–572 (2009)
26. Kahneman, D., Tversky, A.: Prospect theory: An analysis of decision under risk. Econometrica **4**, 263–291 (1979)
27. Smithson, M.: Episodic and framing effects in reactions to conflictive uncertainty. Unpublished manuscript, The Australian National University, Canberra, Australia (2021)
28. Løhre, E., Sobkow, A., Hohle, S.M., Teigen, K.H.: Framing experts' (dis)agreements about uncertain environmental events. J. Behav. Decis. Mak. **32**(5), 564–578 (2019)
29. Smithson, M., Budescu, D.V., Broomell, S.B., Por, H.H.: Never say "not": Impact of negative wording in probability phrases on imprecise probability judgments. Int. J. Approximate Reasoning **53**(8), 1262–1270 (2012)
30. Smithson, M., Priest, D., Shou, Y., Newell, B.R.: Ambiguity and conflict aversion when uncertainty is in the outcomes. Front. Psychol. **10**, 539 (2019)
31. United States Senate: Senate Report No. 98–225 (Senate Judiciary Committee) to Accompany S. 1762, the Comprehensive Crime Control Act of 1983, September 14, 1983. Washington, U.S.: Govt. Print. Off (1983)
32. Stuart, J. O. R., Windschitl, P. D., Miller, J. E., Smith, A. R., Zikmund-Fisher, B. J., Scherer, L. D.: Attributions for ambiguity in a treatment-decision context can create ambiguity aversion or seeking. Journal of Behavioral Decision Making, https://doi.org/10.1002/bdm.2249 (2021)
33. Du, N., Budescu, D.V., Shelly, M.K., Omer, T.C.: The appeal of vague financial forecasts. Organ. Behav. Hum. Decis. Process. **114**(2), 179–189 (2011)
34. Visschers, V.H.: Judgments under uncertainty: evaluations of univocal, ambiguous and conflicting probability information. J. Risk Res. **20**(2), 237–255 (2017)
35. Baillon, A., Cabantous, L., Wakker, P.P.: Aggregating imprecise or conflicting beliefs: An experimental investigation using modern ambiguity theories. J. Risk Uncertain. **44**(2), 115–147 (2012)
36. Viscusi, W.K., Chesson, H.: Hopes and fears: the conflicting effects of risk ambiguity. Theor. Decis. **47**(2), 157–184 (1999)
37. Smithson, M.: Conflict and ambiguity: Preliminary models and empirical tests. In: Proceedings of the Eighth International Symposium on Imprecise Probability: Theories and Applications, Compiegne, France, 2–5 July 2013: pp. 303–310 (2013)
38. Gajdos, T., Vergnaud, J.C.: Decisions with conflicting and imprecise information. Soc. Choice Welfare **41**(2), 427–452 (2013)
39. Lewandowsky, S., Oreskes, N., Risbey, J.S., Newell, B.R., Smithson, M.: Seepage: Climate change denial and its effect on the scientific community. Glob. Environ. Chang. **33**, 1–13 (2015)
40. Gustafson, A., Rice, R.E.: A review of the effects of uncertainty in public science communication. Public Underst. Sci. **29**(6), 614–633 (2020)
41. Löfstedt, R.E.: Science communication and the Swedish acrylamide 'alarm.' J. Health Commun. **8**, 407–432 (2003)
42. Regan, Á., McConnon, Á., Kuttschreuter, M., Rutsaert, P., Shan, L., Pieniak, Z., Barnett, J., Verbeke, W., Wall, P.: The impact of communicating conflicting risk and benefit messages: An experimental study on red meat information. Food Qual. Prefer. **38**, 107–114 (2014)
43. Gustafson, A., Rice, R.E.: The effects of uncertainty frames in three science communication topics. Sci. Commun. **41**(6), 679–706 (2019)
44. Jensen, J.D., Hurley, R.J.: Conflicting stories about public scientific controversies: effects of news convergence and divergence on scientists' credibility. Public Underst. Sci. **21**, 689–704 (2012)

45. Dean, M., Shepherd, R.: Effects of information from sources in conflict and in consensus on perceptions of genetically modified food. Food Qual. Prefer. **18**(2), 460–469 (2007)
46. Nagler, R.H.: Adverse outcomes associated with media exposure to contradictory nutrition messages. J. Health Commun. **19**, 24–40 (2014)
47. McCright, A.M., Dunlap, R.E.: The politicization of climate change and polarization in the American public's views of global warming, 2001–2010. Sociol. Q. **52**(2), 155–194 (2011)
48. Rich, M.D.: Truth decay: An initial exploration of the diminishing role of facts and analysis in American public life. Rand Corporation, Santa Monica, California (2018)
49. Stocking, S.H.: How journalists deal with scientific uncertainty. In: Dunwoody, S., Rogers, C.L. (eds.) Communicating uncertainty: Media coverage of new and controversial science, pp. 23–41. Routledge, New York (1999)
50. Proctor, R.N.: Cancer wars: How politics shapes what we know and don't know about cancer. Basic Books, New York (1995)
51. Oreskes, N., Conway, E.M.: Defeating the merchants of doubt. Nature **465**(7299), 686–687 (2010)
52. Rabinovich, A., Morton, T.A.: Unquestioned answers or unanswered questions: Beliefs about science guide responses to uncertainty in climate change risk communication. Risk Analysis: An International Journal **32**(6), 992–1002 (2012)
53. Joslyn, S.L., LeClerc, J.E.: Climate projections and uncertainty communication. Top. Cogn. Sci. **8**(1), 222–241 (2016)
54. Salem, M., Lakatos, G., Amirabdollahian, F., Dautenhahn, K.: Would you trust a (faulty) robot?: Effects of error, task type and personality on human-robot cooperation and trust. In: Proceedings of the Tenth Annual ACM/IEEE International Conference on Human-Robot Interaction ACM, New York, 141–148 (2015)
55. Smithson, M., Ben-Haim, Y.: Reasoned decision making without math? Adaptability and robustness in response to surprise. Risk Anal. **35**, 1911–1918 (2015)
56. Valentino-Devries, J., Singer-Vine, J., Soltani, A.: Websites vary prices, deals based on users' information. Wall Street J. **10**, 60–68 (2012)
57. Romei, A., Ruggieri, S.: A multidisciplinary survey on discrimination analysis. Knowl. Eng. Rev. **29**(5), 582–638 (2014)
58. Chouldechova, A.: Fair prediction with disparate impact: A study of bias in recidivism prediction instruments. Big Data **5**(2), 153–163 (2017)
59. Rice, R.E., Gustafson, A., Hoffman, Z.: Frequent but accurate: A closer look at uncertainty and opinion divergence in climate change print news. Environ. Commun. **12**(3), 301–320 (2018)
60. Ruhrmann, G., Guenther, L., Kessler, S.H., Milde, J.: Frames of scientific evidence: How journalists represent the (un)certainty of molecular medicine in science television programs. Public Underst. Sci. **24**(6), 681–696 (2015)
61. Kunda, Z.: The case for motivated reasoning. Psychol. Bull. **108**, 480–498 (1990)
62. Chang, C.: Motivated processing: How people perceive news covering novel or contradictory health research findings. Sci. Commun. **37**(5), 602–634 (2015)
63. Dzindolet, M.T., Peterson, S.A., Pomranky, R.A., Pierce, L.G., Beck, H.P.: The role of trust in automation reliance. Int. J. Hum Comput Stud. **58**, 697–718 (2003)
64. Eshete, B.: Making machine learning trustworthy. Science **373**(6556), 743–744 (2021)

How Multi-view Techniques Can Help in Processing Uncertainty

Olga Kosheleva and Vladik Kreinovich

Abstract Multi-view techniques help us reconstruct a 3-D object and its properties from its 2-D (or even 1-D) projections. It turns out that similar techniques can be used in processing uncertainty—where many problems can reduced to a similar task of reconstructing properties of a multi-D object from its 1-D projections. In this chapter, we provide an overview of these techniques on the examples of probabilistic, interval, and fuzzy uncertainty, and of combinations of these three types of uncertainty.

Keywords Multi-view techniques · Processing uncertainty · Probabilistic uncertainty · Interval uncertainty · Fuzzy uncertainty

1 Introduction

What are multi-view techniques: a brief reminder. Our world is 3-dimensional. However, in most practical situations, we only see 2-D projections of the real-world objects, and we need to reconstruct the properties of the 3-D object based on these multi-view 2-D projections.

Because of the ubiquity of this problem, many advanced and efficient multi-view techniques have been developed.

It is advantageous to apply ideas behind multi-view techniques in other problems as well. Because this area is well advanced, it can be advantageous to use its techniques to solve other problems—problems which are less ubiquitous, more recent, and which are, therefore, somewhat behind multi-view research areas—at least in terms of the existence of efficient techniques.

O. Kosheleva · V. Kreinovich (✉)
University of Texas at El Paso, El Paso, TX, USA
e-mail: vladik@utep.edu

O. Kosheleva
e-mail: olgak@utep.edu

W. Pedrycz and S. Chen (eds.), *Recent Advancements in Multi-View Data Analytics*, Studies in Big Data 106, https://doi.org/10.1007/978-3-030-95239-6_2

What we do in this chapter. In this chapter, we show that multi-view techniques can be used in uncertainty quantification (UQ)—which, by the way, is important to multi-view analysis as well.

Specifically, we show that many instances of the uncertainty quantification problem—including estimating the standard deviation and/or the upper bound on the error of the result of data processing—can be equivalently reformulated as the problems of reconstructing the appropriate norm of a multi-dimensional vector from its projections. In this sense, the UQ problems are similar to traditional multi-view problems, where we need to reconstruct a 3-D object from its 2-D (or 1-D) projections.

Historical comment. The idea of relating projections and uncertainty quantification has been described before, see, e.g., [1] and references therein. In this chapter, we further expand on this idea and on its relation to multi-view methods.

Comment about possible future work. In effect, we show that what is useful for UQ is a simplified analog of the usual multi-view problem. This usefulness make us conjecture that more realistic multi-view methods may also be able to help with uncertainty quantification and related problems.

What is uncertainty quantification and why it is important. The main need for uncertainty quantification comes from the fact that, in general, data for processing come from measurements, and measurements are never absolutely accurate, there is always measurement error—the difference between the measurement result and the actual (unknown) value of the corresponding quantity; see, e.g., [2].

Because of this, the value that we obtain by processing measurement results is, in general, different from what we would have got if we processed the actual values of the corresponding quantities. In many practical situations, it is very important to know how big the resulting inaccuracy can be.

For example, if we are prospecting for oil, and we found out that a certain field contains 150 million tons of oil, then our actions depend on the accuracy of this estimate. If it is 150 plus minus 20, we should start exploiting this field right away. However, if it is 150 plus minus 200, maybe there is no oil at all, so we should perform more measurements before investing a lot of money in production.

Challenges of uncertainty quantification. The usual techniques for uncertainty quantification are based on the idea of sensitivity analysis: since we do not know the values of the measurement errors, we simulate different possible combinations of such errors and analyze how it affects the result of data processing. The question is what is the best way to simulate these errors and what is the best way to process the results of this simulation.

How multi-view techniques can help. It turns out that, under reasonable assumptions, the sensitivity of the data processing algorithm can be described by a multi-D vector. In this description, simulation results are 1-D projections of this vector, so the UQ problem means analyzing the property of the multi-D vector based on its 1-D projections. This problem is similar to the usual multi-view analysis.

On the one hand, the uncertainty-related problem is somewhat easier that the usual multi-view analysis, since we have 1-D (and not 2-D) projections, and since, as we will show, the object whose properties we want to reconstruct from these projections is just a vector. On the other hand, the uncertainty-related problem is somewhat more complex that the usual multi-view analysis, since the object of interest is now multi-D (and not just 3-D). We show that multi-view reformulation of UQ problems can be useful for solving these problems.

2 Need for Uncertainty Quantification

What do we want. What do we the humanity want? In a nutshell, we want to predict what will happen in the future—this is one of the main objectives of science, and we want to know what we can do to make the future better—which actions to perform, which actions to avoid, what gadgets to use and how; this is one of the main objectives of engineering.

Of course, these tasks go beyond the narrowly understood engineering. For example, when we go to a doctor because of a cough, we want to know when this cough will stop, and what we need to do to stop it faster. When we teach students, we want to know whether they will learn the required material, and if the prediction is that many of them will fail the class—how to change out teaching strategy to make sure that more students succeed.

Need for data processing. Our knowledge about the state of the world, about the states of all its objects, about the actions and gadgets—all this comes from measurements and expert estimates, and the results of these measurements and expert estimates are usually described by numbers, by the observed or estimated numerical values of the corresponding quantities: numerical values $x_1, \ldots, x_n$ describe the current state of the world, numerical values $y, \ldots,$ describe the future state of the world and the necessary actions.

In these terms, our objective is to determine all the desired values y based on the available data $x_1, \ldots, x_n$. Once the algorithm $y = f(x_1, \ldots, x_n)$ for this determination is found, the computations become straightforward: we plug in the known values x_i into this algorithm, and we get the value $y = f(x_1, \ldots, x_n)$ of the desired quantity.

This is called *data processing*. This is what computers were designed to do in the first place, this is what computers still do a lot.

Need for uncertainty quantification. The above description—in which we implicitly assumed that we know the exact values of the quantities $x_1, \ldots, x_n$ and that we know how to find the desired value y based on the known values $x_1, \ldots, x_n$—was somewhat oversimplified. In many practical situations, the available data processing algorithm provides only an approximate value of the desired quantity. It is therefore important to know how accurate is the resulting algorithm. These exist many statistical methods for estimating this accuracy; see, e.g., [3].

However, in addition to this uncertainty—which can be estimated by usual statistical techniques—there is an additional uncertainty caused by the fact that the values $\widetilde{x}_i$ that we use for data processing come either from measurements, or from expert estimates, and neither of these two procedures produces exact values. There is always a difference $\Delta x_i \stackrel{\text{def}}{=} \widetilde{x}_i - x_i$ between the estimate $\widetilde{x}_i$ and the actual (unknown) value x_i of the corresponding quantity.

Because of this difference, even in the ideal case, when the algorithm $y = f(x_1, \ldots, x_n)$ reflects the exact relation between the quantities x_i and y, the value $\widetilde{y} = f(\widetilde{x}_1, \ldots, \widetilde{x}_n)$ that we obtained by processing the estimates $\widetilde{x}_i$ is, in general, different from the value $y = f(x_1, \ldots, x_n)$ that we would have obtained if we had access to the exact value x_i. Estimating this difference

$$\Delta y = \widetilde{y} - y = f(\widetilde{x}_1, \ldots, \widetilde{x}_n) - f(x_1, \ldots, x_n)$$

is the main task of this paper.

In most practical situations, it is important to understand how big can be the corresponding difference $\Delta y = \widetilde{y} - y$. For example, if we program the trajectory of a self-driving car in a tunnel, and we conclude that in the next second, it will be at a distance $\widetilde{y} = 1$ m from the wall, then the car's reaction should depend on the accuracy of this estimate: if this is $1\,\text{m} \pm 0.5\,\text{m}$, we are safe; however, if this is $1\,\text{m} \pm 2\,\text{m}$, then we need to do something, since otherwise, the car may get too close to the wall and crash.

Finding out what values Δy are possible is known as *uncertainty quantification* (UQ). Informally, uncertainty quantification solves the following task: *given:* an algorithm $y = f(x_1, \ldots, x_n)$, the measurement results $\widetilde{x}_i$, and some information about the uncertainties $\Delta x_i = \widetilde{x}_i - x_i$, *find out* what are the possible values of the difference

$$\Delta y = f(\widetilde{x}_1, \ldots, \widetilde{x}_n) - f(x_1, \ldots, x_n).$$

3 What Makes Uncertainty Quantification Easier and What Makes It More Complex

Linearization. The uncertainty quantification problem is made easier by the fact that the estimation errors are usually small. Therefore, taking into account that $x_i = \widetilde{x}_i - \Delta x_i$, we can expand the dependence

$$\Delta y = f(\widetilde{x}_1, \ldots, \widetilde{x}_n) - f(x_1, \ldots, x_n) = f(\widetilde{x}_1, \ldots, \widetilde{x}_n) - f(\widetilde{x}_1 - \Delta x_1, \ldots, \widetilde{x}_n - \Delta x_n)$$

in Taylor series in Δx_i and ignore terms which are quadratic (or higher order) in Δx_i. As a result, we get the following formula

$$\Delta y = \sum_{i=1}^{n} c_i \cdot \Delta x_i, \tag{1}$$

where we denoted

$$c_i \stackrel{\text{def}}{=} \frac{\partial f}{\partial x_i}(\widetilde{x}_1, \ldots, \widetilde{x}_n). \tag{2}$$

The fact that we can consider linear dependence on Δx_i makes the corresponding computations easier; see, e.g., [2, 4–6].

Additional complexity. In many simple examples, we know all the steps of the algorithm, and we can use this knowledge when solving the corresponding uncertainty quantification problems. This happens when someone wrote the corresponding program "from scratch", then the lines of this program provide a clear idea of what exactly is being done.

However, for complex tasks, it is not possible for one person to write the whole code from scratch. When people write the code for solving such problems, they try to use off-the-shelf packages as much as possible. Many of these packages are proprietary, and it makes sense: to design a huge-size software, it is necessary to employ many programmers, they all need to be paid—and if everything is open access, no one will pay. This makes sense from the economic viewpoint, but from the viewpoint of uncertainty quantification, it makes life more complicated—since now the algorithm $f(x_1, \ldots, x_n)$ is largely a "black box" in the following sense: we can plug in different values x_i and get the result of applying the algorithm f, but we do not know what exactly steps this algorithm went through to produce these results.

4 How Uncertainty Quantification Is Related to Multi-view Techniques

Let us start with the traditional case of uncertainty quantification. Let us start our narrative with the traditional approach to uncertainty quantification (see, e.g., [2, 3]), where we assume that all estimation errors Δx_i are independent, and that each estimation error is normally (Gaussian) distributed with 0 mean and known standard deviation σ_i.

As we will emphasize in the following text, there are many real-life situations in which 'these assumptions are not true: e.g., when we do not know the exact values of the standard deviations and/or the probability distribution of the measurement errors is not Gaussian. In the following text, we will show how methods corresponding to the traditional case of UQ can be extended and modified to cover other possible real-life situations. But first, let us deal with the traditional case of UQ.

In this case, the expression Δy—as described by the formula (1)—is also normally distributed, since it is known that a linear combination of several independent normally distributed random variables is also normally distributed; see, e.g., [3]. The mean of this linear combination is equal to the linear combination of the means, i.e., to 0, and the variance σ^2 of the linear combination (1) is determined by the formula

$$\sigma^2 = \sum_{i=1}^{n} c_i^2 \cdot \sigma_i^2. \tag{3}$$

It is known that a normal distribution is uniquely determined by its mean and its standard deviation σ (or, equivalently, its variance σ^2). Thus, in this case, the uncertainty quantification problem is reduced to the problem of computing the expression (3).

What we can do to estimate the desired variance. Since the data processing algorithm $y = f(x_1, \ldots, x_n)$ is given as a black box, its expression is not known, we cannot differentiate the corresponding function and find the actual values c_i. What we *can* do, once we have computed the result $\widetilde{y} = f(\widetilde{x}_1, \ldots, \widetilde{x}_n)$ of data processing, is to try different tuples $(\Delta x_1, \ldots, \Delta x_n)$; for each such tuple:

- first, we plug in the values $x_i = \widetilde{x}_i - \Delta x_i$ into the algorithm $f(x_1, \ldots, x_n)$, resulting in the value $y = f(\widetilde{x}_1 - \Delta x_1, \ldots, \widetilde{x}_n - \Delta x_n)$, and
- then, we compute the difference $\Delta y = \widetilde{y} - y$ (and we know that this difference is equal to $\sum_{i=1}^{n} c_i \cdot \Delta x_i$).

A typical way to select the values Δx_i is to use Monte-Carlo simulations, i.e., to select the values Δx_i which are normally distributed with zero mean and standard deviation σ_i. Usually, programming languages and simulation packages have methods for simulating the "standard" normal distribution, with 0 mean and standard deviation 1. The resulting desired random variable can be obtained from the result ξ_i of the standard normal random number generator if we multiply this result by σ_i.

From this viewpoint, instead of directly generating the values Δx_i, it makes sense to first generate the values ξ_i, and then take $\Delta x_i = \sigma_i \cdot \xi_i$. In terms of ξ_i, the procedure of generating the corresponding value Δy takes the following form:

- first generate the values ξ_i;
- then, we plug in the values $x_i = \widetilde{x}_i - \sigma_i \cdot \xi_i$ into the algorithm $f(x_1, \ldots, x_n)$, resulting in the value $y = f(\widetilde{x}_1 - \sigma_1 \cdot \xi_1, \ldots, \widetilde{x}_n - \sigma_n \cdot \xi_n)$, and
- finally, we compute the difference $\Delta y = \widetilde{y} - y$ (which we know to be equal to $\sum_{i=1}^{n} c_i \cdot \sigma_i \cdot \xi_i$).

In terms of the auxiliary quantities ξ_i, the expression (1) takes the form

$$\Delta y = \sum_{i=1}^{n} c_i \cdot \sigma_i \cdot \xi_i. \tag{4}$$

This expression can be simplified if we denote $a_i \stackrel{\text{def}}{=} c_i \cdot \sigma_i$, then the expression (4) takes the form

$$\Delta y = \sum_{i=1}^{n} a_i \cdot \xi_i. \tag{5}$$

Interestingly, in terms of a_i, the desired expression (3) also gets a simplified form:

$$\sigma^2 = \sum_{i=1}^{n} a_i^2. \tag{6}$$

Now, we are ready to describe the relation to multi-view techniques.

Geometric formulation of the problem and its relation to multi-view techniques. In geometric terms, the values a_i form a vector $\vec{a} = (a_1, \ldots, a_n)$, and the values ξ_i form a vector $\vec{\xi} = (\xi_1, \ldots, \xi_n)$.

In terms of these vectors, the value Δy—as described by the expression (5)—is simply a scalar (dot) product $\vec{a} \cdot \vec{\xi}$ of these two vectors, and the value σ^2—as describes by the formula (3)—is simply the square $\|\vec{a}\|^2$ of the length $\|\vec{a}\| = \sqrt{\sum_{i=1}^{n} a_i^2}$ of the vector $\vec{a}$. So, the standard deviation $\sigma = \sqrt{\sigma^2}$ is simply equal to the length $\|\vec{a}\|$ of the vector $\vec{a}$.

Thus, in these geometric terms, the newly reformulated problem takes the following form: there is a vector $\vec{a} = (a_1, \ldots, a_n)$ that we do not know. We want to find the length $\|\vec{a}\|$ of this vector. For this purpose, for different vectors $\vec{\xi}$, we can compute the scalar product $\vec{a} \cdot \vec{\xi}$.

Knowing the scalar product is equivalent to knowing the projection

$$\pi_{\vec{\xi}}(\vec{a}) = \frac{\vec{a} \cdot \vec{\xi}}{\|\vec{\xi}\|} \tag{7}$$

of the unknown vector $\vec{a}$ on the 1-D space generated by the vector $\vec{\xi}$. Thus, the problem takes the following form: we want to estimate some characteristic of the unknown multi-D object $\vec{a}$ by studying its projection on different 1-D spaces. In this form, this is clearly a particular cases of the general multi-view reconstruction problem. The main difference from the usual cases of the multi-view problem is that usually, we reconstruct a 3-D object from its 2-D projections, but now, we reconstruct a multi-D object from its 1-D projections.

Comment. We explained the relation between uncertainty quantification and multi-view techniques on the example of the traditional approach to uncertainty quantification; however, as we will see in this paper, the same relation can be traced for all other types of uncertainty.

5 Straightforward ("Naive") Approach and Its Limitations

Straightforward approach: main idea. In terms of the multi-view reformulation of our problem, our goal is to find the length of the vector $\vec{a}$. According to the formula for the length, a straightforward way to compute this length is to find all the components $a_1, \ldots, a_n$ of this vector, and then, to compute $\|\vec{a}\|$ as $\sqrt{\sum_{i=1}^{n} a_i^2}$.

A straightforward way to compute each component a_i is to take into account that this component is simply a scalar product of the vector $\vec{a}$ and the vector $\vec{e}^{\,(i)} = (0, \ldots, 0, 1, 0, \ldots, 0)$ that has all components equal to 0 with the exception of the i-th component $e_i^{(i)}$ which is equal to 1: $a_i = \vec{a} \cdot \vec{e}^{\,(i)}$.

Thus, we arrive at the following straightforward algorithm for computing the desired length $\|\vec{a}\|$—and thus, for solving the corresponding uncertainty quantification problem.

Resulting algorithm.

- for $i = 1, \ldots, n$, we take $\vec{\xi} = \vec{e}^{\,(i)}$ and compute the corresponding value Δy; we will denote the resulting value of Δy by a_i;
- then, we compute the desired length as $\|\vec{a}\| = \sqrt{\sum_{i=1}^{n} a_i^2}$.

Resulting algorithm: a detailed description. If we explicitly describe how Δy is computed, we arrive at the following detailed description of the above algorithm:

- for each i from 1 to n, we prepare the values $\xi_1^{(i)}, \ldots, \xi_n^{(i)}$ for which $\xi_i^{(i)} = 1$ and $\xi_j^{(i)} = 0$ for all $j \neq i$;
- then, we apply the algorithm f to the values

$$x_1 = \widetilde{x}_1 - \sigma_1 \cdot \xi_1^{(i)}, \ldots, x_n = \widetilde{x}_n - \sigma_n \cdot \xi_n^{(i)},$$

  thus computing the value $y^{(i)} = f(\widetilde{x}_1, \ldots, \widetilde{x}_{i-1}, \widetilde{x}_i - \sigma_i, \widetilde{x}_{i+1}, \ldots, \widetilde{x}_n)$, and compute $a_i = \widetilde{y} - y^{(i)}$;
- finally, we compute $\sigma = \sqrt{\sum_{i=1}^{n} a_i^2}$.

Limitations of the straightforward approach. The above straightforward algorithm requires n computations of the corresponding quantity Δy. As we have mentioned earlier, each of these computations means applying the algorithm $f(x_1, \ldots, x_n)$ to appropriate values x_i. Thus, in addition to the original call to the algorithm f—to compute the original result $\widetilde{y} = f(\widetilde{x}_1, \ldots, \widetilde{x}_n)$, we need to call this algorithm n more times.

We have also mentioned that in many practical situations, this algorithm is complicated, each call may require hours on a high performance computer, and the number of inputs n may be huge—for problems like weather predictions, we process thousands of inputs.

Thus, using the straightforward approach would mean spending thousands times more computation time that the actual computation—months instead of hours. This is clearly not realistic. That is why the above straightforward approach is sometimes called “naive” approach.

6 Monte-Carlo Approach: Traditional Probabilistic Case

Main idea: reminder. If the values ξ_i are independent normally distributed random variables with 0 means and standard deviations 1, then their linear combination $\vec{a} \cdot \vec{\xi} = \sum_{i=1}^{n} a_i \cdot \xi_i$ is also normally distributed, with 0 mean and standard deviation $\sigma = \sqrt{\sum_{i=1}^{n} a_i^2}$.

So, if we simply simulate all these ξ_i, then the resulting value $\Delta y = \vec{a} \cdot \vec{\xi}$ will be normally distributed with mean 0 and standard deviation equal to the desired value $\|\vec{a}\|$. If we repeat this simulation several (N) times, we get a sample of values $\Delta y^{(k)}$, based on which we can compute the sample standard deviation

$$\sqrt{\frac{1}{N} \cdot \sum_{k=1}^{N} \left(\Delta y^{(k)}\right)^2}.$$

Comment. Note that the formula we use is different from the usual statistical method for estimating standard deviation based on the sample:

$$\sqrt{\frac{1}{N-1} \cdot \sum_{k=1}^{N} \left(\Delta y^{(k)} - a\right)^2}, \text{ where } a = \frac{1}{N} \cdot \sum_{k=1}^{N} \Delta y^{(k)}.$$

The usual formula assumes that we do not know the actual mean, so we subtract the sample mean a from all the sample values $\Delta y^{(k)}$. In this case, to avoid the bias, we need to divide the sum of the squares of the differences by $N - 1$. In our case, we know that the mean is 0. So, instead of subtracting the sample mean—which is only approximately equal to the actual mean—we can subtract the actual mean, i.e., 0. In this case, the expected value of each term $\left(\Delta y^{(k)}\right)^2$ is exactly σ^2, and thus, the expected value of the sum $\sum_{k=1}^{N} \left(\Delta y^{(k)}\right)^2$ is exactly $N \cdot \sigma^2$. Thus, to get the unbiased estimate for σ^2—i.e., an estimate whose expected value is exactly σ^2—we need to divide this sum by N—and not by $N - 1$ as in the usual estimation.

It is known (see, e.g., [3]), that the accuracy with which the sample standard deviation approximates the actual one is proportional to $1/\sqrt{N}$. So, e.g., if we want to compute the desired value σ with the relative accuracy of 20%, it is sufficient to run just $N = 25$ simulations—since in this case, $1/\sqrt{N} = 20\%$. This is clearly much faster than thousands of calls to f needed for the straightforward approach.

By the way, 20% relative accuracy in determining the uncertainty is very good—it means, e.g., distinguishing between 10% and 12% accuracy. Usually, an even lower accuracy is sufficient: we say that the accuracy is $\pm 10\%$ or $\pm 15\%$, but rarely $\pm 12\%$.

Let us describe the above idea in precise terms.

Resulting algorithm. For each $k = 1, \ldots, N$ (where N is determined by the desired relative accuracy), we do the following:

- first, for each i from 1 to n, we use the random number generator (generating standard normal distribution), and get the corresponding values $\xi_i^{(k)}$;
- then, we use the resulting vector $\vec{\xi}^{(k)}$ to compute the corresponding value $\Delta y^{(k)}$.

Once we have all these values, we compute $\|\vec{a}\| = \sqrt{\frac{1}{N} \cdot \sum_{k=1}^{N} \left(\Delta y^{(k)}\right)^2}$.

Let us describe this idea in detail.

7 Algorithm for the Probabilistic Case

What is given. We are *given* the values $\widetilde{x}_1, \ldots, \widetilde{x}_n$, the standard deviations $\sigma_1, \ldots, \sigma_n$, and an algorithm $f(x_1, \ldots, x_n)$ given as a black box. We also know the value $\widetilde{y} = f(\widetilde{x}_1, \ldots, \widetilde{x}_n)$.

What we want. Our *goal* is to compute the standard deviation σ of the difference $\Delta y = \widetilde{y} - f(\widetilde{x}_1 - \Delta x_1, \ldots, \widetilde{x}_n - \Delta x_n)$, where Δx_i are independent random variables with 0 mean and standard deviation σ_i.

Algorithm. First, we select the number of iterations N based on the desired relative accuracy $\varepsilon > 0$: we want $1/\sqrt{N} \approx \varepsilon$, so we take $N \approx \varepsilon^{-2}$.

Then, for each $k = 1, \ldots, N$, we do the following:

- first, for each i from 1 to n, we use the random number generator (generating standard normal distribution), and get the corresponding values $\xi_i^{(k)}$;
- then, we plug in the values $x_i = \widetilde{x}_i - \sigma_i \cdot \xi_i^{(k)}$ into the algorithm f, thus computing the difference

$$\Delta y^{(k)} = \widetilde{y} - f\left(\widetilde{x}_1 - \sigma_1 \cdot \xi_1^{(k)}, \ldots, \widetilde{x}_n - \sigma_n \cdot \xi_n^{(k)}\right);$$

- once we have all these values, we compute $\sigma = \sqrt{\frac{1}{N} \cdot \sum_{k=1}^{N} \left(\Delta y^{(k)}\right)^2}$.

8 Need to Go Beyond the Traditional Probabilistic Case

Traditional probabilistic case: reminder. The above algorithms are intended for the case when we know the probability distributions of all the approximation errors Δx_i, all these distributions are normal, with 0 mean and known standard deviation σ_i, and all approximation errors are independent.

This case does frequently occur in practice. In practice, there are indeed many situations where all these three conditions are satisfied.

Indeed, factors affecting the inaccuracy of different measurements—such as noise at the location of different sensors—are indeed reasonably independent.

Also, if we know the probability distributions, then the requirement that the mean is 0 makes sense. Indeed, if the mean is not 0, i.e., in probabilistic terms, if we have a *bias*, then we can simply subtract this bias from all the measurement results and thus, end up with measurements for which the mean error is 0.

Even normality makes sense. Indeed, usually, each measurement error is the joint effect of many different independent factors, and, according to the Central Limit Theorem (see, e.g., [3]), the distribution of such a joint effect is close to normal. Indeed, empirical studies [7, 8] show that for about 60% measuring instruments, the probability distribution of measurement errors is close to normal.

Need to go beyond normal distributions. On the other hand, the same statistics means that for about 40% of the measuring instruments the probability distribution is *not* close to normal.

So, the first thing we need to do is to extend out methods to this case—when we *know* the probability distributions, and we know that at least some of them are not normal.

But do we always know the probability distributions? To determine the probability distribution of the measurement error, we need to thoroughly study and test each individual sensor, each individual measuring instrument. Such a testing involves comparing the results of these measurements with the results of some much more accurate ("standard") measuring instrument. This is usually possible, but it is a very expensive procedure—especially taking into account that many sensors are now very cheap.

This procedure makes sense if we are planning a manned space flight or a control system for a nuclear power station, where a wrong decision can lead to catastrophic consequences. However, if, for the purpose of weather prediction, we design a network of reasonably cheap temperature, wind, and humidity sensors, the corresponding expenses are not worth the effort.

What do we know: possible information about uncertainty. Since we do not always perform a thorough study of each sensor, in many situations, instead of knowing the exact probability distribution, we only have partial information about the corresponding probabilities.

In all cases, we should know a guaranteed upper bound Δ on the absolute value $|\Delta x|$ of the measurement error: $|\Delta x| \leq \Delta$. Such an upper bound is needed, since otherwise, if there is no guaranteed upper bound, this would mean that the actual value can be as far from the measurement result as mathematically possible: this is not a measurement, this is a wild guess.

Once we know the upper bound Δ, then, once we have the measurement result $\widetilde{x}$, the only thing that we can conclude about the actual (unknown) value x is that this value belongs to the interval $[\widetilde{x} - \Delta, \widetilde{x} + \Delta]$, and we have no information about the probability of different values from this interval. This case is known as *interval uncertainty*; see, e.g., [2, 9–11].

In some cases, we have partial information about the probabilities: e.g., we know the upper bound $\widetilde{\Delta}$ on the absolute value of the mean $E[\Delta x]$—this mean is known as the *systematic error component*, and we know the upper bound $\widetilde{\sigma}$ on the standard deviation, i.e., the mean square value of the difference $\Delta x - E[\Delta x]$ between the measurement error and its mean; this difference is known as the *random error component*. This is probably the most frequent type of information about the measurement accuracy [2].

There is also a case when some (or even all) estimates $\widetilde{x}_i$ come not from measurements, but from an expert estimates. In this case, the only information that we have about the accuracy of this estimate also comes from the expert, and the expert describes this information in imprecise ("fuzzy") terms of natural language, e.g., by saying that the accuracy is "most probably plus minus 0.5". Such situations are known as situations of *fuzzy uncertainty*; see, e.g., [12–17].

Finally, there are cases when for some measurements, we know the probability distributions of the measurement errors, for some other measurements, we only know interval bounds, and for yet other values, we only have fuzzy descriptions of their accuracy.

In this chapter, we show that in all these cases, multi-view ideas can be helpful. Let us describe how these ideas can be helpful—by considering all different types of uncertainty in the same order in which we have just listed them.

9 Case When We Know the Probability Distributions but They Are Not Necessarily Normal

Description of the case. In this section, we consider the case when we know the probability distributions of all the measurement errors Δx_i, when for each of these distributions, the mean value is 0, and when measurement errors corresponding to different measurements are independent.

Analysis of the problem. For large n, the value Δy is the sum of a large number small independent random variables $c_i \cdot \Delta x_i$. Thus, due to the same Central Limit Theorem that explains the ubiquity of normal distributions, the distribution of Δy is close to normal.

As we have mentioned earlier, a 1-D normal distribution is uniquely determined by its mean and standard deviation. Similarly to the normal cases, we can conclude that the mean is 0, and that the standard deviation is determined by the formula (3).

Thus, to compute σ, we can simply ignore all the information about the known distributions and only take into account the variances σ_i. Once we have found these values, we get the exact same mathematical problem as in the normal case—and the exact same multi-view reformulation of this problem. Thus, we can follow the same algorithm as for the normal case.

In other words, we arrive at the following algorithm.

Resulting algorithm: detailed description. First, for each i, we use our knowledge of the probability distribution of the corresponding measurement error Δx_i to compute the corresponding standard deviation σ_i.

Then, for each $k = 1, \dots, N$ (where N is determined by the desired relative accuracy), we do the following:

- first, for each i from 1 to n, we use the random number generator generating standard normal distribution, and get the corresponding values $\xi_i^{(k)}$;
- then, we plug in the values $x_i = \widetilde{x}_i - \sigma_i \cdot \xi_i^{(k)}$ into the algorithm f, thus computing the difference

$$\Delta y^{(k)} = \widetilde{y} - f\left(\widetilde{x}_1 - \sigma_1 \cdot \xi_1^{(k)}, \dots, \widetilde{x}_n - \sigma_n \cdot \xi_n^{(k)}\right);$$

- once we have all these values, we compute $\sigma = \sqrt{\frac{1}{N} \cdot \sum_{k=1}^{N} \left(\Delta y^{(k)}\right)^2}$.

Important comment. In the algorithm for the normal case, we adequately simulated the measurement errors, by emulating the exact same distribution as we know they have. In this case, however, the actual distributions are *not* normal, but we still use normal distributions.

In other words, in this case, the Monte-Carlo method that we use is not a realistic simulation of measurement errors, it is a computational trick which leads to the same result as the actual simulation but which is computationally much faster.

This trick makes computations faster since, with this trick, there is no need to spend computation time emulating details of complex distributions: we can simply use standard (and fast) random number generator for normally distributed variables.

We make this comment because we will observe the same phenomenon in other algorithms as well: many of our Monte-Carlo simulations will *not* be adequately representing the actual distributions, they will serve as mathematical tricks helping us to compute the desired solution.

A reader should not be surprised by this: our main idea—reduction to a multi-view problem—is also, in effect, a mathematical trick and not a direct representation of the corresponding uncertainty quantification problem.

10 Case of Interval Uncertainty

Formulation of the problem. We know that the measurement errors Δx_i can take any values from the interval $[-\Delta_i, \Delta_i]$. Under this condition, we need to find the range

$$\left\{\Delta y = \sum_{i=1}^{n} c_i \cdot \Delta x_i : |\Delta x_i| \le \Delta_i \text{ for all } i\right\}$$

of possible values of their linear combination $\Delta y = \sum\limits_{i=1}^{n} c_i \cdot \Delta x_i$.

Analysis of the problem. The largest possible value of the sum is attained when each of the terms $c_i \cdot \Delta x_i$ is the largest. Let us consider two possible cases: $c_i \ge 0$ and $c_i \le 0$.

If $c_i \ge 0$, then $c_i \cdot \Delta x_i$ is an increasing function of Δx_i. Thus, its largest value is attained when the variable Δx_i is the largest possible, i.e., when $\Delta x_i = \Delta_i$. For this value Δx_i, the term is equal to $c_i \cdot \Delta_i$.

If $c_i \le 0$, then $c_i \cdot \Delta x_i$ is a decreasing function of Δx_i. Thus, its largest value is attained when the variable Δx_i is the smallest possible, i.e., when $\Delta x_i = -\Delta_i$. For this value Δx_i, the term is equal to $c_i \cdot (-\Delta_i) = -c_i \cdot \Delta_i$.

In both cases, the largest possible value of each term is equal to $|c_i| \cdot \Delta_i$. Thus, the largest possible value Δ of Δy is equal to the sum of these values

$$\Delta = \sum_{i=1}^{n} |c_i| \cdot \Delta_i. \tag{8}$$

Similarly, we can prove that the smallest possible value of Δy is equal to $-\Delta$.

Thus, in the interval case, uncertainty quantification is reduced to the problem of computing the value (8).

What we can do to estimate Δ. Similarly to the probabilistic case, since the data processing algorithm $y = f(x_1, \ldots, x_n)$ is given as a black box, the only thing we can do is to try different tuples $(\Delta x_1, \ldots, \Delta x_n)$; for each such tuple:

- first, we plug in the values $x_i = \widetilde{x}_i - \Delta x_i$ into the algorithm $f(x_1, \ldots, x_n)$, resulting in the value $y = f(\widetilde{x}_1 - \Delta x_1, \ldots, \widetilde{x}_n - \Delta x_n)$, and
- then, we compute the difference $\Delta y = \widetilde{y} - y$ (and we know that this difference is equal to $\sum\limits_{i=1}^{n} c_i \cdot \Delta x_i$).

Reformulating the problem. Similarly to the probabilistic case, instead of directly generating the value Δx_i, let us generate the auxiliary values ξ_i, and then take $\Delta x_i = \Delta_i \cdot \xi_i$.

In terms of ξ_i, the procedure of generating the corresponding value Δy takes the following form:

- first generate the values ξ_i;
- then, we plug in the values $x_i = \widetilde{x}_i - \Delta_i \cdot \xi_i$ into the algorithm $f(x_1, \ldots, x_n)$, resulting in the value $y = f(\widetilde{x}_1 - \Delta_1 \cdot \xi_1, \ldots, \widetilde{x}_n - \Delta_n \cdot \xi_n)$, and
- finally, we compute the difference $\Delta y = \widetilde{y} - y$ (which we know to be equal to $\sum\limits_{i=1}^{n} c_i \cdot \Delta_i \cdot \xi_i$).

In terms of the auxiliary quantities ξ_i, the expression (1) takes the form

$$\Delta y = \sum_{i=1}^{n} c_i \cdot \Delta_i \cdot \xi_i. \tag{9}$$

This expression can be simplified if we denote $a_i \stackrel{\text{def}}{=} c_i \cdot \Delta_i$, then the expression (9) takes the form

$$\Delta y = \sum_{i=1}^{n} a_i \cdot \xi_i. \tag{10}$$

Interestingly, in terms of a_i, the desired expression (8) also gets a simplified form:

$$\Delta = \sum_{i=1}^{n} |a_i|. \tag{11}$$

Now, we are ready to describe the relation to multi-view techniques.

Relation to multi-view techniques. Similarly to the probabilistic case, let us consider the two vectors: a vector $\vec{a} = (a_1, \ldots, a_n)$ formed by the values a_i, and a vector $\vec{\xi} = (\xi_1, \ldots, \xi_n)$ formed by the values ξ_i. In terms of these vectors, the value Δy, as we have mentioned earlier, is simply a scalar (dot) product $\vec{a} \cdot \vec{\xi}$ of these two vectors, and the value Δ—as describes by the formula (11)—is simply the ℓ^1-norm $\|\vec{a}\|_1 \stackrel{\text{def}}{=} \sum\limits_{i=1}^{n} |a_i|$ of the vector $\vec{a}$.

Thus, in these geometric terms, the newly reformulated problem takes the following: there is a vector $\vec{a} = (a_1, \ldots, a_n)$ that we do not know, and we want to find the ℓ^1-norm $\|\vec{a}\|_1$ of this vector. For this purpose, for different vectors $\vec{\xi}$, we can compute the scalar product $\vec{a} \cdot \vec{\xi}$.

Similarly to the probabilistic case, knowing the scalar product is equivalent to knowing the projection

$$\pi_{\vec{\xi}}(\vec{a}) = \frac{\vec{a} \cdot \vec{\xi}}{\|\vec{\xi}\|}$$

of the unknown vector $\vec{a}$ on the 1-D space generated by the vector $\vec{\xi}$. Thus, the problem takes the following form: we want to estimate some characteristic of the unknown multi-D object $\vec{a}$ by studying its projection on different 1-D spaces. In this form, this is clearly a particular cases of the general multi-view reconstruction problem.

Straightforward algorithm. In principle, similarly to the probabilistic case, we can use the following straightforward algorithm to compute the desired value Δ:

- for each i from 1 to n, we prepare the values $\xi_1^{(i)}, \ldots, \xi_n^{(i)}$ for which $\xi_i^{(i)} = 1$ and $\xi_j^{(i)} = 0$ for all $j \neq i$;
- then, we apply the algorithm f to the values

$$x_1 = \widetilde{x}_1 - \Delta_1 \cdot \xi_1^{(i)}, \ldots, x_n = \widetilde{x}_n - \Delta_n \cdot \xi_n^{(i)},$$

  thus computing the value $y^{(i)} = f(\widetilde{x}_1, \ldots, \widetilde{x}_{i-1}, \widetilde{x}_i - \Delta_i, \widetilde{x}_{i+1}, \ldots, \widetilde{x}_n)$, and compute $a_i = \widetilde{y} - y^{(i)}$;
- finally, we compute $\Delta = \sum_{i=1}^{n} |a_i|$.

However, this algorithm has the same main limitation as in the probabilistic case: it requires too much computation time.

Monte-Carlo algorithm for interval computations: towards the main idea. As we have mentioned, for the situation when measurement errors $\Delta x_i = \widetilde{x}_i - x_i$ are independent and have mean 0 and standard deviation σ_i, the standard deviation σ of the measurement error

$$\Delta y = \widetilde{y} - f(\widetilde{x}_1 - \Delta x_1, \ldots, \widetilde{x}_n - \Delta x_n) = \sum_{i=1}^{n} a_i \cdot \Delta x_i$$

of the result of data processing is equal to $\sigma = \sqrt{\sum_{i=1}^{n} a_i^2 \cdot \sigma_i^2}$.

In general, in simulations, we:

- simulate independent random values ξ_i,
- subtract the simulated values ξ_i from the measurement results $\widetilde{x}_i$, plug in the differences into the data processing algorithm $f(x_1, \ldots, x_n)$, and compute the difference

$$\xi = \widetilde{y} - f(\widetilde{x}_1 - \xi_1, \ldots, \widetilde{x}_n - \xi_n) = \sum_{i=1}^{n} c_i \cdot \xi_i.$$

For normal distributions, if each ξ_i is normally distributed with 0 mean and standard deviation σ_i, then their linear combination ξ is also normally distributed, with 0 mean and the desired standard deviation σ. Thus, to estimate σ, it makes sense:

- to simulate ξ_i as having normal distribution with 0 mean and standard deviation σ_i—this can be done by multiplying σ_i and the random variable r_i normally distributed with 0 means and standard deviation 1,

- to subtract the simulated values ξ_i from the measurement results $\widetilde{x}_i$, plug in the differences into the data processing algorithm $f(x_1, \ldots, x_n)$, and compute the difference $\xi = \sum\limits_{i=1}^{n} a_i \cdot \xi_i$.

This way, we get a sample of the values ξ normally distributed with 0 mean and standard deviation σ. We can then use the (slightly modified) usual statistical formulas to estimate the standard deviation based on this sample.

In the interval case, the uncertainty of inputs is characterized by the upper bounds Δ_i on the absolute values of the measurement errors. As we have mentioned, the corresponding upper bound Δ for $|\Delta y|$ is equal to $\Delta = \sum\limits_{i=1}^{n} |c_i| \cdot \Delta_i$. To speed up computation of this value, it is therefore desirable to find a probability distribution for which, if we have random variables ξ_i distributed with parameters Δ_i—i.e., obtained by multiplying some "standard" random variable by Δ_i—then the linear combination $\sum c_i \cdot \xi_i$ should be distributed according to same law, but with the parameter $\Delta = \sum\limits_{i=1}^{n} |c_i| \cdot \Delta_i$. If we find such a distribution, then, to estimate Δ, it makes sense:

- to simulate ξ_i—this can be done by multiplying Δ_i and the appropriately distributed "standard" random variable,
- to subtract the simulated values ξ_i from the measurement results $\widetilde{x}_i$, plug in the differences into the data processing algorithm $f(x_1, \ldots, x_n)$, and compute the difference $\xi = \sum\limits_{i=1}^{n} c_i \cdot \xi_i$.

This way, we get a sample of the values ξ distributed with parameter Δ. We can then use the standard statistical techniques to estimate this parameter based on the sample values.

Good news is that such a distribution exists: namely, the Cauchy distribution has the desired property. In general, Cauchy distribution is characterised by the following probability density function:

$$f(x) = \frac{1}{\pi \cdot \Delta} \cdot \frac{1}{1 + \dfrac{(x-a)^2}{\Delta^2}}, \tag{12a}$$

where a is known as the *location parameter* and Δ is known as the *dispersion parameter*. The standard case is when $a = 0$ and $\Delta = 1$.

In our algorithm, we will only use the case when $a = 0$, i.e., when

$$f(x) = \frac{1}{\pi \cdot \Delta} \cdot \frac{1}{1 + \dfrac{x^2}{\Delta^2}}. \tag{12b}$$

Thus distributed random variable can be obtained if we multiply Δ and a random variable r_i which is Cauchy distributed with $\Delta = 1$.

Comment. For Gaussian distribution, its dispersion parameter—standard deviation—is defined by the distribution's moments. However, the Cauchy distribution has no defined moments—this is how this distribution is often presented in statistics textbooks, as a distribution for which the standard deviation is infinite. As a result, in contrast to the Gaussian distribution, the dispersion parameter Δ is not based on moments.

Why Cauchy distribution. Let us show that Cauchy distribution has the desired property—and that it is the only distribution that has this property. (Readers interested only in the resulting algorithm can skip this section.)

Indeed, each random variable r with probability density function $f(x)$ is uniquely determined by its Fourier transform—known as *characteristic function*:

$$\chi_r(\omega) = E_r[\exp(\mathrm{i} \cdot \omega \cdot r)] = \int f(x) \cdot \exp(\mathrm{i} \cdot \omega \cdot x)\, dx,$$

where we denoted $\mathrm{i} \stackrel{\text{def}}{=} \sqrt{-1}$.

For any constant c, for the variable $c \cdot r$, the characteristic function is equal to

$$\chi_{c \cdot r}(\omega) = E_r[\exp(\mathrm{i} \cdot \omega \cdot (c \cdot r))] = E_r[\exp(\mathrm{i} \cdot (\omega \cdot c) \cdot r)] = \chi_r(c \cdot \omega).$$

In particular, since $\xi_i = \Delta_i \cdot r_i$, we have $a_i \cdot \xi_i = (a_i \cdot \Delta_i) \cdot r_i$ and thus,

$$\chi_{a_i \cdot \xi_i}(\omega) = \chi_r((a_i \cdot \Delta_i) \cdot \omega).$$

The variables $a_i \cdot \xi_i$ are independent, and for the sum $v + v'$ of independent random variables, the characteristic function is equal to the product of the corresponding characteristic functions:

$$\chi_{v+v'}(\omega) = E[\exp(\mathrm{i} \cdot \omega \cdot (v + v'))] = E[\exp(\mathrm{i} \cdot \omega \cdot v) \cdot \exp(\mathrm{i} \cdot \omega \cdot v')] =$$

$$E[\exp(\mathrm{i} \cdot \omega \cdot v)] \cdot E[\exp(\mathrm{i} \cdot \omega \cdot v')] = \chi_v(\omega) \cdot \chi_{v'}(\omega).$$

Thus, the desired property—that the linear combination $\sum c_i \cdot \xi_i$ has the distribution with parameter $\Delta = \sum |c_i| \cdot \Delta_i$—when described as equality of characteristic functions, takes the following form:

$$\prod_{i=1}^{n} \chi_r(c_i \cdot \Delta_i \cdot \omega) = \chi_r\left(\omega \cdot \sum_{i=1}^{n} |c_i| \cdot \Delta_i\right).$$

In particular, for $n = 1$, $c_1 = -1$, and $\Delta_1 = 1$, we get $\chi_i(-\omega) = \chi_r(\omega)$, i.e., $\chi_r(\omega) = \chi_r(|\omega|)$.

For $n = 2$, $\omega = \Delta_1 = \Delta_2 = 1$, and $c_i > 0$, we get $\chi_r(c_1 + c_2) = \chi_r(c_1) \cdot \chi_r(c_2)$. It is known that every measurable function with this property has the form $\chi_r(\omega) = \exp(-k \cdot \omega)$ for $\omega > 0$; see, e.g., [18]. Thus, for all ω, we have $\chi_r(\omega) = \exp(-k \cdot |\omega|)$. By applying the inverse Fourier transform to this characteristic function, we get exactly the Cauchy distribution. So, we have proven that if a distribution has the desired property, then it is the Cauchy distribution.

It is also easy to see that for the function $\chi_r(\omega) = \exp(-k \cdot |\omega|)$, the above equality holds—and thus, the Cauchy distribution indeed has the desired property.

Resulting fact about Cauchy distributions. The above result about Cauchy distribution—the result that we will use in our estimation—is that if n independent random variables $\xi_1, \ldots, \xi_n$ are distributed according to the standard Cauchy distribution, then their linear combination $\sum\limits_{i=1}^{n} a_i \cdot \xi_i$ is distributed according to the Cauchy distribution with parameter $\Delta = \sum\limits_{i=1}^{n} |a_i|$; see, e.g., [19].

Monte-Carlo algorithm: main idea. Thus, similarly to how we use normal distributions to estimate σ in the probabilistic case, we can use Cauchy distribution in the interval case.

There are two additional computational problems here that we did not encounter in the probabilistic case. First, in contrast to the probabilistic case, we do not have a ready random number generator generating Cauchy distribution; this problem can be easily solved: we can take a random variable u uniformly distributed on the interval $[0, 1]$ (for which the random number generator exists), and take

$$\xi = \tan(\pi \cdot (u - 0.5)); \tag{13}$$

Second, once we get a Cauchy distributed sample of values $\Delta y^{(k)}$, there is no standard way to estimate the parameter Δ; for this purpose, we can use the Maximum Likelihood approach of finding the most probable value Δ; this leads to the need to solve the following system of equations:

$$\sum_{k=1}^{N} \frac{1}{1 + \frac{\left(\Delta y^{(k)}\right)^2}{\Delta^2}} - \frac{N}{2} = 0. \tag{14}$$

As a result, we arrive at the following algorithm (see, e.g., [1, 19–25] for details):

Resulting algorithm: first approximation. For each $k = 1, \ldots, N$ (where N is determined by the desired relative accuracy), we do the following:

- first, for each i from 1 to n, we use the formula (13) and get the corresponding values $\xi_i^{(k)}$;
- then, we use the resulting vector $\vec{\xi}^{(k)}$ to compute the corresponding value $\Delta y^{(k)}$.

Once we have all these values, we compute Δ by solving the Eq. (14).

First approximation: detailed description. For each $k = 1, \ldots, N$ (where N is determined by the desired relative accuracy), we do the following:

- first, for each i from 1 to n, we use the random number generator for generating a random number $u_i^{(k)}$ which is uniformly distributed on the interval [0, 1], and then compute $\xi_i^{(k)} = \tan\left(\pi \cdot \left(u_i^{(k)} - 0.5\right)\right)$;
- then, we plug in the values $x_i = \widetilde{x}_i - \Delta_i \cdot \xi_i^{(k)}$ into the algorithm f, thus computing the difference

$$\Delta y^{(k)} = \widetilde{y} - f\left(\widetilde{x}_1 - \Delta_1 \cdot \xi_1^{(k)}, \ldots, \widetilde{x}_n - \Delta_n \cdot \xi_n^{(k)}\right);$$

- once we have all these values, we compute Δ by solving the Eq. (14).

Additional details: how to solve the Eq. (14). To find Δ, we can use, e.g., a bisection algorithm starting with the interval $\left[0, \max\limits_{k=1,\ldots,N} \left|\Delta y^{(k)}\right|\right]$. For the value Δ corresponding to the left endpoint of this interval, the left-hand side of the Eq. (14) is negative, while for the right endpoint, the left-hand side is positive. At each step, we take a midpoint of the previous interval, and select either the left or the right half-interval, so that in the endpoints of the selected half-interval, the left-hand side of (14) has different signs.

Important computational comment. Cauchy distributed values Δx_i can be large, while linearization is only possible for small deviations Δx_i. To make sure that all deviations are within the linearity range, we need to normalize all the simulated measurement errors by dividing them by a sufficiently large value M, and then to re-scale the resulting values $\Delta y^{(k)}$ back, by multiplying them by the same value M.

A natural way to make sure that all simulated values Δx_i are within the range $[-\Delta_i, \Delta_i]$—or, equivalently, that all the values ξ are within the interval $[-1, 1]$—is to divide all the simulated values $\xi_i^{(k}$ by the largest of their absolute values.

As a result, the actual algorithm has the following modified form:

11 Final Algorithm for the Interval Case

What is given. We are *given* the values $\widetilde{x}_1, \ldots, \widetilde{x}_n$, the values $\Delta_1, \ldots, \Delta_n$, and an algorithm $f(x_1, \ldots, x_n)$ given as a black box. We also know the value $\widetilde{y} = f(\widetilde{x}_1, \ldots, \widetilde{x}_n)$.

What we want. Our *goal* is to compute the range

$$\{y = f(x_1, \ldots, x_n) : x_i \in [\widetilde{x}_i - \Delta_i, \widetilde{x}_i + \Delta_i] \text{ for all } i\}.$$

Algorithm. To compute the desired range:

- first, for each i from 1 to n, we use the random number generator for generating a random number $u_i^{(k)}$ which is uniformly distributed on the interval [0, 1], and then compute $\xi_i^{(k)} = \tan\left(\pi \cdot \left(u_i^{(k)} - 0.5\right)\right)$;
- we then compute $M = \max\limits_{i,k} \left|\xi_i^{(k)}\right|$;
- then, we plug in the values

$$x_i = \widetilde{x}_i - \Delta_i \cdot \frac{\xi_i^{(k)}}{M}$$

into the algorithm f, thus computing the value

$$\Delta y^{(k)} = M \cdot \left(\widetilde{y} - f\left(\widetilde{x}_1 - \Delta_1 \cdot \frac{\xi_1^{(k)}}{M}, \ldots, \widetilde{x}_n - \Delta_n \cdot \frac{\xi_n^{(k)}}{M}\right)\right);$$

- once we have all these values, we compute Δ by solving the equation

$$\sum_{k=1}^{N} \frac{1}{1 + \frac{\left(\Delta y^{(k)}\right)^2}{\Delta^2}} - \frac{N}{2} = 0. \qquad (13)$$

To find Δ, we can use a bisection algorithm starting with the interval $\left[0, \max\limits_{k=1,\ldots,N} \left|\Delta y^{(k)}\right|\right]$:

- for the value Δ corresponding to the left endpoint of this interval, the left-hand side of the Eq. (14) is negative, while
- for the right endpoint, the left-hand side is positive.

At each step, we take a midpoint of the previous interval, and select either the left or the right half-interval, so that in the endpoints of the selected half-interval, the left-hand side of (14) has different signs.

The resulting range is equal to $[\widetilde{x} - \Delta, \widetilde{x} + \Delta]$.

How many iterations do we need. To determine the parameter Δ, we used the maximum likelihood method. It is known that the error of this method is asymptotically normally distributed, with 0 average and standard deviation $1/\sqrt{N \cdot I}$, where I is Fisher's information:

$$I = \int_{-\infty}^{\infty} \frac{1}{\rho} \cdot \left(\frac{\partial \rho}{\partial \Delta}\right)^2 dz.$$

For Cauchy probability density $\rho(z)$, we have $I = 1/(2\Delta^2)$, so the error of the above randomized algorithm is asymptotically normally distributed, with a standard deviation $\sigma_e \sim \Delta \cdot \sqrt{2/N}$. Thus, if we use a "two sigma" bound, we conclude that with probability 95%, this algorithm leads to an estimate for Δ which differs from the actual value of Δ by $\le 2\sigma_e = 2\Delta \cdot \sqrt{2/N}$. So, if we want to achieve a 20% accuracy in the error estimation, we must use the smallest N for which $2\sigma_e = 2\Delta \cdot \sqrt{2/N} \leq 0.2 \cdot \Delta$, i.e., to select $N_f = N = 200$.

When it is sufficient to have a standard deviation of 20% (i.e., to have a "two sigma" guarantee of 40%), we need only $N = 50$ calls to f. For $n \approx 10^3$, both values N_f are much smaller than $N_f = n$ required for numerical differentiation.

So, if we have to choose between the straightforward approach and the Monte-Carlo approach, we must select:

- straightforward approach when the number of variables n satisfies the inequality $n \leq N_0$ (where $N_0 \approx 200$), and
- the Monte-Carlo approach if $n \geq N_0$.

Comment. If we use fewer than N_0 simulations, then we still get an approximate value of the range, but with worse accuracy—and the accuracy can be easily computed by using the above formulas.

An example of using this method. As an example, let us consider a multi-dimensional version of oscillator problem that was proposed in [26] as a test for different uncertainty quantification techniques. In the original 1-dimensional oscillator (=spring) problem, the task is to find the range of possible values of the stable-state amplitude y obtained when we apply, to this oscillator, a force with frequency ω and an amplitude that would, in the static case, lead to a unit displacement.

The differential equation describing this simple oscillator is

$$m \cdot \ddot{x} = -k \cdot x - c \cdot \dot{x} + k \cdot \cos(\omega \cdot t),$$

i.e., equivalently,

$$m \cdot \ddot{x} + c \cdot \dot{x} + k \cdot x = k \cdot \cos(\omega \cdot t),$$

where $x(t)$ is the position of the oscillator at moment t, m is the oscillator's mass, k is the spring constant, and c is the damping coefficient. The steady-state solution to this equation also oscillates with the same frequency ω. Any such oscillation can be described, in complex form, as $x(t) = \text{Re}(X(t))$, where $X(t) = Y \cdot \exp(\text{i} \cdot \omega \cdot t)$ for some complex value Y, and we denoted $\text{i} \stackrel{\text{def}}{=} \sqrt{-1}$. For the function $X(t)$, the above equation takes the form

$$m \cdot \ddot{X} + c \cdot \dot{X} + k \cdot X = k \cdot \exp(\text{i} \cdot \omega \cdot t).$$

Substituting the expression $X(t) = Y \cdot \exp(\text{i} \cdot \omega \cdot t)$ into this formula, we conclude that

$$Y \cdot (-m \cdot \omega^2 + c \cdot \text{i} \cdot \omega + k) \cdot \exp(\text{i} \cdot \omega \cdot t) = k \cdot \exp(\text{i} \cdot \omega \cdot t).$$

Thus,

$$Y = \frac{k}{(k - m \cdot \omega^2) + c \cdot \text{i} \cdot \omega}.$$

So, the amplitude $y = |Y|$ of the resulting steady-state oscillations has the form

$$y = \frac{k}{\sqrt{(k - m \cdot \omega^2)^2 + c^2 \cdot \omega^2}}.$$

As a test problem, we consider the problem of computing the sum y of amplitudes of several ($N_{\rm osc}$) oscillators:

$$y = \sum_{i=1}^{N_{\rm osc}} \frac{k_i}{\sqrt{(k_i - m_i \cdot \omega^2)^2 + c_i^2 \cdot \omega^2}}.$$

Specifically, we are interested in computing the parameter Δ that describes the range $[\widetilde{y} - \Delta, \widetilde{y} + \Delta]$ of possible values of y.

To fully describe the problem, we need to select $N_{\rm osc}$ values m_i, $N_{\rm osc}$ values k_i, $N_{\rm osc}$ values c_i, and one value ω—to the total of $3 \cdot N_{\rm osc} + 1$ parameters.

In our simulations, we used $N_{\rm osc} = 400$ oscillators, so we have $3 \cdot 400 + 1 = 1\,201$ parameters.

For each of the parameters k_i, m_i, and c_i, we selected the corresponding range of possible values by dividing the original range from [26] into $N_{\rm osc}$ equal subintervals. For example, we divide the original interval $[\underline{m}, \overline{m}] = [10, 12]$ for m into $N_{\rm osc}$ equal subintervals:

- the interval $\mathbf{m}_1$ of possible values of m_1 is

$$\left[\underline{m}, \underline{m} + \frac{1}{N_{\rm osc}} \cdot (\overline{m} - \underline{m})\right];$$

- the interval $\mathbf{m}_2$ of possible values of m_2 is

$$\left[\underline{m} + \frac{1}{N_{\rm osc}} \cdot (\overline{m} - \underline{m}), \underline{m} + \frac{2}{N_{\rm osc}} \cdot (\overline{m} - \underline{m})\right];$$

- ...
- the interval $\mathbf{m}_i$ of possible values of m_i is

$$\left[\underline{m} + \frac{i-1}{N_{\rm osc}} \cdot (\overline{m} - \underline{m}), \underline{m} + \frac{i}{N_{\rm osc}} \cdot (\overline{m} - \underline{m})\right];$$

- ...
- the interval $\mathbf{m}_{N_{\rm osc}}$ of possible values of $m_{N_{\rm osc}}$ is

$$\left[\overline{m} - \frac{1}{N_{\rm osc}} \cdot (\overline{m} - \underline{m}), \overline{m}\right];$$

For the frequency ω, we used the same interval [2.0, 3.5] as in the original oscillator problem.

Both the straightforward approach and the Monte-Carlo approach work well when the problem is linearizable, i.e., when we can safely ignore terms which are quadratic and of higher order in terms of the deviations Δx_i. From this viewpoint, intervals of possible values of m_i, k_i, and c_i are sufficiently narrow, so that we can indeed ignore terms quadratic in terms of Δm_i, Δk_i, and Δc_i. However, the dependence on ω is far from linear—the corresponding interval is too woide for that. To make it closer to linear, we divided the original interval [2.0, 3.5] into two equal subintervals [2.0, 2.75] and [2.75, 3.5]. In the first subinterval, the dependence is close to linear, but on the second one, not yet, so we divided the second subinterval into two more subintervals [2.75, 3.125] and [3.125, 3.5]. On each of the three resulting intervals, we used both straightforward approach and Monte-Carlo approach to estimate the corresponding value Δ. Here are the results:

Interval	Actual value of Δ	Straightforward approach	Monte-Carlo approach
[2.0, 2.75]	161	151	184
[2.75, 3.125]	23	5	16
[3.125, 3.5]	37	39	42
Number of calls to f		1200	200

As expected, the results of the Monte-Carlo method with 200 iterations are approximately within 20% of the actual value.

By using the Monte-Carlo approach for a problem with $n \approx 1\,200$ variables, we cut the number of calls to f ("gold-plated" calls) 6 times: from 1200 to 200. We them repeated the experiment with $N_{\text{osc}} = 400\,000$ oscillators. In this case, we have $n \approx 1\,200\,000$ variables, and we cut the number of calls to f by a factor of 6 000.

This decrease in computation time is caused by the fact that in the interval-related Monte-Carlo approach, the number of calls to f is always 200 (for 20% accuracy), no matter how many variables we have. In situations when we are satisfied with 40% accuracy, we need an even fewer number of calls—only 50.

Comment. In general, experimental and theoretical confirmation of the Cauchy-based Monte-Carlo algorithm can be found in [19, 27–51]; see also [52].

Methodological comment. Here, as we warned, we have another example when a Monte-Carlo approach is not based on truthful simulation: we do not know the actual probability distribution, but we select a certain distribution for simulation—which is most probably different from the actual (unknown) probability distribution. Similarly to the case of non-normal distributions, here too this use of Monte-Carlo simulations is a mathematical trick helping us to compute the result fast.

12 What if We Have Information About Systematic and Random Error Components

Description of the case: reminder. Let us consider the case which is most common in measurement practice, when for each measurement error Δx_i, we know the upper bound $\widetilde{\Delta}_i$ on the absolute value $|E[\Delta x_i]|$ of its mean value, and the upper bound $\widetilde{\sigma}_i$ on its standard deviation. What can we then conclude about the mean m and the standard deviation s of the value $\Delta y = \sum\limits_{i=1}^{n} c_i \cdot \Delta x_i$?

Analysis of the problem. From the formula (1), we conclude that

$$m = E[\Delta y] = \sum_{i=1}^{n} c_i \cdot E[\Delta x_i]. \tag{15}$$

We only information that we have about each value $E[\Delta x_i]$ is that this value is somewhere in the interval $\left[-\widetilde{\Delta}_i, \widetilde{\Delta}_i\right]$. Thus, from the mathematical viewpoint, this is exactly the problem of uncertainty quantification under interval uncertainty—so we can use the above-described interval algorithm to find the largest possible value $\widetilde{\Delta}$ of the absolute value $|m|$ of the mean m.

For the standard deviation, we have the formula (3), i.e., we have $\sigma^2 = \sum\limits_{i=1}^{n} c_i^2 \cdot \sigma_i^2$. However, here, in contrast to the previously described probabilistic case, we do not know the standard deviations σ_i, we only know the upper bounds $\widetilde{\sigma}_i$ for which $\sigma_i \le \widetilde{\sigma}_i$.

The expression $\sigma^2 = \sum\limits_{i=1}^{n} \sigma_i^2$ is increasing with respect to each of the unknowns σ_i. Thus, its largest possible value is attained when each of the values σ_i is the largest possible, i.e., when $\sigma_i = \widetilde{\sigma}_i$. For $\sigma_i = \widetilde{\sigma}_i$, the resulting largest possible value $\widetilde{\sigma}$ has the form $(\widetilde{\sigma})^2 = \sum\limits_{i=1}^{n} c_i^2 \cdot (\widetilde{\sigma}_i)^2$. This is the same formula as for the probabilistic case—so, to compute $\widetilde{\sigma}$, we can use the above-described probabilistic algorithm.

Thus, we arrive at the following algorithm.

Algorithm. First, we apply the interval algorithm to transform:

- the bounds $\widetilde{\Delta}_i$ on the absolute value of the means of the measurement errors into
- the bound $\widetilde{\Delta}$ on the absolute value of the mean of the resulting approximation error Δy.

Then, we apply the probabilistic-case algorithm to transform:

- the bounds $\widetilde{\sigma}_i$ on the standard deviations of the measurement errors into
- the bound $\widetilde{\sigma}$ for the standard deviation of the resulting approximation error Δy.

13 Fuzzy Case

Formulation of the problem. Suppose now that for each i, we only have fuzzy information about each estimate $\widetilde{x}_i$, i.e., that we have a fuzzy number $\mu_i(x_i)$ that describes the corresponding uncertainty. We want to find the membership function $\mu(y)$ that describes the result of applying the algorithm $y = f(x_1, \ldots, x_n)$ to these uncertain inputs.

Analysis of the problem. In fuzzy techniques, the usual way to transform the initial uncertainty into the uncertainty of the result of data processing is to use the so-called Zadeh's extension principle, which basically means that for each $\alpha \in (0, 1]$, the corresponding "α-cut" $\mathbf{y}(\alpha) \stackrel{\text{def}}{=} \{y : \mu(y) \geq \alpha\}$ is obtained from the α-cuts of the inputs $\mathbf{x}_i(\alpha) = \{x_i : \mu_i(x_i) \geq \alpha\}$ by the usual interval formula

$$\mathbf{y}(\alpha) = \{f(x_1, \ldots, x_n) : x_i \in \mathbf{x}_i(\alpha) \text{ for all } i\}.$$

Thus, for each α, we can apply the above interval algorithm to the corresponding α-cuts.

For this purpose, each α-cut interval needs to be represented in the center-radius form, i.e., as $[\widetilde{x}_i - \Delta_i, \widetilde{x}_i + \Delta_i]$, where $\widetilde{x}_i$ is the interval's center and Δ_i its half-width ("radius"). Thus, we arrive at the following algorithm.

Algorithm. For each α, we:

- represent each α-cut $\mathbf{x}_i(\alpha)$ in the center-radius form, and
- then use the interval algorithm to compute the range.

This range will be the desired α-cut $\mathbf{y}(\alpha)$ for y.

14 General Case: What if We Know Different Inputs with Different Uncertainty

Formulation of the problem. Let us consider the most general case, when we have inputs of all possible types. For the following three types of uncertainty, we know the measurement result $\widetilde{x}_i$, and we also have the following additional information. For some inputs i, we know that the mean value of the measurement error is 0, and we know the standard deviation σ_i of the measurement error. We will denote the set of all such inputs by I_{p}, where p stands for "probabilistic".

For other inputs i, we know the upper bound Δ_i on the absolute value of the mean of the measurement error, and we know the upper bound σ_i on the standard deviation of the measurement error. We will denote the set of all such inputs by I_{m}, where m stands for "measurements". For some inputs i, we only know the upper bound Δ_i on the absolute value of the measurement error. We will denote the set of all such inputs by I_{i}, where i stands for "interval". Finally, for some inputs, instead of the measurement result, we know only the fuzzy number $\mu_i(x_i)$. We will denote the set of all such inputs by $I_{\rm f}$, where f stands for "fuzzy".

Analysis of the problem and the resulting algorithm. If the set I_f is non-empty, let us pick some values $\alpha \in (0, 1]$.

Then, we extend the definitions of Δ_i, σ_i, and $\widetilde{x}_i$ to all indices i:

- for $i \in I_{\rm p}$, we take $\Delta_i = 0$;
- for $i \in I_{\rm i}$, we take $\sigma_i = 0$; and
- for $i \in I_{\rm f}$, we take $\sigma_i = 0$; as Δ_i, we take the radius of the α-cut $\mathbf{x}_i(\alpha)$, and as $\widetilde{x}_i$, we take the center of the α-cut.

Under this definition, for each i, the corresponding value $\Delta x_i \stackrel{\rm def}{=} \widetilde{x}_i - x_i$ can be represented as the sum

$$\Delta x_i = \Delta x_{si} + \Delta x_{ri}, \quad (16)$$

of "systematic" and "random" components, where the only thing we know about Δx_{si} is that $|\Delta_{si}| \le \Delta_i$, and the only thing we know about Δx_{ri} is that it is a random variable with 0 mean and standard deviation not exceeding σ_i.

Substituting the expression (16) into the formula (1), we conclude that:

$$\Delta y = \Delta y_r + \Delta y_s,$$

where we denoted $\Delta y_r \stackrel{\rm def}{=} \sum\limits_{i=1}^{n} c_i \cdot \Delta x_{ri}$ and $\Delta y_s \stackrel{\rm def}{=} \sum\limits_{i=1}^{n} c_i \cdot \Delta x_{si}$.

We can therefore conclude that the only thing we know about Δy_s is that $|\Delta y_s| \le \Delta$, and the only thing we know about Δy_r is that it is a random variable with 0 mean and standard deviation not exceeding σ, where Δ is obtained by applying the interval algorithm to the values Δ_i, and σ is obtained by applying the probabilistic algorithm to the values σ_i.

If some of the inputs are described by fuzzy uncertainty, the procedure of estimating Δ needs to be repeated for several different values α (e.g., for $\alpha = 0.1, 0.2, \ldots, 1.0$), so that Δ becomes a fuzzy number.

Comment. Applications of this idea can be found, e.g., in [40–42, 47, 48].

Important comment. Instead of treating systematic and random components separately (as we did), a seemingly reasonable idea is to transform them into a single type of uncertainty, and then to combine the transformed uncertainties.

Such a transformation is, in principle, possible. For example, if all we know is that the measurement error is located on the interval $[-\Delta_i, \Delta_i]$, but we have no reason to believe that some values on this intervals are more probable than others, then it is reasonable to assume that they are equally probable—i.e., to consider a uniform distribution on this interval; see, e.g., [53]. For the uniform distribution, the mean is 0, and the standard deviation is equal to $\dfrac{1}{\sqrt{3}} \cdot \Delta_i$. On the other hand, if we have a normally distributed random variable Δx_i with mean 0 and standard deviation σ_i, then, with high certainty, we can conclude that this value is located within the 3σ interval $[-3\sigma_i, 3\sigma_i]$; see, e.g., [3].

The problem is that these seemingly reasonable transformations may drastically change the result of uncertainty quantification. Let us show this on the simplest example when $f(x_1, \ldots, x_n) = x_1 + \cdots + x_n$, so that $c_1 = \cdots = c_n = 1$, and all the initial values Δ_i and σ_i are equal to 1.

In this case, for interval uncertainty we get $\Delta = \sum_{i=1}^{n} |c_i| \cdot \Delta_i = n$. However, if we transform it into the probabilistic uncertainty, with $\sigma_i = \frac{1}{\sqrt{3}} \cdot \Delta_i = \frac{1}{\sqrt{3}}$ and process this probabilistic information, then we will get $\sigma^2 = \sum_{i=1}^{n} c_i^2 \cdot \sigma_i^2 = \frac{1}{3} \cdot n$, so $\sigma = \frac{1}{3} \cdot \sqrt{n}$. If we now form an interval bound based on this σ, we will get $\Delta = 3\sigma = \sqrt{3} \cdot \sqrt{n}$—a value which is much smaller than $\Delta = n$. So, if we use this transformation, we will drastically underestimate the uncertainty—which, in many practical situations, can lead to a disaster.

Similarly, in the case of probabilistic uncertainty, we get $\sigma = \sqrt{\sum_{i=1}^{n} c_i^2 \cdot \sigma_i^2} = \sqrt{n}$. However, if we first transform it into interval uncertainty, with $\Delta_i = 3\sigma_i = 3$, and then apply interval estimate to this new uncertainty, we will get $\Delta = \sum_{i=1}^{n} |c_i| \cdot \Delta_i = 3n$. If we transform this value back into standard deviations, we get $\sigma = \frac{1}{3} \cdot \Delta = \sqrt{3} \cdot n$—a value which is much larger than $\sigma = \sqrt{n}$. So, if we use this transformation, we will drastically overestimate the uncertainty—and thus, fail to make conclusions about y which could have made if we estimated the uncertainty correctly.

Bottom line: let 100 flowers bloom, do not try to reduce all uncertainties to a single one.

Acknowledgements This work was supported in part by the National Science Foundation grants 1623190 (A Model of Change for Preparing a New Generation for Professional Practice in Computer Science), and HRD-1834620 and HRD-2034030 (CAHSI Includes), and by the AT&T Fellowship in Information Technology.

It was also supported by the program of the development of the Scientific-Educational Mathematical Center of Volga Federal District No. 075-02-2020-1478.

The authors are thankful to the anonymous reviewers for their valuable suggestions.

References

1. Trejo, R., Kreinovich, V.: Error estimations for indirect measurements: randomized vs. deterministic algorithms for 'black-box' programs. In: Rajasekaran, S., Pardalos, P., Reif, J., Rolim, J. (eds.) Handbook on Randomized Computing, pp. 673–729. Kluwer, Boston, Dordrecht (2001)

2. Rabinovich, S.G.: Measurement Errors and Uncertainties: Theory and Practice. Springer, New York (2005)
3. Sheskin, D.J.: Handbook of Parametric and Non-Parametric Statistical Procedures. Chapman & Hall/CRC, London (2011)
4. Kreinovich, V., Lakeyev, A., Rohn, J., Kahl, P.: Computational Complexity and Feasibility of Data Processing and Interval Computations. Kluwer, Dordrecht (1998)
5. Feynman, R., Leighton, R., Sands, M.: The Feynman Lectures on Physics. Addison Wesley, Boston (2005)
6. Thorne, K.S., Blandford, R.D.: Modern Classical Physics: Optics, Fluids, Plasmas, Elasticity, Relativity, and Statistical Physics. Princeton University Press, Princeton (2017)
7. Novitskii, P.V., Zograph, I.A.: Estimating the measurement errors. Energoatomizdat, Leningrad (1991) in Russian
8. Orlov, A.I.: How often are the observations normal? Ind. Lab. **57**(7), 770–772 (1991)
9. Jaulin, L., Kiefer, M., Didrit, O., Walter, E.: Applied Interval Analysis, with Examples in Parameter and State Estimation, Robust Control, and Robotics. Springer, London (2001)
10. Moore, R.E., Kearfott, R.B., Cloud, M.J.: Introduction to Interval Analysis. SIAM, Philadelphia (2009)
11. Mayer, G.: Interval Analysis and Automatic Result Verification. de Gruyter, Berlin (2017)
12. Zadeh, L.A.: Fuzzy sets. Inf. Control **8**, 338–353 (1965)
13. Klir, G., Yuan, B.: Fuzzy Sets and Fuzzy Logic. Prentice Hall, Upper Saddle River (1995)
14. Novák, V., Perfilieva, I., Močkoř, J.: Mathematical Principles of Fuzzy Logic. Kluwer, Boston (1999)
15. Belohlavek, R., Dauben, J.W., Klir, G.J.: Fuzzy Logic and Mathematics: A Historical Perspective. Oxford University Press, New York (2017)
16. Mendel, J.M.: Uncertain Rule-Based Fuzzy Systems: Introduction and New Directions. Springer, Cham (2017)
17. Nguyen, H.T., Walker, C.L., Walker, E.A.: A First Course in Fuzzy Logic. Chapman and Hall/CRC, Boca Raton (2019)
18. Aczél, J., Dhombres, J.: Functional Equations in Several Variables. Cambridge University Press (2008)
19. Kreinovich, V.: Application-motivated combinations of fuzzy, interval, and probability approaches, with application to geoinformatics, bioinformatics, and engineering. In: Proceedings of the International Conference on Information Technology InTech'07, Sydney, Australia, December 12–14, 2007, pp. 11–20 (2007)
20. Kreinovich, V., Ferson, S.: A new Cauchy-based black-box technique for uncertainty in risk analysis. Reliab. Eng. Syst. Saf. **85**(1–3), 267–279 (2004)
21. Kreinovich, V.: Application-motivated combinations of fuzzy, interval, and probability approaches, and their use in geoinformatics, bioinformatics, and engineering. Int. J. Autom. Control **2**(2/3), 317–339 (2008)
22. Nguyen, H.T., Kreinovich, V., Wu, B., Xiang, G.: Computing Statistics Under Interval and Fuzzy Uncertainty. Springer, Berlin (2012)
23. Kreinovich, V.: Decision making under interval uncertainty (and beyond). In: Guo, P., Pedrycz, W. (eds.) Human-Centric Decision-Making Models for Social Sciences, pp. 163–193. Springer (2014)
24. Kreinovich, V.: Interval computations and interval-related statistical techniques: estimating uncertainty of the results of data processing and indirect measurements. In: Pavese, F., Bremser, W., Chunovkina, A., Fisher, N., Forbes, A.B. (eds.) Advanced Mathematical and Computational Tools in Metrology and Testing AMTCM'X, pp. 38–49. World Scientific, Singapore (2015)
25. Kreinovich, V.: How to deal with uncertainties in computing: from probabilistic and interval uncertainty to combination of different approaches, with applications to engineering and bioinformatics. In: Mizera-Pietraszko, J., Rodriguez Jorge, R., Almazo Pérez, D., Pichappan, P. (eds.) Advances in Digital Technologies. Proceedings of the Eighth International Conference on the Applications of Digital Information and Web Technologies ICADIWT'2017, Ciudad Juarez, Chihuahua, Mexico, March 29–31, 2017, pp. 3–15. IOS Press, Amsterdam (2017)

26. Oberkampf, W.L., Helton, J.C., Joslyn, C.A., Wojtkiewicz, S.F., Ferson, S.: Challenge problems: uncertainty in system response given uncertain parameters. Reliab. Eng. Syst. Saf. **85**(1–3), 11–19 (2004)
27. Kreinovich, V., Pavlovich, M.I.: Error estimate of the result of indirect measurements by using a calculational experiment. Izmeritelnaya Tekhnika (3), 11–13 (in Russian); English translation: Meas. Tech. **28**(3), 201–205 (1985)
28. Ferson, S., Joslyn, C.A., Helton, J.C., Oberkampf, W.L., Sentz, K.: Summary from the epistemic uncertainty workshop: consensus amid diversity. Reliab. Eng. Syst. Saf. **85**(1–3), 355–369 (2004)
29. Bruns, M., Paredis, C.J.J.: Numerical methods for propagating imprecise uncertainty. In: Proceedings of the 2006 ASME Design Engineering Technical Conference (2006)
30. Oberguggenberger, M., King, J., Schmelzer, B.: Imprecise probability methods for sensitivity analysis in engineering. In: de Cooman, G., Vejnarova, J., Zaffalon, M. (eds.) Proceedings of the Fifth International Symposium on Imprecise Probability: Theory and Applications ISIPTA'07, Prague, July 16–19, 2007, pp. 155–164 (2007)
31. Kleb, B., Johnston, C.O.: Uncertainty analysis of air radiation for Lunar return shock layers. In: Proceedings of the AIAA Atmospheric Flight Mechanics Conference, Honolulu, Hawaii, August 18–21, 2008, paper AIAA, pp. 2008–6388 (2008)
32. Pownuk, A., Cerveny, J., Brady, J.J.: Fast algorithms for uncertainty propagation, and their applications to structural integrity. In: Proceedings of the 27th International Conference of the North American Fuzzy Information Processing Society NAFIPS'2008, New York, May 19–22 (2008)
33. Fuchs, M.: Simulation based uncertainty handling with polyhedral clouds. In: Beer, M., Muhanna, R. L., Mullen, R. L. (eds.) Proceedings of the 4th International Workshop on Reliable Engineering Computing REC'2010, National University of Singapore, March 2–5, 2010, pp. 526–535 (2010)
34. Rico, A., Strauss, O.: Imprecise expectations for imprecise linear filtering. Int. J. Approx. Reason. **51**(8), 933–947 (2010)
35. Fuchs, M.: Simulated polyhedral clouds in robust optimisation. Int. J. Reliab. Saf. **6**(1–3), 65–81 (2012)
36. Johnston, C.O., Mazaheri, A., Gnoffo, P., Kleb, B., Bose, D.: Radiative heating uncertainty for hyperbolic Earth entry, part 1: flight simulation modeling and uncertainty. J. Spacecr. Rocket. **50**(1), 19–38 (2013)
37. Rebner, G., Beer, M., Auer, E., Stein, M.: Verified stochastic methods - Markov set-chains and dependency modeling of mean and standard deviation. Soft Comput. **17**, 1415–1423 (2013)
38. Morio, J., Balesdent, M., Jacquemart, D., Vergé, C.: A survey of rare event simulation methods for static input-output models. Simul. Model. Pract. Theory **49**, 287–304 (2015)
39. Calder, A.C., Hoffman, M.M., Willcox, D.E., Katz, M.P., Swesty, F.D., Ferson, S.: Quantification of incertitude in black box simulation codes. J. Phys.: Conf. Ser. **1031**(1), Paper 012016 (2018)
40. McClarren, R.G.: Uncertainty Quantification and Predictive Computational Science: A Foundation for Physical Scientists and Engineers. Springer (2018)
41. Pownuk, A., Kreinovich, V.: Combining Interval, Probabilistic, and Other Types of Uncertainty in Engineering Applications. Springer, Cham (2018)
42. Ceberio, M., Kosheleva, O., Kreinovich, V., Longpré, L.: Between dog and wolf: a continuous transition from fuzzy to probabilistic estimates. In: Proceedings of the IEEE International Conference on Fuzzy Systems FUZZ-IEEE'2019, New Orleans, Louisiana, June 23–26, 2019, pp. 906–910 (2019)
43. de Angelis, M., Ferson, S., Patelli, E., Kreinovich, V.: Black-box propagation of failure probabilities under epistemic uncertainty. In: Papadrakakis, M., Papadopoulos, V., Stefanou, G. (eds.) Proceedings of the 3rd International Conference on Uncertainty Quantification in Computational Sciences and Engineering UNCECOMP'2019, Cre,e Greece, June 24–26, 2019, pp. 713–723 (2019)

44. Kreinovich, V.: Global independence, possible local dependence: towards more realistic error estimates for indirect measurements. In: Proceedings of the XXII International Conference on Soft Computing and Measurements SCM'2019, St. Petersburg, Russia, May 23–25, 2019, pp. 4–8 (2019)
45. Brevault, L., Morio, J., Balesdent, M.: Aerospace System Analysis and Optimization in Uncertainty. Springer (2020)
46. Callens, R., Faes, M.G.R., Moens, D.: Local explicit interval fields for non-stationary uncertainty modelling in finite element models. Comput. Methods Appl. Mech. Eng. **379**, Paper 113735 (2021)
47. Faes, M.G.R., Daub, M., Marelli, S., Patelli, E., Beer, M.: Engineering analysis with probability boxes: a review on computational methods. Struct. Saf. **93**, Paper 102092 (2021)
48. Faes, M.G.R., Valdebenito, M.A., Moens, D., Beer, M.: Operator norm theory as an efficient tool to propagate hybrid uncertainties and calculate imprecise probabilities. Mech. Syst. Signal Process. **152**, Paper 107482 (2021)
49. Kosheleva, O., Kreinovich, V.: Low-complexity zonotopes can enhance uncertainty quantification (UQ). In: Proceedings of the 4th International Conference on Uncertainty Quantification in Computational Sciences and Engineering UNCECOMP'2021, Athens, Greece, June 28–30 (2021)
50. Kosheleva, O., Kreinovich, V.: Limit theorems as blessing of dimensionality: neural-oriented overview. Entropy **23**(5), Paper 501 (2021)
51. Semenov, K., Tselischeva, A.: The interval method of bisection for solving the nonlinear equations with interval-valued parameters. In: Voinov, N., Schreck, T., Khan, S. (eds.) Proceedings of the International Scientific Conference on Telecommunications, Computing, and Control Teleccon 2019, November 18–21, pp. 373–384. Springer (2021)
52. Pownuk, A., Kreinovich, V.: (Hypothetical) negative probabilities can speed up uncertainty propagation algorithms". In: Hassanien, A.E., Elhoseny, M., Farouk, A., Kacprzyk, J. (eds.) Quantum Computing: An Environment for Intelligent Large Scale Real Application, pp. 251–271. Springer (2018)
53. Jaynes, E.T., Bretthorst, G.L.: Probability Theory: The Logic of Science. Cambridge University Press, Cambridge (2003)

Multi-view Clustering and Multi-view Models

Nha Pham Van, Long Ngo Thanh, and Long Pham The

Abstract Over the years, the development of information technology applications, in particular, Artificial Intelligence has spurred the need to collect multi-view data on a large scale. Multi-view data, in general, is large, heterogeneous and uncertain, but also contains a lot of knowledge to mine and apply. Some of the single-view data clustering techniques have been improved to analyze multi-view data by extending the structure of the objective function or building associative models. Currently, multi-view clustering methods are quite rich and diverse, however, each method is usually only effective for a specific group of problems. In order to provide an overview of multi-view data clustering and select suitable methods for specific applications, in this chapter we will review some methods of multi-view clustering. Simultaneously, we also select a number of multi-view data clustering algorithms to present and analyze some potential knowledge extraction mechanisms in the data views. Some experimental results on the benchmark datasets are also analyzed to evaluate the performance and scalability of different multi-view clustering algorithms.

Keywords Multi-view data · Multi-source · Multi-view clustering · Machine learning

---

N. Pham Van (✉)
Department of Mathematical Assurance, Institute of Information Technology, Hanoi, Vietnam
e-mail: nhapv@mod.gov.vn

L. Ngo Thanh · L. Pham The
Department of Information Systems, Le Quy Don Technical University, Hanoi, Vietnam
e-mail: ngotlong@mta.edu.vn

L. Pham The
e-mail: longpt@mta.edu.vn

W. Pedrycz and S. Chen (eds.), *Recent Advancements in Multi-View Data Analytics*,
Studies in Big Data 106, https://doi.org/10.1007/978-3-030-95239-6_3

1 Introduction

1.1 Clustering

Clustering is an unsupervised learning technique used to discover latent structure in data sets. Data objects grouped into the same cluster have similar properties based on their features and characteristics using different measurement [1]. Depending on the relationship of each data object to the clusters, clustering algorithms can also be divided into hard clustering [2] and fuzzy clustering [3, 4].

Fuzzy clustering algorithms often achieve better clustering quality than some hard clustering algorithms. For example, Bezdek et al. designed the fuzzy c-means clustering (FCM) [5] by proposing the concept of overlapping clustering and fuzzy membership. Experimental results have shown that FCM achieves better performance than K-mean hard clustering technique. M. Hanmandlua et al. [6] proposed fuzzy co-clustering (FCCI) by co-clustering pixels and R, B, G color components to improve the quality of color image segmentation. Experimental results have shown that FCCI achieves better performance than FCM and several proposed clustering algorithms. Ngo Thanh Long et al. [7] proposed the interval-value fuzzy co-clustering (IVFCoC) by integrating the advanced techniques of the type-2 fuzzy system to identify uncertainties in data and improve the clustering quality. In works [8, 9] the authors have proposed feature reduction schemes by integrating feature-weighted entropy into the fuzzy clustering objective function to compute and remove small-weighted features. The experiment results were conducted on many-features data sets and hyperspectral images that have demonstrated the outstanding performance of the proposed algorithm compared with some previously proposed algorithms.

Clustering is commonly used in classification, diagnostics, and prediction problems such as pattern recognition, decision support, image analysis, information retrieval, etc. [10]. The outstanding advantage of clustering algorithms compared to traditional data analysis methods is that the computational complexity is not large but has higher accuracy on uncertain data sets. Recently, some clustering techniques have been improved to specialize in multi-feature data analysis [6, 7]. However, like other data processing methods, traditional clustering methods do not perform well on large datasets of this type [11]. In fact, with the exponential growth rate and availability of data collected from different sources, analyzing and organizing data has become an important challenge in big data processing.

1.2 Multi-view Clustering

The need to collect data from multiple sources and multimedia is increasing due to the development of information technology applications, especially artificial intelligence. The one-way, single-source data processing techniques are facing a "bottleneck" situation due to limited responsiveness. In practice, multi-source data is often collected from transceiver stations in different areas or areas of observation. These transceivers can receive part of the data or all the data, but in different dimensions. As a result, the multi-source data is becoming larger, more diverse, heterogeneous, and uncertain, but potentially rich in useful knowledge that is needed for complex data analysis. Therefore, during the past four decades, clustering techniques have been constantly evolving. Depending on the method of data collection, there are two types of multi-source data such as multi-view data and multi-subspace data. Thus multi-source data is a general concept, multi-view data and multi-subspace data are separate cases of multi-source data [12–14]. According to such a concept, multi-source data becomes more complicated than single-source data because of diversity and heterogeneity.

In multi-view data, the data views reflect the same data set, but the data views often have a different number of features. For example, in-ground observation applications we can use multi-spectral, hyper-spectral, and satellite radar sensors. These sensors collect observed information from similar objects, but the observed data have a different number of features.

Multi-subspace data are collected from transceiver stations located in different areas. These transceivers usually have the same structure, which means that the received data has the same number of features. For example, in remote sensing applications, a multi-spectral sensor cannot observe all of the earth's surface; we must use many sensors, each of which collects a subarea's data. These sensors are the same type so that the data obtained have the same number of features.

To analyze multi-source data, there are two branches of clustering techniques corresponding to two types of multi-source data, including multi-view clustering and multi-subspace clustering [15–36].

In multi-view clustering problems, the objects are distributed equally in the data subsets. Data subsets have different weights and feature numbers. The multi-view clustering algorithm finds functional matrices that belong to the optimal object common to all data subsets by learning similar matrices between data subsets. Some methods of multi-view clustering include multi-view k-means [14], multi-view fuzzy c-means clustering [15, 16], multi-view clustering via deep concept factorization [17], multi-view spectral clustering via sparse graph learning [18].

In multi-subspace clustering problems, the objects are scattered in subsets of data. Subsets of data have the same weights and feature numbers. The multi-subspace clustering algorithm finds function matrices that belong to the optimal objects of all data subsets by learning a matrix of characteristic functions between subset data. Some methods of clustering multi-subspace include simultaneously learning feature-wise weights and local structures for multi-view subspace clustering [17], adaptive multi-view subspace clustering for high-dimensional data [18], feature concatenation multi-view subspace clustering [19].

In this chapter, we focus on an overview of multi-view data clustering methods and algorithms. Simultaneously, four typical multi-view data clustering algorithms are presented in detail and analyzed the hidden knowledge of mining mechanisms in data views. Experimental results on the benchmark dataset are also analyzed to evaluate the performance and scalability of different multi-view clustering algorithms.

The remainder of the chapter is organized as follows. Section 2, related works include some concepts of multi-view data and some basic clustering algorithms. In Sect. 3, four typical multi-view clustering algorithms are presented in detail. Section 4 presents some research directions related to the multi-view data analysis problem. Section 5 is the conclusion of the chapter.

2 Related Works

In this section, some basic concepts related to multi-view data and traditional multi-view clustering algorithms are presented. In addition, some benchmark datasets for the experiment and some indexes for evaluating the performance of multi-view clustering techniques are also introduced in this section.

2.1 *Some Related Concepts*

Most of the traditional clustering methods are only dealing with single data coming from a single source. In fact, many real-world applications involve collections of data coming from different sources, which are then pre-processed and synthesized into a single data set. Remote sensing image data, for example, is collected from many different sensors. Each sensor is deployed to collect data in different regions and frequency bands, then preprocessed to aggregate it into a data set in a standard format. In this case, we can call it multi-area data or multi-subspace data. Data generated from various viewpoints or multiple sensors that reflect objects of the same size but different number of features. Such data is called multi-view data.

For example, depth sensors use visible light sensors and near-infrared sensors to determine the depth of water; autopilot systems use both visual sensors and active radar to observe real-time obstacles in the roadway; face analysis algorithms use face images obtained from different capture modes for high fidelity reconstruction and recognition [37]. However, such data tends to increase in size, not heterogeneous and uncertain, which will lead to a great challenge such as difficulty in reducing to the same dimension for comparison, knowledge mining. Thus, there is an urgent need to mitigate the source divergence when facing specific problems by either fusing the knowledge across multiple sources or adapting knowledge from some sources to others [38]. Since there are different terms associated with "multi-source" data analysis. We first give a few different definitions and separate the research focus of what distinguishes it from other related works but in different lines.

Definition 1 Multi-source data include M data subsets that are collected from M different sources. Let $X = \{X_1, X_2, \ldots, X_M\}$, $N = \|X_1 \cup X_2 \cup \ldots \cup X_M\|$, $N_m = \|X_m\|$, $X_m = \left\{x_{m,1}, x_{m,2}, \ldots, x_{m,Nm}\right\}$, $m = 1, .., M$, $qm = 1, .., N_m$, $x_{m,qm} \in R^{Dm}$. Where, N is the number of data objects of X, M is the number of data sources, X_m is the subset of data obtained from the mth source, N_m is the number of data objects in X_m, D_m is the number of features in X_m.

Definition 2 Multi-view data includes M data subsets that are collected from M different views. Let $X = \{X_1, X_2, \ldots, X_M\}$, $N = \|X_1\| = \|X_2\| = \ldots = \|X_M\|$ $X_m = \left\{x_{m,1}, x_{m,2}, \ldots, x_{m,Nm}\right\}$, $x_{m,qm} \in R^{Dm}$, $m = 1, .., M$, $qm = 1, .., N_m$, $D_i \neq D_j$, $x_{iq} \equiv x_{jq}$, $\forall i \neq j; i, j = 1, .., M$, $q = 1, .., N$. Multi-view data is a special case of multi-source data.

Definition 3 Multi-subspace data is multi-source data consisting of N data objects are collected from M subspaces with the same number of dimensions. That means,$X = \{X_1, X_2, \ldots, X_M\}$, $N_m = \|X_m\|$, $N_r \neq N_i$, $N = \sum_{m=1}^{M} N_m$, $D_i = D_j$, $x_{iq} \neq x_{rq}$, $\forall i \neq r, i, r = 1, .., M$, $q = 1, .., N_r$. Multi-subspace data is a special case of multi-source data.

2.2 *Traditional Clustering Algorithms*

2.2.1 Fuzzy C-Means Clustering Algorithm

The fuzzy clustering algorithm introduced by Dunn [39] and later modified by Bezdek [3] is called the Fuzzy C-Means (FCM) algorithm. FCM algorithm is the original fuzzy clustering algorithm of many modern improved fuzzy clustering algorithms and is also the original clustering algorithm of the multi-view clustering algorithms presented in this chapter. Given a data set with N samples $X = \{x_i\}$, $i = \overline{1, N}$. FCM algorithm group X into C different clusters based on minimum criteria the J_{FCM} objective function is as follows:

$$J_{FCM}(U, X, P) = \sum_{c=1}^{C} \sum_{i=1}^{N} u_{ci}^{m} d_{ci}^{2} \tag{1}$$

Let U be a membership degree matrix partition matrix—use this terminology, where u_{ci} denotes the degree of membership of the i data to the c cluster; clustering center matrix $P = \{p_1, p_2, \ldots, p_C\}$, where p_c indicates the c clustering center; $X = \{x_1, x_2, \ldots, x_N\}$, where x_i is the i data points; m is a fuzzification coefficient; u_{ci} is constrained by condition (2)

$$\sum_{c=1}^{C} u_{ci} = 1, \ \textit{with} \ \forall \mathrm{i} = 1, 2, \ldots, \mathrm{N} \tag{2}$$

The components of the objective function J_{FCM} are calculated by (3)–(5).

$$u_{ci} = \frac{1}{\sum_{k=1}^{C} \left(\frac{d_{ci}}{d_{ki}}\right)^{2/m-1}} \tag{3}$$

and p_c is defined by the formula (4).

$$p_c = \frac{\sum_{i=1}^{N} u_{ci}^m x_i}{\sum_{i=1}^{N} u_{ci}^m} \tag{4}$$

d_{ci} is the Euclidean distance between the connection? data record x_i and the cluster center p_c, which is defined as follows.

$$d_{ci} = \sqrt{\sum_{j=1}^{D} \left(x_{ij} - p_{cj}\right)^2} \tag{5}$$

The m index used to adjust the fuzzy weighting of the membership functions, m weighted values increases the opacity of the J_{FCM} objective function, m is usually chosen by 2.

The FCM algorithm is described in the following steps.

Algorithm 1. Fuzzy clustering algorithm FCM

Input: Data set $X = \left\{x_i, x_i \in R^D\right\}$, $i = 1..N$, number of clusters C ($1<C<N$), fuzzy index m ($1 < m < +\infty$) and error coefficient ε.

Output: Clustering results.

1. Initial the centroids matrix $P = \left[p_{cj}\right], P^{(0)} \in R^{CxD}$, partition matrix $U=[u_{ci}]$, $\tau=1$;

2. REPEAT

3. Calculate p_c using (4);

4. Update u_{ci} using (3) and (5);

5. Calculate objective function J_{FCM} (1);

6. $\tau=\tau+1$;

7. UNTIL (($\left|J^{(\tau)} - J^{(\tau-1)}\right| \le \varepsilon$) or ($\tau \ge \tau_{max}$))

The computational complexity of the FCM algorithm is $O(CND\tau_{max})$. Where N is the number of data samples, C is the number of clusters, D is the number of features or dimensions of the data space, and τ_{max} is the maximum number of iterations allowed.

2.2.2 Multi-view Clustering Algorithm

A FCM-based multi-view clustering algorithm was first proposed by W. Pedrycz [28] what the author called the collaborative fuzzy clustering algorithm (MVFCM). MVFCM is built by FCM-based clustering on M data views with collaboration between base clusterings. The J_{MVFCM} objective function is expressed as follows:

$$J_{MVFCM} = \sum_{m=1}^{M}\sum_{c=1}^{C}\sum_{i=1}^{N} u_{ci,m}^{2} d_{ci,m}^{2} + \sum_{m'=1}^{M\,(m'\neq m)} \alpha_{m,m'} \sum_{c=1}^{C}\sum_{i=1}^{N} \left(u_{ci,m'} - u_{ci,m}\right) d_{ci,m}^{2} \quad (6)$$

The J_{MVFCM} objective function (6) is minimized subject to the following constraints (7),

$$\begin{cases} U = \{U_m\}_{1xM},\, U_m = \left\{u_{ci,m}\right\}_{CxN_m} \\ u_{ci,m} \in [0,1] \left| \sum_{c=1}^{C} u_{ci,m} = 1 \quad \forall i, m \text{ and} \right. \\ 0 < \sum_{i=1}^{N} u_{ci,m} < N \quad \text{for} \quad \forall c, m \end{cases} \quad (7)$$

where, N is the number of data objects, M is the number of data views, $d_{ci,m} = \sqrt{\sum_{j=1}^{D_m} \left(x_{ij,m} - p_{cj,m}\right)^2}$ is the Euclidean distance between the ith data object and the cth cluster center in the mth data view, $U_m = \left\{u_{ci,m}\right\}_{CxN}$ is the object partition matrix of the mth data view, $\alpha_{m,m'}$ is the collaborative learning coefficient between the mth data view and the m'th data view.

The U and P components of the J_{MVFCM} objective function are calculated according to the following formulas (8) and (9).

$$p_{cj,m} = \frac{\sum_{k=1}^{N} u_{ck,m}^{2} x_{kj,m} + \sum_{\substack{m'=1 \\ m'\neq m}}^{M} \alpha_{mm'} \sum_{k=1}^{N} \left(u_{ck,m} - u_{ck,m'}\right)^2 x_{kj,m}}{\sum_{k=1}^{N} u_{ck,m}^{2} + \sum_{\substack{m'=1 \\ m'\neq m}}^{M} \alpha_{mm'} \sum_{k=1}^{N} \left(u_{ck,m} - u_{ck,m'}\right)^2} \quad (8)$$

$$u_{ci,m} = 1 - \sum_{j}^{C} \frac{\sum_{\substack{m'=1 \\ m'\neq m}}^{M} \alpha_{mm'} u_{ci,m'}}{1 + \sum_{\substack{m'=1 \\ m'\neq m}}^{M} \alpha_{mm'}} + \frac{\left[1 - \sum_{j}^{C} \frac{\sum_{\substack{m'=1 \\ m'\neq m}}^{M} \alpha_{mm'} u_{ci,m'}}{1 + \sum_{\substack{m'=1 \\ m'\neq m}}^{M} \alpha_{mm'}}\right]}{\sum_{j=1}^{C} \frac{d_{ci,m}^{2}}{d_{ji,m}^{2}}} \quad (9)$$

The pseudocode of MVFCM algorithm is given in Algorithm 2.

Algorithm 2. MVFCM algorithm

Given: Data subsets $X_1, X_2, \ldots, X_M$

Select: Distance function, number of clusters (C), termination criterion, and collaboration matrix $\alpha_{mm'}$.

Phase I: Initiate randomly all partition matrices $U_1, U_2, \ldots, U_M$.

For each data view

Calculate $p_{c,m}$ using (4).

Update $u_{ci,m}$ using (3).

End For

Phase II:

For the given matrix of collaborative links $\alpha_{mm'}$.

$\tau=1$;

Repeat: For each data view

Calculate $p_{c,m}$ using (8);

Update $u_{ci,m}$ using (9);

$\tau=\tau+1$;

Until (($Max(|u_{ci,m}[\tau]-u_{ci,m}[\tau-1]|) \leq \varepsilon$) or ($\tau \geq \tau_{max}$)).

The computational complexity of MVFCM algorithm is O($MCND\tau_{max}$). Where M is the number of data views, N is the number of data samples, C is the number of clusters, D is the number of features or dimensions of the data space, and τ_{max} is the maximum number of iterations allowed.

2.3 *Multi-view Data Sets*

In this chapter, nine benchmark multi-view datasets are used to evaluate the performance of clustering algorithms including:

The ALOI (Amsterdam Library of Object Image) dataset was provided by the Informatics Institute of the University of Amsterdam Kruislaan. ALOI includes 1107 images of 10 subjects. ALOI is divided into four feature sets corresponding to four data views.

Multiple Feature (MF) is an image collection of ten handwritten digits extracted from the Dutch collection of utility maps. Each digit type is enhanced by 200 binary images and divided into six feature sets corresponding to six data views.

3-Sources is a dataset collected from three well-known media channels: BBC, Reuters, and Guardian. 3-Sources data includes 168 articles divided into three views.

The MSRC-v1 dataset consists of 210 image objects that are classified into seven classes including tree, building, airplane, cow, face, car, and bicycle. Similar features are extracted from each image to form five different feature views.

The Caltech101 dataset includes 1447 image objects that are classified into seven classes. Similar features are extracted from each image to form six feature views.

The Image segmentation (IS) dataset includes 2310 image objects from the UCI machine learning repository that are classified into seven classes. Similar features are extracted from each image to form two different feature views.

The One-hundred plant species leaves data set (100leaves) from the UCI machine learning repository consists of 1600 image objects that are classified into 100 classes. Similar features are extracted from each image to form three different feature views.

The PEMS-SF (PEMS-SF) dataset from the UCI machine learning repository consists of 440 video recordings from six cameras. Video recordings depict the occupancy rates of the various auto lanes of San Francisco Bay Area freeways Monday through Sunday. Similar features are extracted from each video to form three views.

6Dims: This dataset consists of six datasets from the Computing University of Eastern Finland. Each subset consists of 1024 objects evenly distributed and ordered in 16 clusters and the number of objects is 32, 64, 128, 256, 512, and 1024 respectively. We assume that six data subsets are copies of the original data set by projecting onto six spaces whose dimensions are 32, 64, 128, 256, 512, and 1024 respectively. Therefore, we consider 6Dims as a six-view data set.

Some properties of the data sets are presented in Table 1.

Table 1 Properties of the real-world datasets

Name	Size	#Clusters	#Views	Views structure
ALOI	1107	10	4	8–27–77–13
MF	2000	10	6	76–216–64–240–47–6
3-sources	169	6	3	3560–3631–3068
MSRC-V1	210	7	5	24–576-512–256-254
Caltech 101–7	1474	7	6	48–40-254–1984-512–928
IS	2310	7	2	9–10
100leaves	1600	100	3	64–64-64
PEMS-SF	440	7	3	138,672–4789-8900
6Dims	1024	16	6	32–64-128–256-512–1024

2.4 Validation Indexes

In clustering, evaluating the performance of clustering algorithms is an important step in choosing a suitable algorithm for a particular problem. To evaluate the clustering performance, we can use the cluster quality evaluation indexes and the consumed time for clustering. Currently, there are many cluster estimation indexes, each suitable for a different class of clustering problems, and very few are suitable for all clustering problems. In this section, we will present some evaluation indexes used for the clustering algorithms introduced in this chapter.

2.4.1 Clustering Accuracy

Let $X = \{x_i\}_N$ be a training dataset with N data objects labeled in C clusters. $Y = \{y_i\}_N$ is the set of ground-truth labels. Accuracy index (ACC) and Adjusted Rand Index (AR) [40] measure clustering accuracy. The higher the value of ACC and AR, the better the clustering quality. ACC index is quantified according to formula (10),

$$ACC = \frac{\sum_{i=1}^{N} \delta(x_i, map(y_i))}{N} \tag{10}$$

where $x_i \in X$ is the label learned by the clustering algorithm and $y_i \in Y$ is the ground-truth label of the ith data object. $map(y_i)$ denotes a permutation function [29]. $\delta(.)$ is the Dirac delta function is calculated by the following formula

$$\delta(x_i, y_i) = \begin{cases} 1, & x_i = y_i \\ 0, & otherwise \end{cases}. \tag{11}$$

AR index is calculated as

$$AR = \frac{\sum_{i,j=1}^{C} C_{N_{ij}}^{2} - \frac{\sum_{i=1}^{C} C_{N_i}^{2} \sum_{i=1}^{C} C_{\tilde{N}_i}^{2}}{C_N^2}}{\frac{1}{2}\left(\sum_{i=1}^{C} C_{N_i}^{2} + \sum_{i=1}^{C} C_{\tilde{N}_i}^{2}\right) - \frac{\sum_{i=1}^{C} C_{N_i}^{2} \sum_{i=1}^{C} C_{\tilde{N}_i}^{2}}{C_N^2}} \tag{12}$$

where C_n^m represents the number of selecting m samples from total n samples.

2.4.2 Normalized Mutual Information Index

Normalized mutual information index (NMI) is used to measure the mutual information between the clustering result set and the ground-truth labels. NMI is calculated in the form

$$NMI = \frac{2\sum_{i=1}^{C}\sum_{j=1}^{\tilde{C}} \frac{N_{ij}}{N} \log \frac{N_{ij}N}{\sum_{i=1}^{C} N_i \sum_{j=1}^{\tilde{C}} \tilde{N}_j}}{-\sum_{i=1}^{C} \frac{N_i}{N} \log \frac{N_i}{N} - \sum_{j=1}^{\tilde{C}} \frac{\tilde{N}_j}{N} \log \frac{\tilde{N}_j}{N}} \tag{13}$$

2.4.3 F-Measure

Let TP_i is the number of data objects correctly grouped into the ith cluster, FP_i is the number of data objects incorrectly grouped into the ith cluster, TN_i is the number of data objects grouped in excess of the actual number of labeled objects into the ith cluster, FN_i is the number of missing data objects of the ith cluster. Mean, $TP_i = N_{ii}$, $FP_i = \tilde{N}_i - TP_i$, $TN_i = N_i - \tilde{N}_i$, $FN_i = \tilde{N}_i - N_i$, $TN_i = -FN_i$. F-score is the harmonic mean of the precision and recall is calculated according to the formula (14),

$$F - score = 2\frac{\Pr ecision.\mathrm{Rec}all}{\Pr ecision + \mathrm{Rec}all} \tag{14}$$

where Precision is the clustering accuracy, and Recall is the correct clustering result. Precision and Recall are calculated by the formula (15).

$$Precision = \frac{TP}{TP + FP}, \mathrm{Rec}all = \frac{TP}{FP + FN} \tag{15}$$

2.4.4 Partition Coefficient Index

Bezdek [41] uses the partition coefficient index (PC) to estimate the degree of convergence of the fuzzy object membership function. As the PC takes on a larger value, the fuzzy object membership function moves closer to the optimal value. PC is calculated by the formula (16).

$$PC = \frac{1}{N} \sum_{c=1}^{C} \sum_{i=1}^{N} u_{ci} \text{ not readable?} \tag{16}$$

2.4.5 Davies–Bouldin Index

Let N_c be the number of data objects of the cth cluster, p_c the center of the cth cluster, S_c is the dispersion in the cth cluster. S_c is calculated according to the formula (17).

$$S_c = \sqrt{\frac{1}{N_c} \sum_{i=1}^{N_c} |x_i - p_c|^2} \tag{17}$$

M_{ij} is a measure of separation between cluster i and cluster j. M_{ij} is calculated according to (18).

$$M_{ij} = \sqrt{\sum_{k=1}^{D} \left(p_{ik} - p_{jk}\right)^2} \tag{18}$$

Let R_{ij} be a measure of the goodness of the clustering scheme. R_{ij} is calculated as

$$R_{ij} = \frac{S_i + S_j}{M_{ij}} \tag{19}$$

D.L. Davies et al. [42] introduced the Davies–Bouldin index (DB) for evaluating clustering algorithms. DB is calculated as

$$DB = \frac{1}{C} \sum_{c=1}^{C} D_c \tag{20}$$

where,

$$D_i = \arg \underset{j \neq i}{Max} R_{ij} \tag{21}$$

Please see Table 2 for how to use indices to estimate the performance of clustering algorithms.

Table 2 Cluster validity indices

Name	Denote	Best if	Range
Accuracy index	ACC	High	0, 1
Adjusted Rand index	AR	High	0, 1
Normalized mutual information index	NMI		
Precision index	Pre	High	0, 1
Recall index	Rec	High	0, 1
F-score	F-score	High	0, 1
Partition coefficient index	PC	High	0, 1
Davies–Bouldin index	DB	Low	$-\infty, +\infty$

3 Multi-view Clustering Algorithms and Models

Clustering algorithms are considered typical candidates for multi-view data classification. Currently, there are many multi-view clustering algorithms. Each algorithm has its own advantages and limitations and is suitable for each specific problem. No algorithm is considered to be effective for all real-world problems. In this chapter, we present three multi-view clustering algorithms which are variants of the partition clustering algorithm, i.e. simultaneous weighting on views and features algorithm [16], collaborative feature-weighted multi-view fuzzy c-means clustering algorithm [43], multi-view fuzzy co-clustering algorithm for high-dimensional data classification [44]. In addition, to mine the knowledge latent in different views which we call dark knowledge, we present a multi-objective clustering model for clustering multi-view data [45].

3.1 Simultaneous Weighting on Views and Features Algorithm

3.1.1 Algorithm Description

In multi-feature data, different features have different importance, so we can encode different weights for each feature. Similarly, in multi-view data, we can also encode different weights for each data view. B. Jiang et al. [16] aims to discover latent knowledge in multi-feature and multi-view data by weighting the features and data views. The proposed technique is called simultaneous weighting on views and features algorithm (SWVF). The objective function of the SWVF algorithm is represented as follows:

$$J_{SWVF} = \sum_{m=1}^{M} \sum_{c=1}^{C} \sum_{i=1}^{N_m} \sum_{\substack{j=1 \\ j \in v_m}}^{D_m} u_{ci,m} w_{j,m}^{\alpha} v_m^{\beta} d_{ci,m}^2 \quad (22)$$

The J_{SWVF} objective function (22) is minimized subject to the following constraints (23),

$$\begin{cases} \sum_{c=1}^{C} u_{ci,m} = 1, u_{ci} \in [0, 1], i = 1, .., N_m, m = 1, .., M \\ \sum_{\substack{j=1 \\ j \in v_m}}^{D_m} w_{j,m} = 1, w_{j,m} \in [0, 1], m = 1, .., M \\ \sum_{m=1}^{M} v_m = 1, v_m \in [0, 1] \end{cases} \quad (23)$$

where, N is the number of data objects, M is the number of data views, $d_{ci,m} = \sqrt{\sum_{j=1}^{D_m} w_{j,m}^2 \left(x_{ij,m} - p_{cj,m}\right)^2}$ is the Euclidean distance between the ith data object and the cth cluster center in the mth data view, $U_m = \left\{u_{ci,m}\right\}_{CxN}$ is the object partition matrix of the mth data view, $W_m = \left\{w_{j,m}\right\}_{j=1 \div D_m}$ is the feature weight vector of the mth data view, $V = \{v_m\}_{1xM}$ is the weight vector of the views, D_i is the feature number of the ith data view. α is the collaborative learning coefficient between data views, β is the weight control coefficients of features and data views.

The U, W, V and P components of the J_{SWVF} objective function are calculated according to the following formulas (24)–(27).

$$u_{ci}^{*} = \begin{cases} 1 \text{ if } c = \underset{c \in N_{1,C}}{\arg\min} \left\{ \sum_{m=1}^{M} \sum_{\substack{j=1 \\ j \in v_m}}^{D_m} w_{j,m}^{\alpha} v_m^{\beta} \left\{p_{cj,m} - x_{ij,m}\right\}^2 \right\} \\ 0 \text{ otherwise} \end{cases} \quad (24)$$

$$w_{j,m}^{*} = 1 / \sum_{\substack{j'=1 \\ j' \neq j}}^{D_m} \left(\frac{E_{j,m}}{E_{j',m}}\right)^{1/(\alpha-1)} \text{ if } \alpha > 1 \quad (25)$$

where, $E_{j,m} = \sum_{c=1}^{C} \sum_{i=1}^{N} u_{ci,m} v_m^{\beta} \left(p_{cj,m} - x_{ij,m}\right)^2$.

$$v_m^{*} = 1 / \sum_{m'=1}^{M} \left(\frac{D_m}{D_{m'}}\right)^{1/(\beta-1)} \text{ if } \beta > 1 \quad (26)$$

where, $S_m = \sum_{c=1}^{C} \sum_{i=1}^{N_m} \sum_{\substack{j=1 \\ j \in v_m}}^{D_m} u_{ci,m} w_j^{\alpha} \left(p_{cj,m} - x_{ij,m}\right)^2$.

$$p^{*}_{cj,m} = \frac{\sum_{i=1}^{N_m} u_{ci,m} x_{ij,m}}{\sum_{i=1}^{N_m} u_{ci,m}} \quad (27)$$

The pseudocode of SWVF algorithm is given in Algorithm 3.

Algorithm 3. Simultaneous weighting on view and feature algorithm (SWVF)

Input: Dataset X, the number of clusters C, M views $X = \{X_m\}, m = \overline{1, M}$ and two fuzzy exponents α and β.
Output: The optimal U, P, W and V.

---

1. Randomly generate initial assignment matrix U;
2. For each view $\{v_m\}_{m=1}^{M} = 1/M$;
3. **Repeat**
4. Update P for given U^* according to (26);
5. Update W for given V^*, U^* and P^* according to (24);
6. Update V for given U^*, P^* and W^* according to (25);
7. Update U for given P^*, W^* and V^* according to (23);
8. Until (($\left|J^{(\tau)} - J^{(\tau-1)}\right| \leq \varepsilon$) or ($\tau \geq \tau_{max}$)).

The SWVF algorithm is an improved algorithm of the original K-means algorithm where the features and data views are considered extended by the weights W and V and the coefficients controlling the α and β weights, respectively. Therefore, the SWVF algorithm diagram is also added compared to that of the K-means algorithm. That is, Steps 5 and Step 6 are inserted between Step 4 and Step 7 in the diagram of algorithm 3 above. Thus, according to the diagram of algorithm 3, we can easily determine the computational complexity of the SWVF algorithm as O($\tau MCDN$) where τ is the number of iterations for the SWVF algorithm to achieve the convergence position, M is the number of data views, C is the number of data clusters, D = argmax(D_1, D_2, ..., D_m) is the largest feature number of the data views, and N is the number of data objects.

3.1.2 Experimental Evaluation

The clustering performance of the proposed SWVF algorithm is compared with three single-view clustering algorithms (LAC [46], EWKM [47], WkMeans [48]), and two multi-view clustering algorithms (TWKM [49], MVKKM [50]). Evaluation metrics such as AR and RI are used to quantify the performance of clustering algorithms. The experimental data consists of three multi-view data sets ALOI, MF and 3-Sources which have been introduced in Table 1. To use single-view clustering algorithms on multi-view data, we take an additional step of aggregating data views into one view by expanding the number of data features. That is, the number of features of the new datasets is equal to the total number of features of the data views combined.

The parameters of the algorithms are set according to the set of values listed in Table 3. The results are synthesized from 30 experiments on each data set. The performance of the experimental algorithms was quantified by the Acc and RI indices. The experimental results are summarized in Table 4.

In Table 4, the best values are highlighted in bold, while the worst values are highlighted in italics. According to the analysis results in Table 4, we can see that the clustering accuracy of the SWVF algorithm is quite high (80%). At the same time, the SWVF algorithm outperformed five other algorithms in both AR and RI indexes. Specifically, for the ALOI dataset, the values of the AR and RI indices obtained from the SWVF algorithm are 69.58% and 91.43% respectively, while the highest values of the remaining methods are 60.89% and 89.84%, respectively.he differences are approximately 8.69% and 1.59%. Similarly, for the MF dataset, the approximate differences are 0.17% and 0.0% for the 3-Sources dataset the approximate differences are –0.87% and 24.8%. From these results we can easily see, in Table 4, that the results obtained from the three multi-view clustering algorithms (TWKM [49], MVKKM [50], and SWVF) are better than the single clustering algorithms (LAC [46], EWKM [47], WkMeans [48]) on two datasets ALOI and MF. However, the same result does not occur for the 3-Source dataset, which has a smaller size and number of clusters and views than the two datasets ALOI and MF. In summary, according to the experimental

Table 3 Parameter setup of the six compared algorithms

Algorithms	Candidate parameters	Best parameters		
		ALOI	MF	3-sources
LAC	h = {1, 5, 10, 15, 20, 25, 30, 35, 40, 45, 50}	5	1	45
EWKM	η = {1, 5, 10, 15, 20, 25, 30, 35, 40, 45, 50}	5	5	50
WkMeans	β = {1, 2, 3,..., 20}	7	8	16
TWKM	λ = {1, 5, 10, 15, 20, 25, 30, 35, 40, 45, 50} 25 30 5 η = {1, 2, 3,..., 20}	25 8	30 7	50 10
MVKKM	α = {1.5, 2, 3, 4, 5, 6, 7, 8, 9, 10}	1.5	2	6
SWVF	α = {1, 2, 3,..., 20} β = {1, 5, 10, 15, 20, 25, 30, 35, 40, 45, 50}	5 10	8 30	4 40

Table 4 Clustering quality of the compared algorithms on the three real-world datasets

Algorithms	AR (%)			RI (%)		
	ALOI	MF	3-sources	ALOI	MF	3-sources
LAC	*37.71*	72.72	**46.29**	84.92	93.29	47.89
EWKM	37.76	*48.01*	39.35	84.96	*79.79*	38.38
WkMeans	51.36	78.16	34.93	*80.80*	94.68	38.61
TWKM	60.89	79.93	40.73	89.84	**95.29**	41.81
MVKKM	48.90	61.75	*34.32*	84.24	89.85	*27.02*
SWVF	**69.58**	**80.10**	45.42	**91.43**	**95.29**	**72.69**

results of clustering on three data sets of different sizes, the SWVF algorithm achieves better performance than the five clustering algorithms TWKM [49], MVKKM [50], LAC [46]., EWKM [47], WkMeans [48] on larger sized data sets.

3.2 Collaborative Feature-Weighted Multi-view Fuzzy c-Means Clustering Algorithm

3.2.1 Algorithm Description

In multi-feature data, the performance of the SWVF algorithm is achieved by considering the weights of features and views. However, the assignment of feature weights and views of the SWVF algorithm is done manually, so the efficiency is limited. MS. Yang et al. [43] proposed a method to automatically calculate the weights of features and views, and integrate the process of removing small-weighted features to increase clustering performance. The collaborative feature-weighted multi-view fuzzy c-means clustering algorithm (CFMF) is proposed on the basis of improving the objective function as follows. CFMF as follows:

$$J_{CFMF} = \sum_{m=1}^{M} v_m^{\beta} \left(\sum_{i=1}^{N_m} \sum_{c=1}^{C} u_{ci,m}^2 w_{j,m}^2 d_{ci,m}^2 + \alpha \sum_{\substack{m'=1 \\ m' \neq m}}^{M} \sum_{i=1}^{N_m} \sum_{c=1}^{C} \left(u_{ci,m} - u_{ci,m'} \right)^2 d_{ci,m}^2 \right) \tag{28}$$

The J_{CFMF} objective function (28) is minimized subject to the following constraints (29),

$$\begin{cases} \sum_{c=1}^{C} u_{ci,m} = 1, u_{ci} \in [0, 1], \forall i = 1, .., N, \forall m = 1, .., M \\ \sum_{j=1}^{D_m} w_{j,m} = 1, w_{j,m} \in [0, 1], \forall m = 1, .., M \\ \sum_{m=1}^{M} v_m = 1, v_m \in [0, 1] \end{cases} \tag{29}$$

where, N is the number of data objects, M is the number of data views, $d_{ci,m} = \sqrt{\sum_{j=1}^{D_m} w_{j,m}^2 \left(x_{ij,m} - p_{cj,m} \right)^2}$ is the Euclidean distance between the ith data object and the cth cluster center in the mth data view, $U_m = \left\{ u_{ci,m} \right\}_{CxN}$ is the object partition matrix of the mth data view, $W_m = \left\{ w_{j,m} \right\}_{j=1 \div D_m}$ is the feature weight vector of the mth data view, $V = \{v_m\}_{1xM}$ is the weight vector of the views, D_i is the feature number of the ith data view. α is the collaborative learning coefficient between data views, β is the weight control coefficients of features and data views.

The β and α coefficients are calculated according to the formulas (30) and (31).

$$\beta = \tau / M \tag{30}$$

$$\alpha = \tau / N \tag{31}$$

where, M is the number of data views, N is the number of data objects, τ is the actual number of iterations. The actual number of iterations t during the clustering processes will change the values of β and α after each iteration. The β parameter is used to distribute the weights of the views with their feature components. A larger value of β will make a clear difference between important and unimportant views. Therefore, a good value of β can automatically identify important feature components in each view. The coefficient parameters β and α will start from a small value and continuously increase until convergence.

An important contribution of the CFMF algorithm compared to the previous multi-view clustering algorithms is the feature reduction scheme by determining the unimportant feature components according to the formula (32). Simultaneously remove unimportant features from the data in the clustering process.

$$W_{j,m}^{(\tau)} < 1/\sqrt{N_m D_m} \tag{32}$$

where, $W_{j,m}^{(\tau)}$ is the weight of the jth feature of the mth view at the τth loop. The $W_{j,m}^{(\tau)}$ feature component satisfying the condition (32) is considered to have small or unimportant weight and will be removed from the data.

The components of the objective function are calculated in the form

$$u_{ci,m} = \left(1 - \sum_{l=1}^{C}\left(\frac{\varphi_{ci,m}}{1+\alpha}\right)\right)\left(\sum_{l=1}^{C}\frac{\sum_{j=1}^{M} w_{j,m}^2 \left(p_{ci,m} - x_{ij,m}\right)^2}{\sum_{j=1}^{D_m} w_{j,m}^2 \left(p_{li,m} - x_{ij,m}\right)^2}\right)^{-1} + \frac{\varphi_{ci,m}}{1+\alpha} \tag{33}$$

where, $\varphi_{ci,m} = \alpha \sum_{\substack{m'=1 \\ m' \neq m}}^{M} \sum_{i=1}^{N_m} \sum_{c=1}^{C} \left(u_{ci,m} - u_{ci,m'}\right)^2 \left(d_{ci,m}\right)^2$

$$v_m = \left(\sum_{r=1}^{M}\left(\frac{\sum_{i=1}^{N_m}\sum_{c=1}^{C}\sum_{j=1}^{D_m} u_{ci,m}^2 \left(p_{cj,m} - x_{ij,m}\right)^2 + \varphi_{ci,m}}{\sum_{i=1}^{N_r}\sum_{c=1}^{C}\sum_{j=1}^{Dr} u_{ci,r}^2 \left(p_{cj,r} - x_{ij,r}\right)^2 + \varphi_{ci,r}}\right)^{\frac{1}{\beta-1}}\right)^{-1} \tag{34}$$

$$w_{j,m} = \left(\sum_{j'=1}^{D_m} \frac{v_m^\beta \sum_{i=1}^{N_m} \sum_{c=1}^{C} u_{ci,m}^2 (x_{ij,m} - p_{cj,m})^2 + \varphi_{ci,m}}{v_m^\beta \sum_{i=1}^{N_m} \sum_{c=1}^{C} u_{ci,m}^2 (x_{ij',m} - p_{cj',m})^2 + \varphi_{ci,m'}} \right)^{-1} \tag{35}$$

where, $\varphi_{ci,m'} = \alpha \sum_{\substack{m'=1 \\ m' \neq m}}^{M} \sum_{i=1}^{N_m} \sum_{c=1}^{C} (u_{ci,m} - u_{ci,m'})^2 (x_{ij',m} - p_{cj',m})^2$

$$p_{cj,m} = \frac{v_m^\beta \sum_{i=1}^{N_m} u_{ci,m}^2 w_{j,m}^2 x_{ij,m} + \alpha \sum_{\substack{m'=1 \\ m' \neq m}}^{M} v_m^\beta \sum_{i=1}^{N_m} (u_{ci,m} - u_{ci,m'})^2 w_{j,m}^2 x_{ij,m}}{v_m^\beta \sum_{i=1}^{N_m} u_{ci,m}^2 w_{j,m}^2 + \alpha \sum_{\substack{m'=1 \\ m' \neq m}}^{M} v_m^\beta \sum_{i=1}^{N_m} (u_{ci,m} - u_{ci,m'})^2 w_{j,m}^2} \tag{36}$$

The global object partition matrix is determined by formula (37), that is, the sum of the view weights with the corresponding object partition matrix.

$$U = \sum_{m=1}^{M} v_m U_m \tag{37}$$

The CFMF clustering algorithm can be summarized as follows:

Algorithm 4. The CFMF Algorithm

Input: Dataset $X = \{x_1, x_2, \ldots, x_N\}$ with $x_i = \{x_{i,m}\}_{m=1}^{M}$ and $x_{i,m} = \{x_{ij,m}\}_{j=1}^{D_m}$, number of cluster C, and $\varepsilon > 0$.

Output: $w_{j,m}, v_m, p_{cj,m}, u_{ci,m}$ and $\bar{u}_{ci}$.

---

1. Initialization: Randomly generate initial U_m, Cluster centers P_m, initialize view weight $V^{(\tau)} = [v_m]_{1xM}$ (user may define $v_m = 1/M\, \forall m$), and set $\tau = 1$.

2. Calculate β and α by (29) and (30).

3. Compute the feature weight $W_m^{(\tau)}$ using $P_m^{(\tau)}$, $U_m^{(\tau-1)}$ and $V_m^{(\tau-1)}$ by (34).

4. Discard total D_r number of these j feature components for $W^{(\tau)}$ if $W^{(\tau)} < 1/\sqrt{N_m D_m}$, and set $D_m = D_{new} = D_m - D_r$.

5. Adjust $W_m^{(\tau)}$ by ($w_{j,m}^{(\tau)} = w_{j,m}^{(\tau)} / \sum_{h=1}^{D_{m,new}} w_{h,m}$.

6. Update the view weight $V_m^{(\tau)}$ using $P_m^{(\tau-1)}$, $U_m^{(\tau-1)}$ and $W_m^{(\tau)}$ by (33).

7. Update the cluster centers $P_m^{(\tau)}$ using $V_m^{(\tau)}$ and $U_m^{(\tau-1)}$ by (35).

8. Update membership $U_m^{(\tau)}$ using $W_m^{(\tau)}$, $P_m^{(\tau)}$ and α by (32).

9. If $\left(\left\|\left\|U_m^{(\tau)}\right\| - \left\|U_m^{(\tau-1)}\right\|\right\|\right) / N_m C < \varepsilon$, then stop;

Else set $\tau = \tau + 1$ and go back to 2.

10. Compute the global fuzzy partition matrix U by (36).

The overall computational complexity of the CFMF algorithm is $O(\tau N^2 CDM)$. Where, τ is the number of iterations for the CFMF algorithm to achieve the convergence position, M is the number of data views, C is the number of data clusters, $D = \text{argmax}(D_1, D_2, \ldots, D_m)$ is the largest feature number of the data views, N is the number of data objects.

3.2.2 Experimental Evaluation

To evaluate the performance of clustering algorithms, the clustering results of several multi-view clustering algorithms are compared with each other. In this chapter, we use the clustering results of Co-FKM [51], MultiNMF [52], WV-Co-FCM [53], and minimax-FCM [54] algorithms to compare with the Co-FW-MVFCM algorithm. Clustering performance is quantified by AR, RI, and NMI indexes. The three multi-view datasets MSRCV1, Caltech 101–7, and IS introduced in Table 1 are provided for clustering experiments.

The results are aggregated from 40 experiments on each data set, each experiment providing a set of values of the evaluation indicators. The result sets obtained on each data set are then analyzed to derive each set of minimum, maximum, and average values of each metric. The results of this analysis are presented in Table 5 for the MSRC-V1 dataset, Table 6 for the Caltech 101–7 dataset, and Table 7 for the IS dataset.

In results Tables 5, 6 and 7, the best values are highlighted in bold, while the worst values are highlighted in italics. The analysis results in Tables 5, 6 and 7 show that

Table 5 Classification results of Co-FKM, MultiNMF, WV-Co-FCM, Minimax FCM, proposed CFMF in MSRC-V1 data set

Index	Co-FKM	MultiNMF	WV-Co-FCM	Minimax-FCM	CFMF
AR	0.60/0.65/0.74	0.42/0.56/0.68	***0.17/0.25/0.31***	0.53/0.58/0.60	**0.61/0.66/0.76**
RI	0.85/0.87/0.89	0.81/0.84/0.87	***0.56/0.67/0.75***	0.85/0.86/0.87	**0.85/0.87/0.89**
NMI	**0.56/0.59/0.64**	0.42/0.50/0.59	***0.11/0.25/0.31***	0.51/0.54/0.55	0.48/0.55/0.63

Table 6 Classification results of Co-FKM, MultiNMF, WV-Co-FCM, Minimax FCM, and proposed CFMF in Caltech 101–7 data set

Index	Co-FKM	MultiNMF	WV-Co-FCM	Minimax-FCM	CFMF
AR	0.37/0.63/0.76	0.34/0.39/0.48	***0.27/0.49/0.75***	0.33/0.41/0.46	**0.64/0.68/0.72**
RI	0.62/0.72/0.79	0.66/0.69/0.73	***0.39/0.57/0.75***	0.60/0.63/0.66	**0.69/0.75/0.78**
NMI	**0.33/0.41/0.59**	0.38/0.43/0.50	***0.01/0.19/0.45***	0.14/0.19/0.27	0.31/0.35/0.40

Table 7 Classification results of Co-FKM, MultiNMF, WV-Co-FCM, Minimax FCM, and proposed CFMF in IS data set

Index	Co-FKM	MultiNMF	WV-Co-FCM	Minimax-FCM	CFMF
AR	0.36/0.51/0.64	0.48/0.59/0.62	***0.12/0.25/0.46***	0.28/0.50/0.52	**0.55/0.58/0.65**
RI	0.79/0.84/0.86	0.83/0.86/0.88	***0.14/0.61/0.79***	0.78/0.84/0.85	**0.84/0.86/0.87**
NMI	0.41/0.53/0.60	**0.53/0.57/0.63**	***0.00/0.16/0.57***	0.44/0.56/0.57	0.48/0.53/0.57

the clustering accuracy of the CFMF algorithm is quite high (76%). Simultaneously, the CFMF algorithm outperformed four other algorithms in both AR and RI indexes. However, it is easy to see that the values of the NMI index of the CFMF algorithm are not the highest compared to the other four algorithms.

To evaluate the performance of the CFMF algorithm according to the consumption time, the authors compared the execution time of the CoFW-MVFCM algorithm with four algorithms Co-FKM, MultiNMF, WV-Co-FCM, minimax-FCM on three data sets MSRC-V1, Caltech 101–7 and IS. In this experiment, the authors use optimal parameter sets determined by these algorithms. The results of the time consumed by the clustering algorithms are shown in Table 8.

According to the experimental results presented in Table 8, we easily see that the minimax-FCM algorithm consumes the least time, while the CFMF algorithm consumes the most time. However, in practice, to determine the optimal set of parameters for the algorithms Co-FKM, MultiNMF, WV-Co-FCM, Minimax-FCM they must consume a lot of time. That is, these algorithms have to execute many times on each data set to find the optimal set of parameters. Meanwhile, the CFMF algorithm only includes the weights of the feature and the data view. That is, the parameters $w_{j,m}$, α, and β which are automatically updated according to the formulas (17), (18), and (21). According to the above commentary, overall, we can evaluate that the CFMF algorithm is more efficient than the algorithms compared in Table 8 in terms of clustering quality and time efficiency.

Table 8 Comparison of total running times (in seconds) of the three algorithms on data sets MSRC-V1, Caltech 101–7 and IS

Data set	Co-FKM	MultiNMF	WV-Co-FCM	Minimax-FCM	CFMF
MSRC-V1	5387.09	430.81	1935.01	**77.47**	***5515.21***
Caltech 101–7	645.53	510.60	2278.89	**70.86**	***9748.34***
IS	3818.15	288.69	6739.38	**141.43**	***8562.76***

3.3 Multi-view Fuzzy Co-Clustering Algorithm

3.3.1 Algorithm Description

In the MVFCM algorithm [28], the authors determined the weights of features and data views by manual tests. That is, the determination of parameters is only relative, passive and depends entirely on the subjective factors of the user. It is easy to see that this is a limitation of the MVFCM algorithm.

To overcome the limitation of MVFCM, the authors [16] have proposed the SWFV algorithm, where they consider adding the parameters of the features of each view and the weight of different views. The fuzzy coefficients of these weights are also considered. In addition, these weighting coefficients are optimized after each learning loop. The authors in [43] not only consider the weights of features and views but also argue that it is possible to remove features with small weights to improve clustering performance. In addition, the α and β fuzzy coefficients of the views are also considered to change with each learning loop. In general, the multi-view clustering enhancements have improved the multi-view clustering performance. However, we believe that the problems that the improved algorithms care about the weight of features and views are only indirect and passive. That is, they get the weighted average as a common measure for all samples. This may cause a loss of important data. In fact, a large data set (many samples) may contain only a few anomalous samples manifesting in some of its features, which if using the averaging method to determine the weight of the feature will Ignore the features of the anomalous sample.

Fuzzy co-clustering (FCoC) is an unsupervised learning technique [4]. The FCoC algorithm considers data simultaneously in terms of objects and features through object membership function u_{ci} and feature membership function v_{cj}. In addition, the Euclidean distance between samples in the FCoC is also decomposed into the distance between their features. This makes the FCoC algorithm suitable for large data clustering, with many features and noise, and uncertainty. The objective function J_{FCoC} is expressed in the following form (38),

$$J_{FCoC} = \sum_{c=1}^{C} \sum_{i=1}^{N} \sum_{j=1}^{D} u_{ci} v_{cj} d_{cij} + T_U \sum_{c=1}^{C} \sum_{i=1}^{N} u_{ci} \log u_{ci} + T_V \sum_{c=1}^{C} \sum_{j=1}^{D} v_{cj} \log v_{cj} \quad (38)$$

The J_{FCoC} objective function (38) is minimized subject to the following constraints (39),

$$\begin{cases} \sum_{c=1}^{C} u_{ci} = 1, u_{ci} \in [0, 1], \forall i = 1, .., N \\ \sum_{j=1}^{D} v_{cj} = 1, v_{cj} \in [0, 1], \forall c = 1, .., C \end{cases} \quad (39)$$

where C is the number of clusters, N is the number of data objects, and D is the number of data features. $P = \{p_1, p_2, \ldots, p_C\}$ is centroids, $U = [u_{ci}]_{CxN}$ is the object membership function and $V = \left[v_{cj}\right]_{CxD}$ is the feature membership function.

d_{cij} is the square of the Euclidean distance between x_{ij} and p_{cj} given by:

$$d_{cij} = \left\|x_{ij}, p_{cjj}\right\|^2 = (x_{ij} - p_{cj})^2 \quad (40)$$

T_u and T_v are the weights that determine the degree of fuzzy. Increase T_u and T_v increase the opacity of the clusters and will be determined experimentally.

The components of the objective function J_{FCM} are calculated by (41)–(43).

$$u_{ci} = \frac{\exp\left(-\sum_{j=1}^{K} \frac{v_{cj} d_{cij}}{T_u}\right)}{\sum_{k=1}^{C} \exp\left(-\sum_{j=1}^{K} \frac{v_{kj} d_{kij}}{T_u}\right)} \quad (41)$$

$$v_{cj} = \frac{\exp\left(-\sum_{i=1}^{N} \frac{u_{ci} d_{cij}}{T_v}\right)}{\sum_{p=1}^{D} \exp\left(-\sum_{i=1}^{N} \frac{u_{ci} d_{cip}}{T_v}\right)} \quad (42)$$

$$p_{cj} = \frac{\sum_{i=1}^{N} u_{ci}^m x_{ij}}{\sum_{i=1}^{N} u_{ci}^m} \quad (43)$$

In order to contribute a new technique for multi-view clustering, motivated by the advanced ideas in the MVFCM [16], Co-KFM [28] and FCoC algorithms, the MVFCoC algorithm has been proposed for multi-view and high-dimensional data classification. The objective function of the MVFCoC algorithm is expressed in the following form (44),

$$J_{MVFCoC} = \begin{cases} \sum_{m=1}^{M} \sum_{c=1}^{C} \sum_{i=1}^{N_m} \sum_{j=1}^{D_m} u_{ci,m} v_{cj,m} d_{cij,m} \\ +\eta_u \frac{1}{M-1} \sum_{\substack{m'=1 \\ m' \neq m}}^{M} \sum_{c=1}^{C} \sum_{i=1}^{N_m} \sum_{j=1}^{D_m} (u_{ci,m'} - u_{ci,m}) d_{cij,m} + T_u \sum_{m=1}^{M} \sum_{c=1}^{C} \sum_{i=1}^{N_m} u_{ci,m} \log u_{ci,m} \\ +\eta_v \frac{1}{M-1} \sum_{\substack{m'=1 \\ m' \neq m}}^{M} \sum_{c=1}^{C} \sum_{i=1}^{N_m} \sum_{j=1}^{D_m} \left(v_{cj,m'} - v_{cj,m}\right) d_{cij,m} + T_v \sum_{m=1}^{M} \sum_{c=1}^{C} \sum_{j=1}^{D_m} v_{cj,m} \log v_{cj,m} \end{cases} \quad (44)$$

To get optimal clustering results, the objective function (44) is minimized subject to the following constraints (45),

$$\begin{cases} \sum_{c=1}^{C} u_{ci,m} = 1, u_{ci,m} \in [0, 1], \forall i = 1, .., N, \forall m = 1, .., M \\ \sum_{j=1}^{D_m} v_{cj,m} = 1, v_{cj,m} \in [0, 1], \forall c = 1, .., C, \forall m = 1, .., M \end{cases} \tag{45}$$

The components of the J_{MVFCoC} objective function are calculated according to the following formulas (46)–(48),

$$u_{ci,m} = \frac{e^{\frac{-\sum_{j=1}^{D_m} v_{cj,m} d_{cij,m} - \eta_u \frac{1}{M-1} \sum_{m'=1}^{M\,(m' \neq m)} \sum_{j=1}^{D_m} \left(u_{ci,m'} - 1\right) d_{cij,m}}{T_u} - 1}}{\sum_{k=1}^{C} e^{\frac{-\sum_{j=1}^{D_m} v_{kj,m} d_{ki,m} - \eta_u dfrac1M - 1 \sum_{m'=1}^{M\,(m' \neq m)} \sum_{j=1}^{D_m} \left(u_{ki,m'} - 1\right) d_{ki,m}}{T_u} - 1}} \tag{46}$$

$$p_{cj,m} = \frac{\sum_{i=1}^{N_m} u_{ci,m} x_{ij,m}}{\sum_{i=1}^{N_m} u_{ci,m}} \tag{47}$$

$$v_{cj,m} = \frac{e^{\frac{-\sum_{j=1}^{D_m} u_{ci,m} d_{cij,m} - \eta_v \frac{1}{M-1} \sum_{m'=1}^{M\,(m' \neq m)} \sum_{j=1}^{D_m} \left(v_{cj,m'} - 1\right) d_{cij,m}}{T_v} - 1}}{\sum_{k=1}^{C} e^{\frac{-\sum_{j=1}^{D_m} u_{ki,m} d_{ki,m} - \eta_v \frac{1}{M-1} \sum_{m'=1}^{M\,(m' \neq m)} \sum_{j=1}^{D_m} \left(v_{kj,m'} - 1\right) d_{kij,m}}{T_v}} - 1} \tag{48}$$

In MVFCoC algorithm, the idea of fuzzy co-clustering ensemble is adopted to combine individual view fuzzy partitions $u_{ci,m}$, $v_{cj,m}$ and obtain the global clustering result $\overline{u}_{ci}$, $\overline{v}_{cj}$. The consensus function is defined as the geometric mean of $\mathrm{u_{ci,m}}$, $\mathrm{v_{cj,m}}$ for each view and expressed as follows:

$$\overline{u}_{ci} = \sqrt[M]{\prod_{M=1}^{M} u_{ci,m}} \tag{49}$$

$$\overline{v}_{cj} = \sqrt[M]{\prod_{m=1}^{M} v_{cj,m}} \tag{50}$$

The MVFCoC algorithm diagram consists of the learning processes of partition matrixs U, V that are shown as Algorithm 5.

Algorithm 5. MVFCoC algorithm

Input: M data sets $X_m = \{x_{i,m}, x_{i,m} \in R^{D_m}\}$, $i = \overline{1, N_m}$, the number of clusters C.

Output: Clustering result.

1. Initialize parameters T_u, T_v, η_u, η_v, ε, η_1, η_2, ε, the maximum number of iterations τ_{max}.
2. Initialize $u_{ci,m}$ satisfying Eq. (44).
3. τ=1.
4. REPEAT
5. Update $p_{cj,m}$ using (46).
6. Calculate $d_{cij,m}$ using (39).
7. Update $v_{cj,m}$ using (47).
8. Update $u_{ci,m}$ using (45).
9. Update $\overline{u}_{ci}$ using (48).
10. Update $\overline{v}_{cj}$ using (49).
11. $\tau=\tau+1$.
12. UNTIL ($\left\| \left\| \overline{U}(\tau) \right\| - \left\| \overline{U}(\tau+1) \right\| \right\| < \varepsilon$) **or**($\tau > \tau_{max}$)

The overall computational complexity of the MVFCoC algorithm is $O(NCD_{max}M\tau_{max})$, where, N is the number of data objects, C is the number of clusters, D_{max} is the maximum number of features, M is the number of views, and τ_{max} is the maximum number of iterations.

3.3.2 Experimental Evaluation

In this section, we evaluate our proposed algorithm on real-world data. Experiments are conducted on four real-world benchmark datasets, i.e., 100leaves, MF, IS, and PEMS-SF.

The clustering results are evaluated by comparing the obtained label of each instance with the provided label by the dataset. Three metrics, the accuracy and the recall index and precision index are used to measure the clustering performance. To compare the effects of different clustering algorithms on multi-view clustering, MVFCoC is fed with other three clustering algorithms: The multi-view fuzzy clustering algorithm [51], the fuzzy co-clustering - FCoC algorithm [4] and the interval-valued fuzzy co-clustering - IVFCoC algorithm [5]. Table 9 show the comparison results which is quantified by Precision, Recall and Accuracy. Where the values in bold indicate the best performance results among the four algorithms. From Table 9, we observe that the MVFCoC algorithm obtained Precision, Recall and Accuracy indexes higher than Co-KFM, FCoC and IVFCoC algorithms. This means that the MvFCoC algorithm achieves higher classification accuracy than Co-KFM, FCoC and IVFCoC algorithms. In addition, the classification time of the MvFCoC algorithm is lower than Co-KFM, FCoC and IVFCoC algorithms.

To clarify the effectiveness of the MVFCoC algorithm, we have a look at a few comparisons below. For the Co-KFM algorithm, the MVFCoC algorithm is based on

Table 9 Clustering performance comparison of each algorithm in terms of precision, recall, and accuracy on four real-world datasets

Data sets	Algorithms	Pre	Rec	Acc
100leaves	Co-FKM	0.824	0.827	0.832
	FCCI	0.815	0.812	0.828
	IVFCoC	0.835	0.838	0.842
	MVFCoC	**0.848**	**0.843**	**0.852**
MF	Co-FKM	0.913	0.912	0.920
	FCCI	0.933	0.932	0.937
	IVFCoC	0.964	0.964	0.965
	MVFCoC	**0.974**	**0.974**	**0.975**
IS	Co-FKM	0.684	0.682	0.759
	FCCI	0.882	0.881	0.894
	IVFCoC	0.913	0.909	0.920
	MVFCoC	**0.924**	**0.923**	**0.929**
PEMS-SF	Co-FKM	0.737	0.727	0.788
	FCCI	0.946	0.946	0.949
	IVFCoC	0.965	0.964	0.965
	MVFCoC	**0.980**	**0.980**	**0.981**

the original FCoC algorithm which is dedicated to processing data with many features. Therefore, the algorithm MVFCoC will classify more accurately and faster than the Co-KFM algorithm. For the FCoC algorithm which is the original algorithm of the MVFCoC algorithm. There, the FCoC algorithm processes each data view and then aggregates their results into the general result. The MVFCoc algorithm concurrently processes views, then interacts between views using Eqs. (49) and (50). Therefore, the algorithm MvFCoC will get more accurate and faster classification results. For IVFCoC algorithm has higher computational complexity than the FCoC algorithm and the same operating mechanism as the FCoC algorithm. Therefore, the IVFCoC algorithm will process the data more slowly and less accurately than the MvFCoC algorithm.

3.4 Fuzzy Optimization Multi-objective Clustering Ensemble Model for Multi-source Data Analysis

3.4.1 Model Description

In fact, there are several variations of multi-view such as multi-subspace or multi-area, and multi-source is a generalized case of multi-view and multi-area. Meanwhile, multi-area and multi-source data is more complex than multi-view data but also contains a lot of dark knowledge that can be exploited to improve data clustering performance. To expand data clustering capabilities and exploit latent knowledge of complex data [55], a clustering ensemble model (FOMOCE) is proposed [45]. FOMOCE is not simply a multi-view data clustering algorithm but a multi-objective clustering ensemble model that can implement efficient multi-view data clustering. Because FOMOCE does not use a single clustering objective function but uses multiple base clustering objective functions, each base clustering is responsible for processing sub-datasets that come from a view or a subspace or a subarea or a source and exploit latent knowledge on these for the purpose of improving clustering performance.

Definition 4 The fuzzy multi-objective clustering ensemble model denoted Ω, is characterized by data coming from M different sources S_1, S_2, ..., S_M. The data is passed through P filter to classify the data source. The base clustering $\Pi = \{\Pi_1, \Pi_2, \ldots, \Pi_M\}$ is linked by L parallel processing fuzzy clustering techniques. An F classification technique is used to consensus the base clustering results and evaluate the quality of clusters to produce X clustering results of M data sources. The model Ω is represented by the following expression,

$$\Omega = \{S, P, \Pi, L, F, X, V\} \tag{51}$$

where,

(i) *Multi-source data input:* $S = \{S_1, S_2, \ldots, S_M\}$, $S = S_1 \cup S_2 \cup \ldots \cup S_M$.
(ii) *Data classification module:* $P = P(S)$
(iii) *Base clusterings:* $\Pi = \{\Pi_1, \Pi_2, \ldots, \Pi_M\}$, $\exists \Pi_i \neq \Pi_j$, *with* $i \neq j$, *i.e.* $\|\Pi\| \geq 2$
(iv) *Basic clustering link module:* $L = L(\Pi)$, $\Pi_i \infty \Pi_j$, $i, j = 1, M$ *với* $i \neq j$
(v) *Ensemble function of base clustering results:* $\Pi^* = argmin\Pi_1(S_1) \cup argmin\Pi_2(S_2) \cup \ldots \cup argmin\Pi_M(S_M)$.
(vi) *The model's clustering optimization function:* $X = argmin\ F(S, P, \Pi^*, L)$ *with* $X = \{X_1, X_2, \ldots, X_C\}$, $X = X_1 \cup X_2 \cup \ldots \cup X_C$; $X \Leftrightarrow S$: $S_1 \cup S_2 \cup \ldots \cup S_M = X_1 \cup X_2 \cup \ldots \cup X_C$, $||X|| = ||S||$; $\exists x_i \in X$, $i = 1 \div ||X||$, $\exists x_j \in S$, $i = 1 \div ||S||$, $x_i \equiv x_j$.
(vii) *Cluster quality assessment module:* $V = V(X)$

The functional diagram of the FOMOCE model is summarized in Fig. 1.

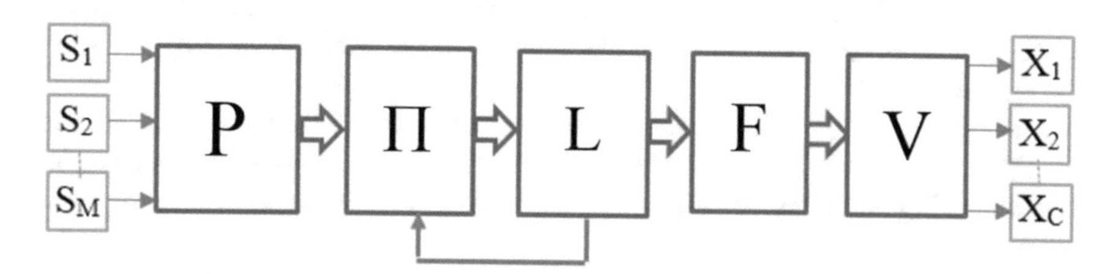

Fig. 1 FOMOCE model modularization diagram

3.4.1.1 Input Data S

The input data set is described by Eq. (52),

$$S = \{M, S, D\} \tag{52}$$

where, M is the number of input data sources, S = {S$_1$, S$_2$, ..., S$_M$}, $S_i = \{x_{i,j}\}$, $x_{i,j} \in R^{Di}$, $i = 1 \div M, j = 1 \div \|S_i\|$, $\|S_i\|$ is the number of data objects from the S_i. $D = \{D_1, D_2, \ldots, D_M\}$ is the number of features of M data sources. When $M = 1$, input data is collected from a single source, $M > 1$, data is collected from multiple sources.

3.4.1.2 Input Data Classifier P

The P input data classifier is described by Eq. (53),

$$P = \{S, R_P, f\} \tag{53}$$

where, S is the input data set described by Eq. (6), R_P is the rule set for classifying the data source, f is the value set that classifies the input data source. R_P is a function of f with parameter S. That is, f with S. That is $f = R_p(S)$ or $f_i = R_P(S_i)$, $i = 1 \div M$.

3.4.1.3 Set of Base Clustering

The set of base clustering methods, which includes the input data for each base cluster, the set of clustering algorithm selection rules, and the termination conditions of the base clustering, is described by:

$$\Pi = \{M, S, f, A, R_\Pi, E, O_\Pi\} \tag{54}$$

where, $\Pi = \{\Pi_1, \Pi_2, \ldots, \Pi_M\}$ includes M base clusterings. No plural $S = \{S_1, S_2, \ldots, S_M\}$ input data sets corresponding to the set of categorical values $f = \{f_1, f_2, \ldots, f_M\}$. $A = \{A_1, A_2, \ldots, A_M\}$ is the set of objective functions used for M base clusterings. R_Π is a set of rules for selecting the objective function for the base clusters based on the f classification value set. Mean, $A = R_\Pi(f)$ or $A_i = R_\Pi(f_i)$, $i = 1 \div M$.

$E = \{E_1, E_2, \ldots, E_M\}$ is the set of stop conditions for the base clusters. The stop condition E_i depends on the A_i clustering algorithm to determine whether the *ith* base clustering continues or ends the learning loop. That mean, $E = E(A)$ or $E_i = E(A_i)$, $i = 1 \div M$.

$O_\Pi = \{O_{\Pi 1}, O_{\Pi 2}, \ldots, O_{\Pi M}\}$ are the clustering results of the base clustering with $O_{\Pi 1} \cup O_{\Pi 2} \cup \ldots \cup O_{\Pi M} = S_1 \cup S_2 \cup \ldots \cup S_M$. Where, $O_{\Pi i} = \{O_{\Pi i,1}, O_{\Pi i,2}, \ldots, O_{\Pi i,C}\}$, $O_{\Pi i,1} \cup O_{\Pi i,2} \cup \ldots \cup O_{\Pi i,C} = S_i$ with $i = 1 \div M$ is the output of the ith base clustering.

3.4.1.4 The Link Module of Base Clusterings L

The link module of base clusterings L, which includes the clustering results of the base clusters in the learning loops, the global best clustering results, and the knowledge exchange links from and to the base clusterings, is described by:

$$L = \{O_{\Pi}, O_{\Pi G}, R_L\} \tag{55}$$

where O_{Π} is the set of clustering results obtained from the base clusterings in the learning loops. Clustering results include the membership function of data objects, list of objects and cluster center of each cluster in each data set. O_{Π} is the result of globally optimal clustering at the iteration steps of the base clusterings. O_{Π} is called dark knowledge in the base clusterings. R_L is the rule for defining and exchanging dark knowledge between the base clusterings.

3.4.1.5 Consensus Module F

The consensus module of the FOMOCE model, which includes input data, consensus algorithm, and final clustering result, is described by:

$$F = \{O_{\Pi}, A^*, X^*) \tag{56}$$

where, O_{Π} is the set of output clusters of base clusterings, that is $O_{\Pi} = \{O_{\Pi 1}, O_{\Pi 2}, \ldots, O_{\Pi M}\}$ which is the input data of consensus function. A^* is the objective function of the consensus algorithm. X^* is the output result of the consensus module.

Let $O_{\Pi i} = \{O_{\Pi i,1}, O_{\Pi i,2}, \ldots, O_{\Pi i,C}\}$ is the result of the base clustering O_{Π}. Then, $O_{\Pi i,j} = \{x_{i,j,1}, x_{i,j,2}, \ldots, x_{i,j,Mij}\}, j = 1 \div C$ is the jth cluster of the base cluster O_{Π}, that is $O_{\Pi} = \{O_{\Pi ij}\}_{MxC}$ includes MxC result clusters from M the base clusterings. Each cluster $O_{\Pi ij} = \{\overline{x}_1, \overline{x}_2, \ldots s, \overline{x}_{M_{ij}}\}, j = \overline{1, C}$ consists of objects derived from S_i which are represented by a cluster center $\overline{C_{ij}} \in R^{D_i}$. So, O_{Π} can be seen as a super-object. We perform $O_{\Pi} = \{O_1, O_2, \ldots, O_{MxC}\}_{MxC}$ is a set of super-objects that are represented by $\overline{C} = \{\overline{C}_1, \overline{C}_2, \ldots, \overline{C}_{MxC}\}$ is a set of cluster centers.

3.4.1.6 The Clustering Results Evaluating Module X*

The clustering results evaluation module including data sets of each cluster, cluster center and cluster quality assessment index is described by:

$$V = \{X^*, C^*, I^*\} \tag{57}$$

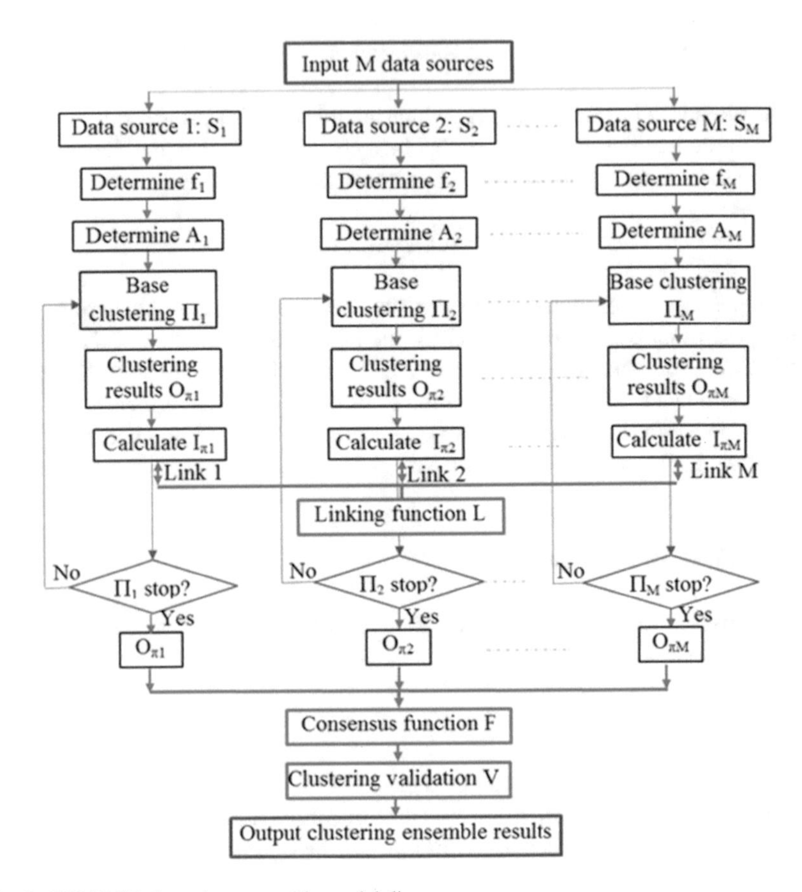

Fig. 2 FOMOCE clustering ensemble model diagram

where $X^* = \{X_1, X_2, \ldots, X_C\}, X_i = \{x_{i,j}\}, x_{i,j} \in R^D, i = 1 \div C, j = 1 \div ||X_i||$ are sets of data objects in result clusters corresponding to cluster $C^* = \{C_1, C_2, \ldots, C_C\}$. $I^* = \{I_1, I_2, \ldots, I_Q\}$ is the set of values of the final cluster quality assessment indicators.

3.4.1.7 Schematic of the FOMOCE Clustering Ensemble Model

Schematic diagram of the FOMOCE clustering ensemble model including S input data, data classification module, base clustering module, clustering consensus module, cluster quality assessment module, and clustering results and their detailed components are depicted in Fig. 2.

3.4.1.8 The Dark Knowledge and Derivation Rules of the FOMOCE Model

The term dark knowledge in machine learning was proposed by Geoffrey Hinton and Wang [55]. Dark knowledge refers to useful information that is often hidden or unused in big data or complex data processing models. If we can fully exploit this dark knowledge to train a ensemble algorithm, the performance of the algorithm will be improved. In the clustering ensemble, the results of the base clusterings, if not exploited, are also a kind of dark knowledge. For example, (i) the fuzzy matrix obtained by fuzzy c-means clustering contains specific knowledge besides the labels, but this knowledge is not used in traditional clustering ensembles; (ii) centroids represent another kind of dark knowledge, but are not used in traditional clustering ensembles.

(a) *The dark knowledge in the FOMOCE model*

Each input data source from receiving stations located in different environments with different characteristics. Therefore, each input data source may have a different uncertainty and number of data dimensions.

Definition 5 The dark knowledge in input data sources is their degree of uncertainty and number of features.

For the FOMOCE model, data were collected from many different sources, each with different characteristics. Therefore, the dark knowledge in data needs to be exploited to distinguish the different sources. Also, because the FOMOCE model is a clustering model of multiple objective functions. Therefore, dark knowledge in clustering objective functions should be exploited to take advantage of each clustering objective function corresponding to each appropriate data set. Here we will present methods of identifying and using dark knowledge in our FOMOCE model.

- Dark knowledge in data

Each different source is often affected by a different working environment, so the obtained data has different properties. In addition, data can be observed and projected by different dimensions. In the FOMOCE model, the dark knowledge in the data is classified into two categories. The first kind of knowledge based on the level of noise and uncertainty in each data source, called the degree of fuzziness. They are divided into three different levels: uncertainty, noise and uncertainty, very noisy and uncertainty. The second kind of knowledge is based on the degree of the feature. They are divided into three different levels: A small number of features, a high number of features and a very high number of features. These types of knowledge will be classified by us to serve the selection of processing techniques suitable for each data source.

- Dark knowledge in clustering techniques

Each algorithm to be accepted must identify the challenges facing and demonstrate the ability to solve those challenges. However, the more and more advanced technology, the demand of people is higher and higher, the more challenges are increasing. Hence an algorithm cannot solve all the problems that arise in the relevant field. In data mining, a clustering algorithm can only solve a few challenging problems. They may be suitable for this data type but not for other data types. In summary, each clustering algorithm has its own advantages and can be applied to solve a specific type of problem.

Definition 6 Dark knowledge in clustering techniques is the ability to solve a certain type of clustering problem.

In the FOMOCE model, we use five different algorithms for the base clustering A = {K-means, FCM, IT2FCM, FCCI, IVFCoC}. Each algorithm has its own strengths and weaknesses and is applied to solve specific problems. These clustering techniques all use a similar quantitative method based on the Euclidean distance measurement. Each algorithm has different advantages and disadvantages in terms of data type, computational complexity, and clustering accuracy. Usually, algorithms with low computational complexity are fast but with low accuracy, and vice versa, algorithms with high complexity are slow but with higher accuracy. In the FOMOCE model, we exploit the capabilities of five algorithms as dark knowledge to determine the appropriate clustering algorithm for each base clustering.

- Dark knowledge in the base clusterings

One of the fundamental differences between the FOMOCE model and the traditional cluster ensemble models is the existence of the L base clusterings linking module. In L, the base clusterings can exchange knowledge with each other during the learning loop. So what is the dark knowledge in the base clusterings, please refer to the definition below.

Definition 7 Dark knowledge in base clusterings is the parameters of clustering algorithms, cluster centers and membership functions obtained in learning loops.

(b) Derivation rules in the FOMOCE model

- The derivation rule of dark knowledge in multi-source data

Dark knowledge in data sources includes the degree of uncertainty and the degree of a feature. Where, the degree of uncertainty includes no noise and uncertainty, noise and uncertainty, a lot of noise and uncertainty; the degree of a feature includes small number of features, high number of features and very high number of features. In the FOMOCE model, the data source type is determined by the RP rule in algorithm 6. In the R_P rule, the two main parameters, the degree of the "noisy and uncertainty" and the "number of features" need to be supplied. In fact, the parameter "number of features" can be specified based on the input data. The "noisy and uncertainty" parameter is very difficult to determine which is usually qualitative and provided by the user.

Algorithm 6. The rule determines the type of data source:

***R_P*: ClassifySource(*S, M*)**

BEGIN

Initialize the $f = \{f_1, f_2, \ldots, f_M\}$.

For (i=1 to M)

If (S_i is no "noisy and uncertainty") and (S_i is "small number of features") then (f_i = "Simple");
If (S_i is "noisy and uncertainty") and (S_i is "small number of features") then (f_i = "Not simple");
If (S_i is much "noisy and uncertainty") and (S_i is "small number of features") then (f_i = "Quite complicated");
If (S_i is "noisy and uncertainty") and (S_i is "high number of features") then (f_i = "Complicated");
If (S_i is much "noisy and uncertainty") and (S_i is "high number of features") then (f_i = "Very complicated").

End For

END

- The selection rule of objective function for the base clusterings

The difference between the FOMOCE model and the traditional clustering ensemble models is the presence of many different objective functions, each base clustering uses a different objective function. The base clusterings instead of randomly selecting the objective function, choose the objective function based on the characteristics of the data. That is, the base clustering is based on dark knowledge in data sources and clustering techniques to choose the appropriate clustering objective function.

Algorithm 7. Rules for determining objective functions for base partitions

R_Π: *SelectObjectiveFunction(f, A)*

BEGIN

For (*i=1 to M*)

Select Case *f_i*:
"simple": return A_i (K-means);
"not simple": Return A_i (FCM);
"quite complicated": Return A_i (IT2FCM);
"complicated": Return A_i (FCCI);
"very complicated": Return A_i (IVFCoC);
End Select Case

End For

END

In the FOMOCE model, the four types of knowledge in the data are $f = \{$simple, not simple, quite complicated, complicated, and very complicated$\}$ and the five clustering objective functions are A = $\{$K-means, FCM, IT2FCM, FCCI, IVFCoC$\}$. Where K-means is is the simplest, FCM is not simple, IT2FCM is quite complicated,

FCCI is complicated and IVFCoC is very complicated. We have designed the rules to determine the objective function for the basis clusterings as in Algorithm 7.

- The dark knowledge derivation rule in the base clusterings

The dark knowledge derivation rule in the base clusterings is designed in the base clustering linking module. This design is seen as a new improvement over traditional clustering ensemble models. Specifically, in Module *L*, we design the rules for determining dark knowledge by quantifying the clustering performance at the learning loops using the cluster quality assessment indicators. Which base clustering has the best performance will extract knowledge to share with the remaining base clusterings for subsequent loops. The rules for defining and exchanging dark knowledge are designed as follows:

$$\begin{aligned} R_L : If\ \ (L(O_{\Pi i}) > L(O_{\Pi G}))\ then\ O_{\Pi G} = O_{\Pi i} \\ Else\ O_{\Pi i} = O_{\Pi G} \end{aligned} \tag{58}$$

where, *L(.)* is a function of clustering performance. $O_{\Pi i}$ is the result of the *i*th base clustering, $L(O_{\Pi i})$ is the performance of the *i*th base clustering. $O_{\Pi G}$ is the best clustering result globally, $L(O_{\Pi G})$ is the best clustering performance. The expression $O_{\Pi G} = O_{\Pi i}$ implies the knowledge sharing process from the best base clustering to the *i*th base clustering. Similarly, the expression $O_{\Pi i} = O_{\Pi G}$ implies the process of knowledge acquiring from the *i*th base clustering.

- The consensus objective function of the base clusterings

To get the final clustering result, a consensus function in Eq. (42) is used to group *M* results of base clusterings into *C* different clusters. Several consensus functions have been developed to produce the final data clustering result. Recently, we have introduced a clustering tendency assessment method SACT [56] applied in hyperspectral image classification. The SACT is viewed as a consensus function based on graph-based approaches. In the FOMOCE model, we use SACT as a consensus function to classify the partitions obtained from the base clusterings into the final clustering result. We first aggregate the partitions obtained from the base clusterings into a set of *MxC* partitions. Next, we represent the partitions as super-objects that are represented by cluster centers and data object lists. Then, the SACT algorithm is used to group the set of *MxC* super objects into *C* clusters which is the final clustering result.

3.4.1.9 The FOMOCE Algorithm

In the format of the FOMOCE mathematical model and the derivation rules described above, the procedure of the FOMOCE algorithm is shown in Algorithm 8.

Algorithm 8. Fuzzy multi-objective clustering ensemble algorithm FOMOCE

Input: M data sources $S_1, S_2, \ldots, S_M$

Output: C clusters.

1. Initialize classifier parameters $f = \{f_1, f_2, \ldots, f_M\}$, $A = \{A_1, A_2, \ldots, A_M\}$

2. Determine the classification of M data sources f_i= ClassifySource(S_i, M) according to Algorithm 4.

3. Initialize M parallel processing threads for M base clustering

4. Select objective function for each basic cluster Ai= SelectObjectiveFunction(f_i, A) according to Algorithm 5.

5. DO: Executing learning loops of base clusterings

  5.1. Update the parameters U, V of each base clustering.

  5.2. Update the cluster centers C of the base clusterings.

  5.3. Update and share knowledge according to R_L rules Eq. (43).

6. WHILE The stop condition of all base clusterings is satisfied.

7. Collect the results of the base clusterings and represent them as super-objects.

8. Consensus on the results of the base clusterings to get the result of final clustering X*.

9. Evaluate the clustering quality of X*.

10. Output the $O_{\Pi i}$ clustering results.

The outstanding features of the FOMOCE algorithm are the multi-source data and the multi-objective clustering function. These are the two basic elements for exploiting latent dark knowledge in data and processing techniques that form the difference and prominence of the FOMOCE model compared to the traditional clustering ensemble models.

The FOMOCE algorithm will achieve better efficiency by implementing flexible rules that allow the algorithm to be flexible in handling different situations without the need for a user. Specifically, in the FOMOCE algorithm, we have designed the stages of determining dark knowledge in different data sources and selecting suitable clustering algorithms to process data coming from the respective data sources. At the same time, we also designed a module that controls the linking and sharing of dark knowledge between the base clusterings. In theory, these are very significant contributions to support reciprocity between the base clusterings to speed up and improve the clustering quality of the FOMOCE ensemble model.

3.4.2 Experimental Evaluation

In this section, we will present some experimental results to simulate the working mechanism of the FOMOCE model and demonstrate the effectiveness of the proposed clustering consensus method. For a fair comparison, we have installed a clustering experiment along with state-of-the-art clustering methods such as MKCE [57], eFCoC [58], and FCME [59]. These are single-source clustering ensemble models. To experiment on multi-source datasets, we implement experimentally multi-view

Table 10 Three single-source data sets from the UCI repository

Dataset	No. of clusters	No. of objects	No. of features
Avila	12	20,867	10
Chess	18	28,056	6
Farm-Ads	2	4143	54,877

clustering methods which is a special case of multi-source data. Multi-view clustering methods include WCoFCM [10], Co-FKM and Co-FCM [60].

To evaluate the clustering performance of the proposed approach, we use three standard evaluation metrics, i.e., clustering accuracy (ACC), Davies–Bouldins index (DBI), and partition coefficient (PC). These indicators have been widely used in the field of clustering. Please note that the larger the values of ACC and PC metrics and the smaller the value of the DBI metric, the better the performance of the algorithm.

3.4.2.1 Single-Source Data Clustering

In this subsection, we collect the results of experiments on three single-source datasets Avila, Chess, and Farm Ads. These data sets come from the UCI Machine Learning Repository. Table 10 provides some of the basic properties of clustering data sets.

Three clustering ensemble algorithms, i.e. MKCE, FCME, eFCoC are involved in the experimental process. To use clustering consensus models on single-source datasets, we divide these datasets into five data subsets. Since the five data subsets come from a single source, in the FOMOCE model we consider them the same. We design five base clusterings corresponding to five algorithms K-mean, FCM, IT2FCM, FCCI, and IVFCoC. Each base clustering corresponds to a single algorithm and a data subset. Experimental results are reported in Table 11.

3.4.2.2 Multi-view Data Clustering

In this subsection, we collect experimental results on two multi-view datasets such as 6Dims and Multiple Features. Where 6Dims includes six data subsets from the Computing University of Eastern Finland. Each subset consists of 1024 objects evenly distributed and ordered in 16 clusters and the number of features is 32, 64, 128, 256, 512, and 1024, respectively. We assume that six data subsets as copies of an original data set by projecting on six spaces having the number of dimensions 32, 64, 128, 256, 512, and 1024 respectively. Therefore, we consider the 6Dims as the multi-view data set. Multiple Features data set comes from the UCI Machine Learning Repository. Table 16 provides some of the basic properties of two multi-view datasets, where D1 $\div$ D6 is the number of features in each source.

Three multi-view clustering methods, i.e., WCoFCM [10], Co-FKM and Co-FCM [12] were involved in the experimental process. To use the FOMOCE model

Table 11 Clustering performance (ACC, PC and DBI) of different clustering algorithms on three single- source datasets

Data sets	Algorithms	ACC	PC	DB
Avila	eFCoC	0.92	0.96	0.62
	MKCE	0.79	0.71	4.23
	FCME	0.82	0.77	3.95
	FOMOCE	**0.98**	**0.98**	**0.45**
Chess	eFCoC	0.93	0.92	0.73
	MKCE	0.80	0.74	3.15
	FCME	0.83	0.81	2.87
	FOMOCE	**0.98**	**0.98**	**0.52**
FarmAds	eFCoC	0.95	0.96	0.72
	MKCE	0.63	0.70	5.96
	FCME	0.66	0.71	5.24
	FOMOCE	**0.97**	**0.98**	**0.59**

Table 12 Clustering performance (ACC, PC and DBI) of different clustering algorithms on two multi-view datasets

Data sets	Algorithms	ACC	PC	DB
6Dims	Co-FKM	0.79	0.65	3.85
	Co-FCM	0.80	0.71	3.53
	WCoFCM	0.84	0.79	3.25
	FOMOCE	**0.97**	**0.98**	**0.55**
MF	Co-FKM	0.83	0.85	2.82
	Co-FCM	0.84	0.85	2.66
	WCoFCM	0.86	0.87	2.56
	FOMOCE	**0.93**	**0.94**	**0.65**

on multi-view datasets, we initialize the number of base clusterings corresponding to the number of views of each data set. Each base clustering chooses an algorithm using the R_{Π} rule. Experimental results are reported in Table 12.

3.4.2.3 Discussion

Tables 11 and 12 report the means of the ACC, PC, and DBI values obtained by the clustering algorithms in fifty runs on single-source and multi-view dataset. These results clearly indicate that the FOMOCE model is the best among all these methods. The experimental results of the proposed algorithm indicate that multi-objective together with dark knowledge is an effective way to enhance the performance of the clustering algorithm.

4 Some Research Directions on Multi-view Data

Multi-view data in particular and multi-source, in general, are concepts that will become increasingly popular. Because in reality, data collection does not depend only on a single receiver or sensor but will be gathered from many different destinations. Each destination is treated as a source or a view, so the data becomes multiple views. For example, in news gather, the same issue is received by many different reporters, who all try to process to get the best article for their magazine. Therefore, the articles will have different accuracy. The crawler will collect all the articles and filter out the main content of interest. With such characteristics, the data becomes more complex and larger in size because it is collected from many different views, the views do not have the same, the same number of features, the certainty. To analyze such data, single clustering techniques cannot be performed, but multi-source, multi-view specialized clustering techniques are needed.

Within the framework of this chapter, we suggest some problems that are oriented to research and develop the application of multi-source, multi-view clustering techniques as follows:

4.1 Information Search Problem

In information retrieval and synthesis, the news is often presented in different formats and languages and conveyed in different media. Therefore, it is necessary to have powerful enough data analysis tools to extract the desired information. However, this is considered a difficult problem because it faces challenges such as multiple sources, heterogeneity, uncertainty, and big data.

4.2 The Problem of Ground Observation Data Analyzing

Today, the problem of ground observation is widely applied in a number of important fields such as forestry, agricultural, environmental and military monitoring applications. Ground observation technology is also increasingly diverse such as super-spectral remote sensing, multi-spectral remote sensing, and radar satellite. In addition, due to the large coverage area, one sensor cannot see all the surfaces of interest, so many different sensors are needed. Therefore, remote sensing data is often aggregated from many different revenue sources and has a large size. In addition, the climate and weather of different observation areas also create uncertainty and uniformity of remote sensing data. To analyze ground observation data, it is necessary to have a powerful enough multi-source data analysis tool for the applications of detecting changes in the earth's surface, anomaly identification, and classification of targets on land, sea, and islands, in the forest, etc.

4.3 The Problem of System Management and Operation

Today, with the development of information and communication technology, the management of production, business, or social organizations has gone beyond the limit of one location to become a global connection. The method of information collection and processing also changes in the direction of multiple sources of revenue and centralized information processing to improve the accuracy of the information obtained. For example, as for pharmaceutical production and business management and administration systems, the system no longer collects information about drug use demand or human resource supply needs in one place or country to Business planning for all different branches. Because, each country has different economic conditions, environment, and human resources, so the need to use different drugs. As well as finished products in different places combined with different consumption needs, leading to different production and business plans. Thus, to solve the problem of data analysis for management and operating systems, we need a powerful enough tool to analyze multi-view data and big data.

4.4 The Problem of Predicting the Status

Recently, due to the development of intelligent computing techniques such as artificial intelligence, cloud computing, big data, and blockchain technology, the demand for status predicting has increased. Status predicting is related to the need for early prediction of equipment failure, human health, weather conditions, natural disasters, and diseases caused by nature. Based on the results of the early forecast of the condition, we can plan for the prevention, cure of diseases, or against possible disasters. However, in order to accurately forecast the status of any type, we must obtain the past and present status of the objects involved. In addition, the application of condition prediction only makes sense if it is possible to predict the condition for many different objects, at least of the same category. Then, we must collect past and present status data of as many subjects as possible. Thus, to solve the problem of status prediction, we need a powerful enough tool to analyze multi-view data and big data.

5 Conclusion and Future Works

Multi-view data has challenges such as large size, heterogeneity, and uncertainty, but there is also a lot of knowledge hidden in the views. The development of multi-view data clustering techniques is inevitable to replace traditional clustering techniques in modern data analysis problems. In this chapter, we investigate the overview of multi-view data clustering algorithms. The survey results show that multi-view data clustering techniques are developing rapidly. Each algorithm has its own advantages and limitations, suitable for each specific problem. Finally, to provide readers with the choice of a suitable algorithm for their multi-view data clustering problem. We focus on presenting four new and most promising multi-view data clustering algorithms. Some experimental results are also presented to demonstrate the effectiveness of clustering algorithms in some specific data analysis problems.

In the future, we will implement parallel multi-view clustering algorithms on GPUs to solve the complex problems that were pointed in this chapter.

References

1. Miyamoto, S., Ichihashi, H., Honda, K.: Algorithms for Fuzzy Clustering. Springer: Studies in Fuzziness and Soft Computing, vol. 229 (2008)
2. Xu, D., Tian, Y.: A comprehensive survey of clustering algorithms. Ann. Data Sci. **2**(2), 165–193 (2015)
3. Bezdek, J.C., Ehrlich, R., Full, W.: The fuzzy C-means clustering algorithm. Comput. Geosci. **10**(2–3), 191–203 (1984)
4. Hanmandlua, M., Verma, O.P., Susan, S., Madasu, V.K.: Color segmentation by fuzzy co-clustering of chrominance color features. Neurocomputing **120**, 235–249 (2013)
5. Van Nha, P., Long, N.T., Pedrycz, W.: Interval-valued fuzzy set approach to fuzzy co-clustering for data classification. Knowl.-Based Syst. **107**, 1–13(2016)
6. Wierzchon, S., Kłopotek, M.: Modern Algorithms of Cluster Analysis, Springer: Studies in Big Data, vol. 34 (2018)
7. Dong, X., Yu, Z., Cao, W., Shi, Y., Ma, Q.: A survey on ensemble learning. Front. Comput. Sci. **14**, 241–258 (2020)
8. Van Nha, P., Long, P.T., Pedrycz, W., Long, N.T.: Feature-reduction fuzzy co-clustering approach for hyperspectral image analysis. Knowl.-Based Syst. **216**, 106549 (2021)
9. Yang, M.S., Nataliani, Y.: A feature-reduction fuzzy clustering algorithm based on feature-weighted entropy. IEEE Trans. Fuzzy Syst. **26**(2), 817–835 (2018)
10. Okun, O., Valentini, G., Re, M.: Ensembles in Machine Learning Applications, Springer: Studies in Computational Intelligence, vol. 373 (2011)
11. Adil, F., Najlaa, A., Zahir, T., Abdullah, A., Ibrahim, K., Zomaya, A.Y., Sebti, F.: A survey of clustering algorithms for big data: taxonomy and empirical analysis. IEEE Trans. Emerg. Top. Comput. **2**(3), 267–279 (2014)
12. Chen, C., Wang, Y., Hu, W., Zheng, Z.: Robust multi-view k-means clustering with outlier removal. Knowl.-Based Syst. **210**, 106518 (2020)
13. Jing, Z., Xijiong, X., Xin, X., Shiliang, S.: Multi-view learning overview: recent progress and new challenges. Inf. Fusion **38**, 43–54 (2017)
14. Fu, L.: An overview of recent multi-view clustering. Neurocomputing **402**, 148–161 (2020)
15. Guillaume, C., Matthieu, E., Lionel, M., Sublemontier, J.H.: CoFKM: a centralized method for multiple-view clustering. In: 2009 Ninth IEEE International Conference on Data Mining (2009)
16. Jiang, B., Qiu, F., Wang, L.: Multi-view clustering via simultaneous weighting on views and features. Appl. Soft Comput. **47**, 304–315 (2016)
17. Chang, S., Hu, J., Li, T., Wang, H., Peng, B.: Multi-view clustering via deep concept factorization. Knowl.-Based Syst. **217**, 106807 (2021)
18. Hu, Z., Nie, F., Chang, W., Hao, S., Wang, R., Li, X.: Multi-view spectral clustering via sparse graph learning. Neurocomputing **384**, 1–10 (2020)
19. Lin, S.X., Zhong, G., Shu, T.: Simultaneously learning feature-wise weights and local structures for multi-view subspace clustering. Knowl.-Based Syst. **205**, 106280 (2020)

20. Yan, F., Wang, X., Zeng, Z., Hong, C.: Adaptive multi-view subspace clustering for high-dimensional data. Pattern Recognit. Lett. **130**, 299–305 (2020)
21. Zheng, Q., Zhu, J., Li, Z., Pang, S., Wang, J., Li, Y.: Feature concatenation multi-view subspace clustering. Neurocomputing **379**, 89–102 (2020)
22. Chaudhuri, K., Kakade, S.M., Livescu, K., Sridharan, K.: Multi-view clustering via canonical correlation analysis. In: Proceedings of the 26th Annual International Conference on Machine Learning, pp. 129–136 (2009)
23. Bickel, S., Scheffer, T.: Multi-view clustering. In: Proceedings of the 4th IEEE International Conference on Data Mining, pp. 19–26 (2004)
24. Sa, V.R.D.: Spectral clustering with two views. In: Proceedings of the 22th IEEE International Conference on Machine Learning, pp. 20–27 (2005)
25. Blaschko, M.B., Lampert, C.H.: Correlational spectral clustering. In: Proceedings of IEEE Conference on Computer Vision and Pattern Recognition, pp. 1–8 (2008)
26. Zhou, D., Burges, C.J.C.: Spectral clustering and transductive learning with multiple views. In: Proceedings of the 12th IEEE International Conference on Data Mining, pp. 675–684 (2012)
27. Bruno, E., Marchand-Maillet, S.: Multiview clustering: a late fusion approach using latent models. In: Proceedings of the 32nd International ACM SIGIR Conference on Research and Development in Information Retrieval, pp. 736–737 (2009)
28. Pedrycz, W.: Collaborative fuzzy clustering. Pattern Recognit. Lett. **23**, 1675–1686 (2002)
29. Zhu, H., Zhou, M.C.: Efficient role transfer based on Kuhn-Munkres algorithm. IEEE Trans. Syst., Man, Cybern.-Part A: Syst. Hum. **42**(2), 491–496 (2012)
30. Jiang, Y.: Collaborative fuzzy clustering from multiple weighted views. IEEE Trans. Cybern. **45**(4), 688–701 (2015)
31. Shudong, H., Zhao, K., Tsang, W., Zenglin, X.: Auto-weighted multi-view clustering via kernelized graph learning. Pattern Recognit. **88**, 174–184 (2019)
32. Shizhe, H., Xiaoqiang, Y., Yangdong, Y.: Dynamic auto-weighted multi-view co-clustering. Pattern Recognit. **99**, 107101 (2020)
33. Nie, F.: Auto-weighted multi-view co-clustering via fast matrix factorization. Pattern Recognit. **102**, 107207 (2020)
34. Yu, H.: Multi-view clustering by exploring complex mapping relationship between views. Pattern Recognit. Lett. **138**, 230–236 (2020)
35. Sharma, K.K., Seal, A.: Outlier-robust multi-view clustering for uncertain data. Knowl.-Based Syst. **211**, 106567 (2021)
36. Zeng, S., Wang, X., Cui, H., Zheng, C., Feng, D.: A unified collaborative multikernel fuzzy clustering for multiview data. IEEE Trans. Fuzzy Syst. 1671–1687 (2018)
37. Ding, Z., Zhao, H., Fu, Y.: Learning Representation for Multi-view Data Analysis, Advanced Information and Knowledge Processing. Springer Nature Switzerland AG (2019)
38. Deepak, P., Jurek-Loughrey, A.: Linking and Mining Heterogeneous and Multi-view Data, Unsupervised and Semi-Supervised Learning. Springer Nature Switzerland AG (2019)
39. Dunn, J.: A fuzzy relative of the ISODATA process and its use in detecting compact well-separated clusters. J. Cybern. **3**, 32–57 (1973)
40. Kun, Z., Chaoxi, N., Changlu, C., Feiping, N., Changqing, Z., Yi, Y.: Graph structure fusion for multiview clustering. IEEE Trans. Knowl. Data Eng. **31**(10), 1984–1993 (2019)
41. Bezdek, J.C.: Cluster validity with fuzzy sets. J. Cybern. **3**, 58–73 (1974)
42. Davies, D.L., Bouldin, D.W.: A cluster separation measure. IEEE Trans. Pattern Anal. Mach. Intell. **1**(2), 224–227 (1979)
43. Yanga, M.S., Sinaga, K.P.: Collaborative feature-weighted multi-view fuzzy c-means clustering. Pattern Recognit. **119**, 108064 (2021)
44. Binh, L.C., Van Nha, P., Long, P.T.: Multi-view fuzzy co-clustering algorithm for high-dimensional data classification. In: The 23th National Symposium of Selected ICT Problems (2020)
45. Binh, L.C., Van Nha, P., Long, N.T.: Fuzzy optimization multi-objective clustering ensemble model for multi-source data analysis. In: IFSA-EUSFLAT 2021 - The 19th World Congress of the International Fuzzy Systems Association (2021)

46. Domeniconi, C., Gunopulos, D., Ma, S., Yan, B., Al-Razgan, M., Papadopoulos, D.: Locally adaptive metrics for clustering high dimensional data. Data Min. Knowl. Discov. **14**(1), 63–97 (2007)
47. Liping, J., Michael, K., Joshua, Z.H.: An entropy weighting k-means algorithm for subspace clustering of high-dimensional sparse data. IEEE Trans. Knowl. Data Eng. **19**(8), 1026–1041 (2007)
48. Huang, J.Z., Rong, H., Li, Z.: Automated variable weighting in k-means type clustering. IEEE Trans. Pattern Anal. Mach. Intell. **27**(5), 657–668 (2005)
49. Chen, X., Xu, X., Huang, J., Ye, Y.: Tw-k-means: automated two-level variable weighting clustering algorithm for multi-view data. IEEE Trans. Knowl. Data Eng. **25**(4), 932–944 (2013)
50. Tzortzis, G., Likas, A.: Kernel-based weighted multi-view clustering. In: IEEE 12th International Conference on Data Mining, pp. 675–684 (2012)
51. Cleuziou, G., Exbrayat, M., Martin, L., Sublemontier, J.H.: CoFKM: a centralized method for multiple-view clustering. In: Proceedings of the IEEE International Conference on Data Mining (ICDM'09), pp. 752–757 (2009)
52. Liu, J., Wang, C., Gao, J., Han, J.: Multi-view clustering via joint nonnegative matrix factorization. In: Proceedings of the SIAM International Conference on Data Mining, SDM, pp. 252–260 (2013)
53. Jiang, Y., Chung, F.L., Wang, S., Deng, Z., Wang, J., Qian, P.: Collaborative fuzzy clustering from multiple weighted views. IEEE Trans. Cybern. **45**, 688–701 (2015)
54. Wang, Y., Chen, L.: Multi-view fuzzy clustering with minimax optimization for effective clustering of data from multiple sources. Expert Syst. Appl. **72**, 457–466 (2017)
55. Wenting, Y., Hongjun, W., Shan, Y., Tianrui, L., Yan, Y.: Nonnegative matrix factorization for clustering ensemble based on dark knowledge. Knowl.-Based Syst. **163**, 624–631 (2019)
56. Van Nha, P., Long, P.T., Thao, N.D., Long, N.T.: A new cluster tendency assessment method for fuzzy co-clustering in hyperspectral image analysis. Neurocomputing **307**, 213–226 (2018)
57. Baia, L., Lianga, J., Cao, F.: A multiple k-means clustering ensemble algorithm to find nonlinearly separable clusters. Inf. Fusion **61**, 36–47 (2020)
58. Binh, L.C., Long, N.T., Van Nha, P., Long, P.T.: A new ensemble approach for hyperspectral image segmentation. In: Conference on Information and Computer Science (NICS) (2018)
59. Baraldi, P., Razavi-Far, R., Zio, E.: Bagged ensemble of fuzzy C-means classifiers for nuclear transient identification. Ann. Nucl. Energy **38**(5), 1161–1171 (2011)
60. Boongoen, T., Iam-On, N.: Cluster ensembles: a survey of approaches with recent extensions and applications. Comput. Sci. Rev. **28**, 1–25 (2018)
61. Kun, Z., Feiping, N., Jing, W., Yi, Y.: Multiview consensus graph clustering. IEEE Trans. Image Process. **28**(3), 1261–1270 (2019)

Rethinking Collaborative Clustering: A Practical and Theoretical Study Within the Realm of Multi-view Clustering

Pierre-Alexandre Murena, Jérémie Sublime, and Basarab Matei

Abstract With distributed and multi-view data being more and more ubiquitous, the last 20 years have seen a surge in the development of new multi-view methods. In unsupervised learning, these are usually classified under the paradigm of multi-view clustering: A broad family of clustering algorithms that tackle data from multiple sources with various goals and constraints. Methods known as collaborative clustering algorithms are also a part of this family. Whereas other multi-view algorithms produce a unique consensus solution based on the properties of the local views, collaborative clustering algorithms aim to adapt the local algorithms so that they can exchange information and improve their local solutions during the multi-view phase, but still produce their own distinct local solutions.

In this chapter, we study the connections that collaborative clustering shares with both multi-view clustering and unsupervised ensemble learning. We do so by addressing both practical and theoretical aspects: First we address the formal definition of what is collaborative clustering as well as its practical applications. By doing so, we demonstrate that pretty much everything called collaborative clustering in the literature is either a specific case of multi-view clustering, or misnamed unsupervised ensemble learning. Then, we address the properties of collaborative clustering methods, and in particular we adapt the notion of clustering stability and propose a bound for collaborative clustering methods. Finally, we discuss how some of the properties of collaborative clustering studied in this chapter can be adapted to broader contexts of multi-view clustering and unsupervised ensemble learning.

Keywords Collaborative clustering · Multi-view clustering · Stability

P.-A. Murena (✉)
Helsinki Institute for Information Technology HIIT, Department of Computer Science, Aalto University, Helsinki, Finland
e-mail: pierre-alexandre.murena@aalto.fi

J. Sublime
ISEP, School of Engineering, 10 rue de Vanves, 92130 Issy-Les-Moulineaux, France
e-mail: jsublime@isep.fr

J. Sublime · B. Matei
LIPN–CNRS UMR 7030, LaMSN–Sorbonne Paris North University, 93210 St Denis, France
e-mail: matei@lipn.univ-paris13.fr

W. Pedrycz and S. Chen (eds.), *Recent Advancements in Multi-View Data Analytics*, Studies in Big Data 106, https://doi.org/10.1007/978-3-030-95239-6_4

1 Introduction

Clustering techniques play a central role in various part of data analysis and are key to finding important clues concerning the structure of a data sets. In fact, clustering is often considered to be the most commonly used tool for unsupervised exploratory data analysis. However due to an explosion both in the number of frequently occurring multi-view data (be it with organic views or with "artificial views" created by different feature extraction algorithms), and also the number and diversity of available clustering methods to tackle them, there has been a surge in the number of clustering methods that are either multi-view, multi-algorithm, or both.

While regular clustering itself presents its own challenges (the most common of which is to find which methods are "best" for a given task), these new paradigms, involving multiples views and sometimes multiple clustering algorithms, make the problem even more complex. Yet, despite an extensive literature on multi-view clustering, unsupervised ensemble learning and collaborative clustering, very little is known about the theoretical fundations of clustering methods belonging to these families of algorithms. Furthermore, while it is easy to tell the difference between unsupervised ensemble learning and multi-view clustering, the third family of algorithm—namely collaborative clustering [23, 33]—which is also the most recent of the three is a lot more problematic in the sense that this notion is ill-defined in the literature and that it shares many similarities with both multi-view clustering and ensemble learning, both in terms of practical applications, but also when it comes to the algorithms used.

To address these issues with a special focus on collaborative clustering, in this chapter we propose the following main contributions:

- First, we propose formalization of the three notions of collaborative clustering, multi-view clustering and unsupervised ensemble learning. And from it, we clearly define what are the foundations of collaborative clustering.
- Second, we show that collaborative clustering is very much related and overlapping with multi-view clustering and ensemble learning. We also show that ultimately multi-view clustering and collaborative clustering are equivalent to regular clustering.
- We define the notion of pure collaborative clustering algorithms, a notion that guaranties the definitive aspect of the results produced by such algorithms.
- We introduce two key notions, namely novelty and consistency, that can be used as quality metric for both collaborative clustering and unsupervised ensemble learning.
- Then, we lay the groundwork for a theory of multi-view and collaborative clustering approaches, by extending the notion of clustering stability originally proposed by Ben-David et al. [2]. We propose three complementary characterizations of stable multi-view clustering algorithms, involving in particular the stability of the local algorithms and some properties of the collaboration.

We tackle these issues all the while trying to keep a theoretical setting as generic as possible.

Chapter organisation This chapter is organized as follows: In Sect. 2 we introduce the state of the art and current situation of collaborative clustering. Section 3 introduces the main notations and concepts that will be used in this chapter, and it formalizes some definitions of collaborative and multi-view clustering. Based on these formal definitions, it introduces the interconnections between all these tasks. Section 4 presents the main contributions of our chapter, presentation and the study of several theoretical aspects and properties of collaborative clustering. Section 5 lists various open questions subsequent to the theoretical framework we presented. We conclude the chapter with a brief discussion on how the formal classification of these fields could help future research.

2 Collaborative Clustering: State of the Art of a Polymorphic Notion with Very Diverse Applications

Before we start to describe the different variations of collaborative clustering, we first present, in Table 1 below, the full spectrum of methods often falling under the umbrella of collaborative clustering in the literature. We detail each sub-category of methods according to their other denominations in the literature, its characteristics and its inputs. Please note that the term "different algorithms" may include cases where the same algorithm is used with different number of cluster, or different parameters. This classification is our personal view of the field, and is in no way fixed. Nevertheless, we needed it to make clear what we were referring to when we mention these different notions throughout the chapter.

Table 1 The spectrum of notions sometimes falling under the term collaborative clustering

Method family	Characteristics	Input	References
Multi-view clustering	Organic views of the same objects under different features	Data only	[48]
Horizontal collaboration	Same object, different features	Data, local partition, local algorithms	[21, 23, 47]
Multi-algorithm collaboration	Same objects, same features, different algorithms	Data, local partition, local algorithms	[31, 39, 42, 44]
Vertical collaboration	Different objects, same features	Data, local partitions, local algorithms	[21, 23, 40]
Partition collaboration	Same objects, different features	Data local partitions	[14, 16, 17]
Ensemble learning	Same objects, same features, different algorithms	Local partitions only, no data, no algorithm	[15, 20, 38]

2.1 Evolution of the Notion of Collaborative Clustering

Collaborative clustering is a term that was first coined by Pedrycz [33] to describe a clustering framework whose aim is to find common structures in several data subsets. It essentially involves an active way of jointly forming or refining clusters through exchanges and information communication between the different data sites with the goal of reconciling as many differences as possible [35]. In its original design collaborative clustering was not aimed at a specific application and was essentially targeted at fuzzy clustering [32, 37] and rough set clustering [30] applications.

As the original idea gained in visibility, the notion of collaborative clustering was re-used by several research teams who thought of various possible applications. This led to a diversification of what can be considered collaborative clustering, but also to the question of its place as a tool or as a field compared with already established problems such as multi-view clustering [3, 48] and unsupervised ensemble learning [38].

A first attempt at formalizing collaborative clustering and categorizing its different uses was made by Grozavu and Bennani [23]. In their work the authors describe collaborative clustering as a two-steps process where partitions are first created locally in each data site, and then refined during a collaborative step which involves information communication between the different sites with a goal of mutual improvement. The full process is illustrated in Fig. 1. In the same paper, the authors also distinguish two types of collaborative clustering: *horizontal collaboration* (previously mentioned but not fully formalized in [33, 34]) which involves the same data with attributes spread over different sites, and *vertical collaboration* where different data with the sames attributes are spread across different sites. The algorithms developed following these ideas are heavily inspired by the original work of Pedrycz et al. and rely on the same principles applied to self-organizing maps [28] and generative topographic mapping [4] instead of the fuzzy C-Means algorithm [10].

It is worth mentioning that the term *horizontal collaborative clustering* had already been used prior to their work to describe a version of collaborative fuzzy C-Means applied to what can reasonably be considered a multi-view clustering application [47].

Following the same idea as *horizontal collaboration*, a large number of methods [7, 12, 21, 22, 24, 42] have since been developed with applications that fall under the umbrella of what is traditionally known as multi-view clustering. Sometimes the multiple views will be organic to the data, and sometimes it will be something more artificial like different features or views artificially created by feature extracting algorithms. The common point of these horizontal collaborative methods is that they perform the clustering of the same data spread across several sites. Differences exist however as some of them aim at a single consensus result, while others highlights that the specificity of collaborative clustering is that a consensus should not be the main goal [21]. In other cases, collaborative clustering is directly referenced as being multi-view clustering [22, 25, 43]. And sometimes the authors use collaborative clustering for multi-view application, but prefer to opt for a neutral name [31].

While horizontal collaborative clustering and multi-view clustering appear to be the same thing, a few authors point out that on the one hand most multi-view clustering methods have access to all the views, while on the other hand collaborative clustering

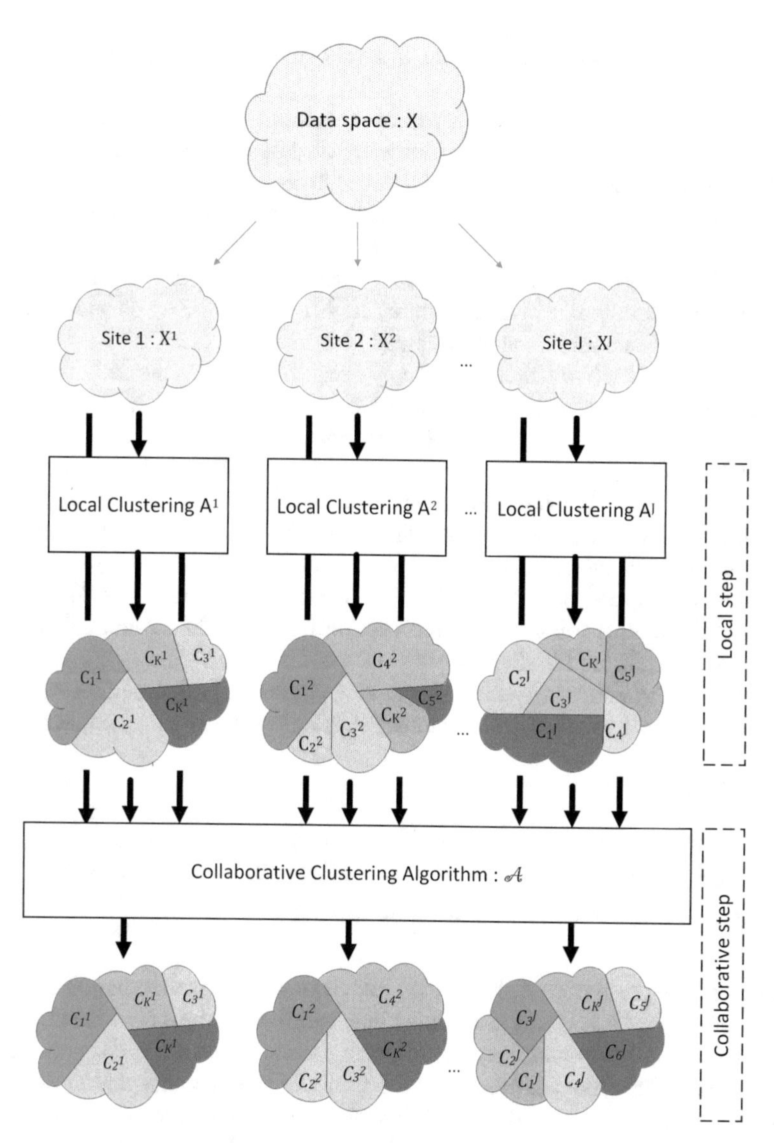

Fig. 1 Graphical definition of collaborative clustering as it is defined by Grozavu and Bennani [23]: local clustering algorithms produce clustering partitions locally during the local step. During the collaborative step, these local partitions are passed alongside the original data and local algorithms to produce improved partitions in each site after the collaboration process. This graphical definition works for both horizontal and vertical collaboration

with its local algorithms and exchanges of information offer more possibilities for data anonymization and the control of privacy [26, 47, 49]. Furthermore, it is worth mentioning that collaborative clustering in its so called *horizontal from* encompasses both real multi-view applications [21, 23, 47], and also cases of multi-algorithms collaboration where several algorithms tackle the same data without any views [31, 39, 42, 44]. In the second case, this is in a way similar to boosting techniques but for unsupervised methods.

Finally, another common recent use of collaborative clustering for multi-view application is its application to the clustering of data sets spread across networks under various constraints [11, 41, 45]. As with the privacy issues, collaborative clustering with its local methods and information exchanges offers more possibilities than classical multi-view clustering framework for this type of applications.

The second form described by Grozavu and Bennani, *vertical collaboration*, appears to be less common in the literature [21, 40]: In this case we consider different samples of the same initial database spread across several sites. It is likely that this term is less common in the literature simply because it matches the definition of *federated learning*, which ironically is sometimes coined under *collaborative learning*. However, unlike vertical collaborative clustering which so far has only been tested on mostly outdated algorithms (K-Means, GTM, and SOM), federated learning is currently researched for deep learning and using block chain technology [1, 5, 9, 36]. Indeed, in the case where we consider different data distributed across several sites and with nearly identical distribution, vertical collaborative clustering can be seen as a form of unsupervised federated learning. On the other hands, if the distributions are too different, this becomes transfer learning for which the current collaborative clustering methods are ill-adapted.

Lastly, we can mention hybrid collaborative clustering [13], a mix between horizontal and vertical collaborative clustering with little to no practical application.

2.2 Remarkable Branches of Collaborative Clustering and Applications that Blur the Lines Between Ensemble Learning and Multi-view Clustering

While we have presented an almost chronological evolution of the notion of collaborative clustering, it is worth mentioning that not all algorithms coined as "collaborative clustering" fall under these definitions. There is indeed a whole spectrum of collaborative clustering algorithm ranging from the mostly multi-view applications that we have seen, to collaborations between algorithms working on the same data subsets, and even some methods discarding the original data and algorithms altogether to have a "collaboration" only between partitions, thus drifting towards what seems to be ensemble learning.

We can for instance mention group of methods and algorithms described by their authors as collaborative clustering, but that differ slightly from both the original idea by Pedrycz and the notion of *horizontal collaboration* coined by Grozavu and Bennani. In [15–17, 20, 46], the authors propose various iterations of the SAMARAH method [19]. Like in Grozavu and Bennani [23], they define collaborative clustering as a two-steps process where results are first produced by local algorithms, and are then refined. For some applications, several algorithms are applied to the same source data [15, 20], and in other multiple views or sources for the same data are considered [14, 16, 17]. In their case, they do not use collaborative clustering for multi-view learning but to merge the results of several and potentially different clustering algorithms applied to the exact same data and attributes. Furthermore, unlike in previous collaborative methods described previously in this step of the art, the algorithms are completely removed from the collaborative step and only the local partitions are kept to search for a consensus. As one can see, this type of collaborative clustering is identical to what is known as unsupervised ensemble learning [38]. It is worth mentioning that the strength of this approach is that is it compatible with any clustering algorithm, and this is due to the removal of the algorithms from the collaborative step.

The collision between collaborative clustering and ensemble learning was further increased by the third attempt at formalizing collaborative clustering by a group researchers from several teams working on the subject [8]. Furthermore, this approach of giving less importance to the local algorithm and to focus more on the partitions appears to be a growing trend too in collaborative clustering for multi-view applications as more and more authors appear to favor it in recently produced collaborative clustering algorithms [18, 31].

As one can see from the state of the art and from Table 1, collaborative clustering is a polymorphic notion whose main applications in the literature range from multi-view clustering to ensemble learning. Yet, it is also obvious that many of the methods under the name *collaborative clustering* share common points that are unique to them. One of the goals of this chapter is to address the overlap and confusion that may exist between the 3 notions and we will do so in Sect. 3 by formally defining what we consider to be the properties specific to each type of method.

The second goal of this chapter, detailed in Sect. 4, is to introduce a formal understanding of collaborative clustering, which appears to be missing in all papers from the state of the art that we have mentioned previously: With dozens of methods described, none of them so far has studied the theoretical properties of collaborative clustering such as the question of its stability, the question of the consistency between the original local partitions and the collaborative result(s), and potential guarantees that it will produce novel solutions compared with the local ones, a property important for both multi-view and ensemble learning applications. We will formalize and address these notions for collaborative clustering in general, and then we will discuss to what degree they can be extended to multi-view clustering and unsupervised ensemble learning.

3 Distinguishing Regular Clustering, Collaborative Clustering, Multi-view Clustering and Unsupervised Ensemble Learning

3.1 Notations

In the remainder of this work, we will use the notations presented in this subsection.

3.1.1 Regular Clustering

Let us consider that all clustering methods—regular, collaborative, multi-view or otherwise—will be applied a data space $\mathbb{X}$ endowed with a probability measure P. If $\mathbb{X}$ is a metric space, let ℓ be its metric. In the following, let $S = \{x_1, \ldots, x_m\}$ be a sample of size m drawn i.i.d. from $(\mathbb{X}, P, \Sigma)$, where Σ is the set of finite partitions of $\mathcal{X} \subseteq \mathbb{X}$.

In this work, we emphasize the difference between a clustering or partition, and the clustering algorithm that produces this partition: In regular clustering, a *clustering* C of a subset $X \subseteq \mathbb{X}$ is a function $C : X \to \mathbb{N}$ which to any of said subset X associates a solution vector in the form of matching clusters $S = C(X)$. Individual *clusters* are then defined by: $C_i = C^{-1}(\{i\}) = \{x \in X; C(x) = i\}$. The *clustering algorithm* A is the function which produces the clustering partition, i.e. a function that computes a clustering of X for any finite sample $\mathcal{S} \subseteq X$, so that $A : X \mapsto C$.

These definitions are graphically explained in Fig. 2. The proposed definition differs from a more standard view of clustering, in that they aim to produce clustering partition for the whole space and not only for the dataset of interest. Note however

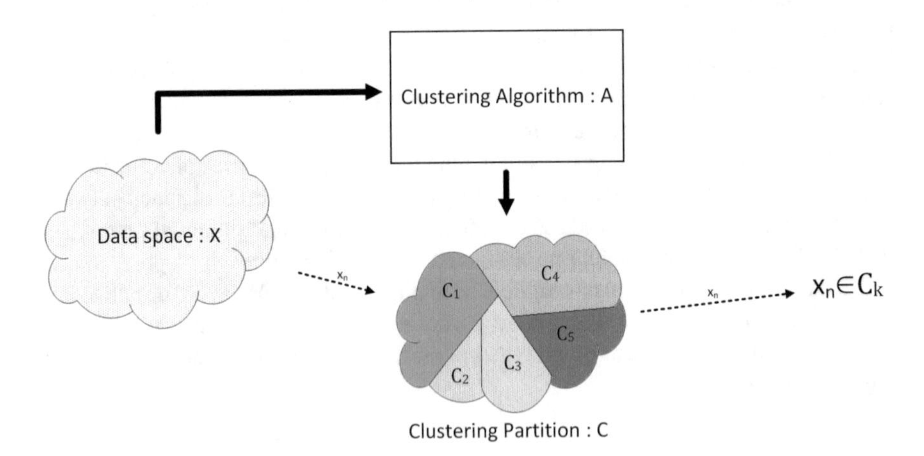

Fig. 2 Graphical definition of the notions of clustering algorithm, clustering partition, and cluster in the case of *regular* clustering

that this specific case can be retrieved easily from the definitions, by defining the total space to correspond to the dataset. This trivial case is weaker, in the sense that the theoretical analysis proposed below does not apply to it.

Example 1 Consider data representing individuals, represented by two features, the height and the weight. Here, the *data space* $\mathbb{X}$ is $\mathbb{X} = \mathbb{R}^2$. Consider a population distribution P over $\mathbb{X}$, and a sample of m individuals drawn from distribution P. The K-means algorithm is a *clustering algorithm* which, given the sample, produces the *partition* defined as a Voronoi diagram associated to some optimal seeds, the *means* computed by the algorithm.

3.1.2 Reminders on Risk Optimization Schemes

A large class of clustering algorithms choose the clustering by optimizing some risk function. The large class of center based algorithms falls into this category, and spectral clustering can also be interpreted in this way. Risk optimization schemes are an important clustering notion discussed by Ben David et al. [2]. We will also use them when discussing clustering stability for both regular, multi-view and collaborative clustering. This subsection reviews some of the basics needed to understand our work.

Definition 1 (*Risk optimization scheme*) A risk optimization scheme is defined by a quadruple $(\mathbb{X}, \Sigma, \mathcal{P}, \mathcal{R})$, where $\mathbb{X}$ is some domain set, Σ is a set of legal clusterings of $\mathbb{X}$, and $\mathcal{P}$ is a set of probability distributions over $\mathbb{X}$, and $\mathcal{R} : \mathcal{P} \times \Sigma \rightarrow [0, \infty)$ is an objective function (or risk) that the clustering algorithm aims to minimize. We denote $opt(\mathcal{P}) := inf_{C \in \Sigma} \mathcal{R}(\mathcal{P}, C)$. For a sample $\mathcal{X} \subseteq \mathbb{X}$, we call $\mathcal{R}(\mathcal{P}_X, C)$ the empirical risk of C, where $\mathcal{P}_X$ is the uniform probability distribution over X. A clustering algorithm A is called R-minimizing, if $\mathcal{R}(\mathcal{P}_X, A(X)) = opt(\mathcal{P}_X)$, for any sample X.

Example 2 Generic examples regarding risk optimization schemes usually use center-based clustering algorithms such as K-means and K-medians, and any K-medoid based algorithm fuzzy or not. Those algorithms pick a set of k center points $c_1, \dots, c_k$ and then assign each data point in the metric space to the closest center point. Such a clustering is a k-cell Voronoi diagram over (X, ℓ), ℓ being the metric on the space X. To choose the centers, the K-Means algorithm minimizes the following risk function:

$$R(P, C) = \mathop{\mathbb{E}}_{x \sim P} \left[\min_{1 \leq i \leq k} (\ell(x, c_i))^2 | \text{Vor}(c_1, c_2, \dots, c_k) \right]$$

while the K-medians algorithm minimizes:

$$R(P, C) = \mathop{\mathbb{E}}_{x \sim P} \left[\min_{1 \leq i \leq k} (\ell(x, c_i)) | \text{Vor}(c_1, c_2, \dots, c_k) \right]$$

where $\text{Vor}(c_1, c_2, \ldots, c_k)$ is the minimization diagram of the k functions $(\ell(x, c_i))^2$ respectively $\ell(x, c_i)$, $1 \leq i \leq k$.

Usually, risk-minimizing algorithms are meant to converge to the true risk as the sample sizes grow to infinity, which is formalized by the notion of *risk convergence*.

Definition 2 (*Risk convergence*) Let A be an R-minimizing clustering algorithm. We say that A is risk converging, if for every $\epsilon > 0$ and every $\delta \in (0, 1)$ there is m_0 such that for all $m > m_0$, $\underset{S \sim P^m}{Pr}[R(P, A(S)) < opt(P) + \epsilon] > 1 - \delta$ for any probability distribution $P \in \mathcal{P}_X$.

For example, Ben-David et al. [2] have shown that, on bounded subset of $\mathbb{R}^d$ with Euclidean metric, both K-means and K-medians minimize risk from samples.

Note that this definition represents, from measure theory point of view, the almost everywhere convergence.

3.1.3 Notations in the Multi-view Context

The problem of interest in this chapter involves clustering algorithms that can be applied to several data sites. This setting includes all applications in Table 1. Therefore, we consider a data space $\mathbb{X}$ which is decomposed into the product $\mathbb{X} = \mathbb{X}^1 \times \cdots \times \mathbb{X}^J$ of J spaces $\mathbb{X}^j$, that may or may not overlap depending on the application. We will call the spaces $\mathbb{X}^j$ *view spaces* or simply *views*. The interdependence between the views is not solely contained in the definition of the different views $\mathbb{X}^j$, but also in the probability distribution P over the whole space $\mathbb{X}$.

For the remainder of this chapter, we will use strict notation conventions. Upper indexes will usually refer to the view or data site index, and lower indexes to individual data or clusters in specific views. For instance, C_k^j would be the kth cluster of data site j, x_n^j the nth data element of site j, etc. For simplicity purposes, we will sometimes use the notation $O^{1:J}$ to designate the tuple $O^1, \ldots, O^J$, where O can be any object distributed among the J views (including algorithms, partitions or data).

Example 3 Consider the data described in Example 1. Consider that now data are available in two different sites, corresponding to two view spaces (i.e. $J = 2$). In the first site, both height and weight are observed ($\mathbb{X}^1 = \mathbb{R}^2$), while only the height is observed in the second site ($\mathbb{X}^2 = \mathbb{R}$). In this multi-view description, the data space $\mathbb{X}$ is then defined as $\mathbb{X} = \mathbb{X}^1 \times \mathbb{X}^2 = \mathbb{R}^2 \times \mathbb{R}$. The total distribution on $\mathbb{X}$ must satisfy the equality of the height between $\mathbb{X}^1$ and $\mathbb{X}^2$ (if $x \sim P$, then feature 0 of x^1 is equal to x^2).

3.2 *Definitions, Context, and Practical Setting*

We now formalize the different tasks presented in Table 1 and show, based on their definitions, how interconnected they are.

3.2.1 Multi-view Clustering Partition

From the definitions above, we define the notion of multi-view clustering partition as follows:

Definition 3 (*Multi-view clustering partition*) A multi-view partition is defined as a combination of local clustering in the following sense: A multi-view clustering $\mathcal{C}$ of the subset $X \subseteq \mathbb{X}$ is a function $\mathcal{C} : X \to \mathbb{N}^q$, where $q \in \{1, J\}$ is called the *index of the partition* and indicates whether the goal is to reach a single consensus solution ($q = 1$) or to keep independent clustering in each view ($q = J$).

As one can see, the very broad definition of a multi-view partition given above covers all cases of collaborative and multi-view clustering, with the different objectives of reaching a consensus between the views ($q = 1$) and refining the partitions produced for each view ($q = J$). Both cases are important depending on the context. When the views describe features of same objects but the goal is to have groups of similar objects, then a consensus is needed: In Example 3, it would be the case if the goal is to group individuals with similar morphological traits. Conversely, refining the results of the views is important when the goal is not to propose a unique group for each object, but one group per view: In Example 3, the joint information of height and weight can provide refined information about the height distribution, but the clustering of heights must still provide a description of the height characteristics only.

A very important observation here is that a multi-view clustering partition can be interpreted as a regular clustering partition. This is clear when $q = 1$, since the definitions of regular and multi-view partitions match completely. When $q = J$, this result is based on the observation that $\mathbb{N}$ and $\mathbb{N}^J$ are equipotent (i.e. there exists a bijection $\nu : \mathbb{N}^J \to \mathbb{N}$, for instance the Cantor pairing function). For instance, saying that a point $x = (x^1, \ldots, x^J) \in \mathbb{X}$ is associated to clusters $(c^1, \ldots, c^J)$ in the multi-view setting, is equivalent to considering that x is associated to cluster $\nu(c^1, \ldots, c^J)$ in a regular clustering of $\mathbb{X}$.

Example 4 Consider $\mathbb{X} = \mathbb{X}^1 \times \mathbb{X}^2$. In a case where both spaces are partitioned into 2 clusters (namely $\mathcal{C}_0^1$ and $\mathcal{C}_1^1$ for $\mathbb{X}^1$ and $\mathcal{C}_0^2$ and $\mathcal{C}_1^2$ for $\mathbb{X}^2$), this can be represented as a partition of $\mathbb{X}$ into four clusters: $\mathcal{C}_0 = \mathcal{C}_0^1 \times \mathcal{C}_0^2$, $\mathcal{C}_1 = \mathcal{C}_0^1 \times \mathcal{C}_1^2$, $\mathcal{C}_2 = \mathcal{C}_1^1 \times \mathcal{C}_0^2$ and $\mathcal{C}_3 = \mathcal{C}_1^1 \times \mathcal{C}_1^2$.

Far from being anecdotal, this observation shows that any multi-view clustering sums up to a regular clustering. This result will be exploited further to extend the main property of stability to collaborative clustering (Theorem 1). The converse is not true though, since any partition of the total space $\mathbb{X} = \mathbb{X}^1 \times \ldots \times \mathbb{X}^J$ does not correspond to a multi-view partition. In order to correspond to a valid multi-view partition, the global partition needs to satisfy another additional property:

Proposition 1 *A global clustering partition $\mathcal{C}$ on $\mathbb{X} = \mathbb{X}^1 \times \ldots \times \mathbb{X}^J$ corresponds to a local multi-view partition $(\mathcal{C}^1, \ldots, \mathcal{C}^J)$ on the views $\mathbb{X}^1, \ldots, \mathbb{X}^J$ if and only if for all $j \in \{0, \ldots, J\}$ and for all $x^j \in \mathbb{X}^j$, all clusters of $\mathcal{C}$ containing a point x' with $x'^j = x^j$ have for projection over $\mathbb{X}^j$ the set $\mathcal{C}^j(x^j)$.*

Proof Suppose first that the global partition $\mathcal{C}$ corresponds to the local multi-view partition $(\mathcal{C}^1, \ldots, \mathcal{C}^J)$, i.e. there exists a bijection $\nu : \mathbb{N}^J \to \mathbb{N}$ such that, for all $c^1, \ldots, c^J \leq 0, \mathcal{C}_{\nu(c^1,\ldots,c^J)} = \mathcal{C}^1_{c^1} \times \ldots \times \mathcal{C}^J_{c^J}$. Consider a view j and a point $x^j \in \mathbb{X}^j$. By construction, the clusters in $\mathcal{C}$ containing points with x^j as their jth component are the clusters of the form $\mathcal{C}_{\nu(c^0,\ldots,\mathcal{C}^j(x^j),\ldots,c^J)}$ which have all $\mathcal{C}^j(x^j)$ as their projection on $\mathbb{X}^j$.

Suppose now the converse, and let us show that $\mathcal{C}$ corresponds to the multi-view partition $(\mathcal{C}^1, \ldots, \mathcal{C}^J)$. From the hypothesis, we see that the projection of each cluster $\mathcal{C}_i$ onto $\mathbb{X}^j$ is the union of some clusters from $\mathcal{C}^j$. However, the clusters being disjoint, the projection of $\mathcal{C}_i$ onto $\mathbb{X}^j$ being equal to one cluster implies that the union contains one single element. Therefore, each cluster $\mathcal{C}_i$ is the Cartesian product of clusters in local views: $\mathcal{C}_i = \mathcal{C}^1_{c^1_i} \times \ldots \times \mathcal{C}^J_{c^J_i}$ for some c^j_i. We must show that the function defined by $\nu(c^1_i, \ldots, c^J_i) = i$ is bijective. It is direct to show that $\nu(n_1, \ldots, n_J)$ is well defined for n_j lower than the number of clusters on $\mathcal{C}^j$ (this can be shown by considering $x = (x^1, \ldots, x^J)$ with $x^j \in \mathcal{C}^j_{n_j}$). Then, $\nu(c^1_i, \ldots, c^J_i) = \nu(c^1_i, \ldots, c^J_i)$ implies $(c^1_i, \ldots, c^J_i) = (c^1_i, \ldots, c^J_i)$ since the clusters are distinct, which concludes the proof.

3.2.2 Multi-view and Collaborative Clustering Algorithms

Based on the notions introduced before, we can now formalize the various notions exposed in Table 1, in particular the notions of collaborative clustering algorithm, multi-view clustering algorithm and unsupervised ensemble learning.

Defining these notions requires understanding the main differences between them. Assessing *multi-view clustering* problems is direct: a multi-view clustering problem simply partitions the data based on observations of samples from the views $\mathbb{X}^j$ (Fig. 3). For collaborative clustering algorithms, we will focus on algorithms that have at least the following property: the collaboration process should include local clustering algorithms exchanging information and must not be limited to only exchanging local partitions. We feel like this definition is the broadest we can have as it includes most algorithms developed by Pedrycz et al., as well as all algorithms falling under the definition of vertical and horizontal collaboration as defined by Grozavu and Bennani [23], and thus only excludes so called collaborative methods that are in fact unsupervised ensemble learning as they deal only with partitions fusion (see Fig. 4).

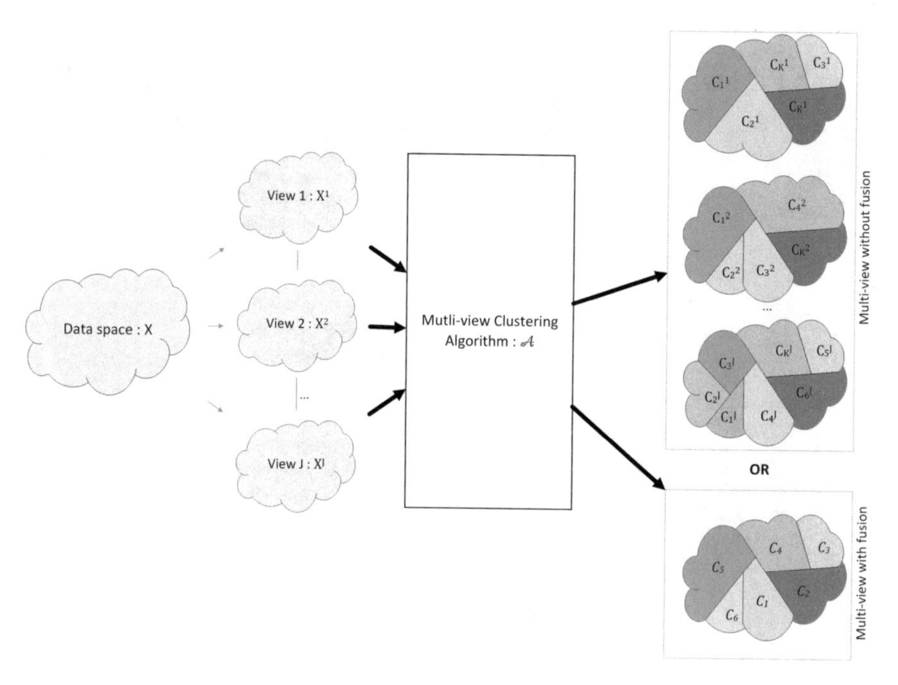

Fig. 3 Graphical definition of multi-view clustering: In this figure we display two possible cases, namely multi-view clustering without fusion, and multi-view clustering leading to a single consensus partition

Definition 4 (*Multi-view and collaborative clustering algorithms*) Consider a total space $\mathbb{X} = \mathbb{X}^1 \times \ldots \times \mathbb{X}^J$ and a subspace $X \subseteq \mathbb{X}$. Let S^j be a sample of X^j. Then a **collaborative clustering algorithm** is a mapping

$$\mathcal{A}^{col} : (S^1, \ldots S^J, A^1, \ldots, A^J, C^1 \ldots, C^J) \mapsto \mathcal{C} \tag{1}$$

and a **multi-view clustering algorithm** is a mapping

$$\mathcal{A}^{MV} : (S^1, \ldots, S^J) \to \mathcal{C} \tag{2}$$

where A^j designates a clustering algorithm over $\mathbb{X}^j$ and C^j is a partition of X^j.

For collaborative clustering algorithms, we denote the local algorithms as index: $\mathcal{A}^{col}_{\langle A^1,\ldots,A^J\rangle}(S^1, \ldots, S^J, C^1, \ldots, C^J)$.

These two definitions describe successfully the multi-view clustering, horizontal collaboration, multi-algorithm collaboration, vertical collaboration and partition collaboration families described in Table 1. We observe that many existing collaborative clustering methods suppose the application of a same clustering algorithm to the different views, which corresponds in essence to having $A^1, \ldots, A^J$ belonging to

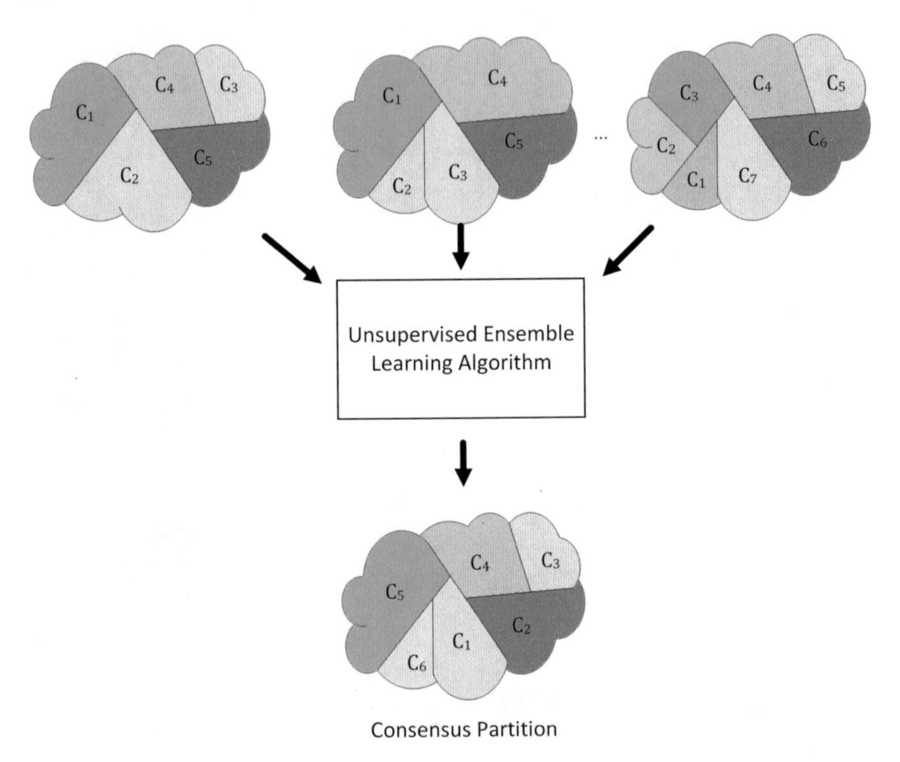

Fig. 4 Graphical definition of unsupervised ensemble learning: Notice that the data themselves are never involved in the process

a same class of algorithms[1]. Please note that Eq. 1 is compatible with both horizontal and vertical collaborative clustering as it makes no assumptions about whether the full data space is cut into sub-sites alongside the features or the data themselves.

It is noticeable that the collaborative clustering algorithms are given local algorithms $A^1, \ldots, A^J$ as input, and could be rewritten as functions of the form: $\mathcal{A}^{col} : (A^1, \ldots, A^J) \mapsto \mathcal{C}$.

The role of local algorithms as inputs in collaborative clustering is twofold: They have an influence over the collaboration between the views. Intuitively, the decision of altering an optimal local partition to incorporate information from other views must be constrained by the biases of the local algorithm. This strategy is explicit in the multi-algorithm collaboration setting [31, 39, 42, 44], where the impact of a local algorithm intervenes as a penalization of a risk minimization objective by the information held in the produced partitions [31, 44], as the core of a risk minimization scheme with a penalty for divergences between views [42], or as a bias in the selection of data for learning the local partitions from one step to another [39].

[1] Formally, it would be incorrect to state that $A^1 = \ldots = A^J$, since the algorithms A^j are defined relatively to different spaces $\mathbb{X}^j$ and are therefore of different natures.

Noticeably, many collaborative clustering algorithms are thought to apply to one single class of local clustering algorithms, such as C-Means, Self-Organizing Maps or Generative Topographic Maps [12, 21, 23, 35]. In this case, the nature of the local algorithms is directly exploited for the collaboration. In addition, following the idea of a 2-step process introduced by [23], where the algorithm is divided into the generation of local partitions and the refinement of these partitions, the local algorithms A^j are naturally involved in the first step. The support of these algorithms (i.e. the space of parameters on which they are properly defined) is then constrained to satisfy $C^j = A^j(S^j)$ for all j. It follows directly from Definition 4 that such collaborative clustering algorithms, once the local clustering algorithms are fixed, are strictly equivalent to multi-view clustering algorithms.

Proposition 2 *Let $\mathcal{A}^{col}$ be a collaborative clustering algorithm the support of which is restricted to satisfy $C^j = A^j(S^j)$. Given $A^1, \ldots, A^J$, J fixed local clustering algorithms, the function*

$$S^1, \ldots, S^J \mapsto \mathcal{A}^{col}_{\langle A^1,\ldots,A^J \rangle}\left(S^1, \ldots, S^J, C^1(X^1), \ldots, C^J(X^J)\right)$$

is a multi-view clustering algorithm.

This proposition is a direct application of Definition 4. A consequence of this result is that, if we follow the definition of collaborative clustering as given by Grozavu and Bennani [23] where partitions should not be merged (See Fig. 1), we have that collaborative clustering algorithms are a specific case of multi-view clustering algorithms and they produce multi-view partition $\mathcal{C}$ of the subset $\mathcal{X} \subseteq \mathbb{X}$ whose mapping follows the form $\mathcal{C} : \mathcal{X} \to \mathbb{N}^J$, given J algorithms collaborating together.

A very straightforward property of collaborative clustering algorithms which, when used in practice, applies quite naturally the constraint $C^j = A^j(S^j)$, is that the produced solution cannot be altered by a second application of the collaboration. We call that property the *purity of the collaboration*:

Definition 5 (*Pure collaborative clustering algorithm*) Let $\mathcal{A}^{col}$ be a collaborative clustering algorithm that outputs a multi-view partition of index J. $\mathcal{A}^{col}$ is said to be pure, if and only if

$$\mathcal{A}^{col}\left(S^{1:J}, A^{1:J}, \mathcal{A}^{col}\left(S^{1:J}, A^{1:J}, C^{1:J}\right)\right) = \mathcal{A}^{col}\left(S^{1:J}, A^{1:J}, C^{1:J}\right). \quad (3)$$

The previous definition entails that a collaborative algorithm is pure if and only if re-applying it to its own output will not change the resulting partitions. As such purity is a desirable property for collaborative algorithms since it ensures the definitive aspect of the results for algorithms that have this property.

3.2.3 Unsupervised Ensemble Learning

The main difference between the aforementioned collaborative and multi-view clustering algorithms, and the unsupervised ensemble learning ones, is that the latter operate at the level of partitions only (Fig. 4) and are therefore slightly difference in essence.

Definition 6 (*Unsupervised ensemble learning*) Consider a total space $\mathbb{X} = \mathbb{X}^1 \times \ldots \times \mathbb{X}^J$ and a subspace $X \subseteq \mathbb{X}$. For all j, let C^j be a partition of X^j. An unsupervised ensemble learning algorithm is defined as a mapping $\mathcal{A}^{ens} : C^1 \times \ldots \times C^J \to C^j$, where the C^j is a partition of a given view X^j.

Although this definition is chosen to be as general as possible, most applications focus on the simplest case where all views are equal ($\mathbb{X}^1 = \ldots = \mathbb{X}^J$ and $X^1 = \ldots = X^J$) with the implicit assumption that the data are the same in all views (which, formally, corresponds to an assumption on the probability distribution P, such as in Example 3).

This very different nature makes it impossible to relate unsupervised ensemble learning algorithms to the other families, such as done for instance for multi-view and collaborative frameworks (Proposition 2). However, we can observe the following relations between the three families of algorithms:

Proposition 3 *Let $\mathcal{X} = \mathbb{X}^1 \times \ldots \times \mathbb{X}^J$ be a total space and $A^1, \ldots, A^J$ local clustering algorithms on the views $\mathbb{X}^j$. Consider a multi-view algorithm $\mathcal{A}^{MV}$ on $\mathbb{X}$. Let $\mathcal{A}^{ens,1}, \ldots, \mathcal{A}^{ens,J}$ be J unsupervised ensemble learning algorithms, where $\mathcal{A}^{ens,j}$ produces a partition in $\mathbb{X}^j$. We define*

$$\mathcal{A}^{ens}(C^1, \ldots, C^J) = (\mathcal{A}^{ens,1}(C^1, \ldots, C^J), \ldots, \mathcal{A}^{ens,J}(C^1, \ldots, C^J))$$

Then the following statements are correct:

1. *The function $S^1, \ldots, S^J, A^1, \ldots, A^J, C^1, \ldots, C^J \mapsto \mathcal{A}^{ens}(C^1, \ldots, C^J)$ is a collaborative clustering algorithm.*
2. *If $\mathcal{A}^{MV}$ produces a multi-view partition of index J, then the function $\mathcal{A}^{ens} \circ \mathcal{A}^{MV}$ is a multi-view clustering algorithm of index J.*
3. *$\mathcal{A}^{MV}$ can be decomposed as the combination $\mathcal{A}^{MV} = \mathcal{A}^{ens,\prime} \circ \mathcal{A}^{MV,loc}$ of local regular clustering algorithms: $\mathcal{A}^{MV,loc} : (S^1, \ldots, S^J) \mapsto (A^1(S^1), \ldots, A^J(S^J))$ and of unsupervised ensemble learning algorithms $\mathcal{A}^{ens,\prime}$ (defined in a similar manner as $\mathcal{A}^{ens}$).*

These statements are direct consequences of the definitions. Point (3) formalizes the decomposition of multi-view algorithms into two steps: applying local algorithms to each view to produce a partition, and applying an unsupervised ensemble learning to exchange the information between the views. The combination of this point and of Proposition 2 formalizes the idea of Grozavu and Bennani [23] of a two-step process (applying local algorithms, then refine). Regarding the notations, the multi-view algorithm $\mathcal{A}^{MV,loc}$ will be called, in the following section, *concatenation* of local algorithms and will be denoted by $\bigoplus_j A^j$. It will play a central role in the study of theoretical properties of a collaboration.

3.3 Summary: Four Interleaving Notions

In this section, we have introduced four interleaving notions: regular clustering, multi-view clustering, collaborative clustering and unsupervised ensemble learning. The definitions we proposed are extremely general, and in particular do not incorporate some classical (sometimes implicit) properties associated to these notion, for instance the independence to the order of arguments (a permutation of S^i and S^j in the arguments of $\mathcal{A}^{MV}$ simply yields a permutation of C^i and C^j in the output). Actually, these properties are not essential to the very nature of these notions and it would be reasonable to think of applications where these do not hold.

A fundamental result we showed is that multi-view clustering and regular clustering are in essence similar, in the sense that multi-view clustering (be it with or without fusion) generates a clustering of the total space and, conversely any regular clustering satisfying some constraints can be understood as a multi-view clustering (Proposition 1). These constraints can be understood as some regularization of the produced clustering. This result however is not meant to lower the importance of multi-view clustering as an independent domain: on contrary, implementing these constraints is a challenge *in se* and is the core motivation of a whole field. It is however of primary importance for a theoretical study of multi-view clustering (and by extension, of collaborative clustering), since it entails that the same tools and results apply to it.

Such as introduced in Definition 4, multi-view and collaborative clustering algorithms differ mostly on the nature of their input arguments. Collaborative clustering algorithms are more general, since they consider the local algorithms and initial partitions in addition to the data points. As a direct result, it is clear that multi-view methods are specific cases of collaborative clustering, but, conversely, the collaborative clustering algorithms inspired by the works of Grozavu and Bennani [23] can be reduced to multi-view algorithms. Indeed, such algorithms use the local algorithms only to constrain the local views.

Unsupervised ensemble learning algorithms are not clustering algorithms in the sense of Sect. 3.1.1, since they do not take data points as input. However, they are strongly involved in collaborative and multi-view clustering: It has been discussed that a multi-view algorithm can be decomposed into a regular clustering algorithm and an unsupervised ensemble learning. This decomposition, which may be only theoretical and does not necessarily reflect how the algorithms work *in practice*, amounts to considering the output of a multi-view algorithm as a correction of local partitions based on the information provided by the other views. In the next section, a measure of the influence of this ensemble algorithm will be used to define the *novelty* and *consistency* of a collaborative/multi-view algorithm.

4 Properties of Collaborative Clustering: Stability, Novelty and Consistency

In this section, we introduce various properties which could be expected from collaborative algorithms. These notions and formal definitions are inspired by the notion of clustering *stability* introduced in the original work of Ben David et al. [2]. We will see how stability can be extended to collaborative clustering and how the stability of a collaborative clustering algorithm depends inherently from a novel notion, called *consistency*.

All these notions focus more precisely on the influence of the local clustering algorithms $A^1, \ldots, A^J$ and of the input data $S^1, \ldots, S^J$, onto the produced partitions. In particular, we will consider the initial partitions $C^1, \ldots, C^J$ as fixed, for instance as $C^j = A^j(S^j)$. We will use the notation $\mathcal{A}_{\langle A^1,\ldots,A^J \rangle}$ (shorten in $\mathcal{A}$ when the local context is explicit) to designate such a collaborative clustering algorithm based on local algorithms $A^1, \ldots, A^J$. Notice that the defined function $\mathcal{A}_{\langle A^1,\ldots,A^J \rangle}$ corresponds to a multi-view algorithm, when the local algorithms are fixed.

4.1 Reminders on Clustering Stability

Stability is a key notion in regular clustering that assesses the ability of a clustering algorithm to find a consistent partitioning of the data space on different subsamples [2, 6, 29]. From its definition it is a neutral quality index that evaluates the noise robustness of clustering algorithms.

The stability of a clustering algorithm is defined as the ability to produce always the same partition, given enough data from a fixed distribution. In order to formalize this idea, it is important to be able to measure how to partitions differ. This is done with the notion of *clustering distance*:

Definition 7 (*Clustering distance*) Let $\mathcal{P}$ be a family of probability distributions over some domain $\mathbb{X}$. Let Σ be a family of clusterings of $\mathbb{X}$. A clustering distance is a function $d : \mathcal{P} \times \Sigma \times \Sigma \to [0, 1]$ that for any $P \in \mathcal{P}$ and any clusterings C, C', C'' satisfies:

1. $d_P(C, C) = 0$
2. $d_P(C, C') = d_P(C', C)$ (*symmetry*)
3. $d_P(C, C'') \leq d_P(C, C') + d_P(C', C'')$ (*triangle inequality*)

Please note that clustering distances as we have defined them are not required to satisfy $d_P(C, C') = 0 \Rightarrow C = C'$, which is not true with most clustering distances that are commonly used.

Example 5 A typical example of a clustering distance (introduced for instance by Ben-David et al. [2]) is the Hamming distance:

$$d_P^H(C, C') = \underset{\substack{x \sim P \\ y \sim P}}{\mathbb{P}} \Big[\big(C(x) = C(y)\big) \oplus \big(C'(x) = C'(y)\big) \Big] \tag{4}$$

where $\oplus$ denotes the logical XOR operation. The Hamming distance measures how much the two partitions group together the same pairs of points. It can be easily checked that d_P^H satisfies the properties of a clustering distance. It is also clear that two partitions C and C' can be different and yet have a 0 distance. For instance, if the space is continuous and C and C' differ only on one point, we still have $d_P^H(C, C')$.

As we have mentioned, clustering stability measures how a perturbation in the data affects the result of a clustering algorithm. Using the proposed definition of a clustering algorithm, the stability of algorithm A can then be formalized as the distance between the produced partitions $A(X_1)$ and $A(X_2)$ for X_1 and X_2 sampled from the same distribution P:

Definition 8 (*Stability of a clustering algorithm*) Let P be a probability distribution over $\mathcal{X}$. Let d be a clustering distance. Let A be a clustering algorithm (a regular one). The stability of the algorithm A for the sample of size m with respect to the probability distribution P is:

$$stab(A, P, m) = \underset{\substack{X_1 \sim P^m \\ X_2 \sim P^m}}{\mathbb{E}} [d_P(A(X_1), A(X_2))] \tag{5}$$

From there, the stability of algorithm A with respect to the probability distribution P is:

$$stab(A, P) = \limsup_{m \to \infty} stab(A, P, m) \tag{6}$$

We say that a regular clustering algorithm A is stable for P, if $stab(A, P) = 0$.

A very strong property of clustering stability, demonstrated by Ben David et al. [2], states that a risk minimizing clustering algorithm (see Definition 1) satisfying a specific property of unicity of the optimal produced partition (such as defined below), is stable.

Definition 9 (*Unique minimizer*) We fix a risk minimization scheme $(\mathbb{X}, \Sigma, \mathcal{P}, \mathcal{R})$. Let d be a clustering distance. We say that a probability distribution P has unique minimizer C^* if:

$$(\forall \eta > 0)(\exists \epsilon > 0)(R(P, C) < opt(P) + \epsilon) \Longrightarrow d_P(C^*, C) < \eta).$$

More generally, we say a probability distribution P has n distinct minimizers, if there exists $C_1^*, C_2^*, \ldots, C_n^*$ such that $d_P(C_i^*, C_j^*) > 0$ for all $i \neq j$, and

$$(\forall \eta > 0)(\exists \epsilon > 0)(R(P, C) < opt(P) + \epsilon \Longrightarrow (\exists 1 \leq i \leq n) \quad d_P(C_i^*, C) < \eta).$$

Note that there is a technical subtlety here: the definition does not require that there is only a single clustering with the minimal cost, but rather that for any two optima

$C_1^*, C_2^*, d_P(C_1^*, C_2^*) = 0$, which does *not* imply that $C_1^* = C_2^*$. Technically, we can overcome this difference by forming equivalence classes of clusterings, saying that two clusterings are equivalent if their clustering distance is zero. Similarly, n distinct optima correspond n such equivalence classes of optimal clusterings.

4.2 *From Regular to Collaborative Clustering: Stability, Novelty and Consistency*

As discussed previously, multi-view clustering, and in particular collaborative clustering, can be interpreted as a specific constrained form of clustering. Following this idea, we show now how the general theoretical notions presented for regular clustering can be formulated for collaborative clustering.

We remind that regular clustering and multi-view partitions are theoretically equivalent because $\mathbb{N}^J$ and $\mathbb{N}$ are equipotent. In the following, we will denote by $\nu : \mathbb{N}^J \to \mathbb{N}$ a bijective application mapping $\mathbb{N}^J$ to $\mathbb{N}$. With this application, the mapping $\nu \circ \mathcal{C}$ is a clustering of $X \subseteq \mathbb{X}$.

4.2.1 Multi-view Clustering Distance and Stability

Any analysis of the theoretical properties of collaborative clustering requires us to firstly define the relevant clustering distance used to measure the discrepancy between the produced clusters on the total space $\mathbb{X} = \mathbb{X}^1 \times \ldots \mathbb{X}^J$. Even though in theory any distance satisfying the conditions of Definition 7 would be applicable, it seems also interesting to consider distances more adapted to the specificity of the space decomposition. In the following proposition, we show that a simple linear combination of local distances is a valid distance for the total space.

Proposition 4 (Canonical multi-view clustering distance) *Let* $\mathbb{X} = \mathbb{X}^1 \times \cdots \times \mathbb{X}^J$ *be a domain, and the* d^j *clustering distance on* $\mathbb{X}^j$*. We define the function* $d : \mathcal{P} \times \mathcal{S} \times \mathcal{S} \to [0, 1]$ *such that* $d_P(\mathcal{C}_1, \mathcal{C}_2) = \frac{1}{J} \sum_{j=1}^{J} d^j_{P_j}(\mathcal{C}_1^j, \mathcal{C}_2^j)$*. Then* d *defines a clustering distance on* $\mathcal{X}$*. We call it the* ***canonical multi-view clustering distance****.*

Proof The clustering distance properties follow directly from the linearity in terms of d^j and from the properties of the local clustering distances.

In this definition, we chose the coefficients of the linear combination to be uniformly equal to $1/J$, with the will to give the same importance to all views, but we would like to emphasize that none of the following results would be altered by choosing non-uniform weights.

Given a distance on the total space $\mathbb{X} = \mathbb{X}^1 \times \ldots \times \mathbb{X}^J$, the definition of clustering stability above (Definition 8) can be applied directly to the case of collaborative and multi-view clustering.

When the used distance is the canonical multi-view clustering distance, the stability of a collaborative or multi-view algorithm on the total space has a simple interpretation. Let $\mathcal{A}$ be the total algorithm and $\mathcal{A}^j : S^1, \ldots, S^J \mapsto \big(\mathcal{A}(S^1, \ldots, S^J)\big)^j$. Algorithm $\mathcal{A}^j$ considers the projection of the multi-view partition produced by $\mathcal{A}$ onto the subspace $\mathbb{X}^j$. Note that, in the case of collaborative clustering, this algorithm $\mathcal{A}^j$ is distinct from the local algorithm A^j. The following characterization of multi-view stability comes directly from the definitions:

Proposition 5 *Multi-view algorithm $\mathcal{A}$ is stable for the canonical multi-view clustering distance if and only if, for all j, the projections $\mathcal{A}^j$ are stable.*

Such a characterization of multi-view stability, despite being intuitive, is actually a consequence of the choice of the canonical distance. For a general clustering multi-view clustering distance, there is no guarantee that this result remains correct. Satisfying Proposition 5 can be an important property that a reasonable multi-view clustering distance should satisfy.

Example 6 Given $\mathbb{X} = \mathbb{X}^1 \times \ldots \mathbb{X}^J$ and local clustering distances d^j, the function $d_P(C_1, C_2) = d_{P^1}(C_1^1, C_2^1)$ defines a proper multi-view clustering distance. With this distance, the equivalence in Proposition 5 is not always satisfied. For instance, any algorithm which would be stable on view 1 but unstable on at least one other view would still be globally stable with respect to the chosen multi-view distance.

4.2.2 Novelty and Consistency

Using Definition 4 and under the assumptions of Proposition 2, we have seen that collaborative clustering algorithms (with fixed local algorithms) are a specific case of multi-view clustering algorithms that aim to refine locally found partitions throughout the collaboration process.

It is worth mentioning that in general, the expected goal of collaborative clustering is that the projection of the clustering obtained by the collaborative algorithm onto one of the views j should be distinct from the original clustering results obtained by the local algorithm $\mathcal{A}^j$ for the same view. This corresponds to the effect of the unsupervised ensemble learning step as discussed in Sect. 3.3 and case 3 of Proposition 3. In other words, if $\mathcal{C} = \mathcal{A}(X)$, then in general we have that $\mathcal{C}^j \neq A^j(X^j)$. It is this property that makes the collaboration interesting and more valuable than a simple concatenation of the local results.

Definition 10 (*Concatenation of local clustering algorithms*) The concatenation of local clustering algorithms A^1 to A^J, denoted by $\bigoplus_{j=1}^J A^j$ is defined as follows: If $\mathcal{C}$ is the global clustering induced by $\mathcal{A} = \bigoplus_{j=1}^J A^j$ on a dataset X, then:

$$\forall x \in \mathbb{X}, \forall j \in \{1, \ldots, J\}, \qquad \mathcal{C}^j(x^j) = \big(A^j(X^j)\big)(x^j) \tag{7}$$

This defines the concatenation of local clustering algorithms as a collaborative algorithm that "does nothing" (i.e. in which there is no exchange of information between the various views), and produces the exact same results as the ones obtain by the local algorithms A^j. Such a degenerate algorithm had been already used in Proposition 3 under the name of $\mathcal{A}^{MV,loc}$.

We now introduce the notion of *novelty*, the property of any collaborative clustering algorithm to do more than just concatenating the local solutions. This represents the ability of a collaborative algorithm to produce solutions that could not have been found locally.

Definition 11 (*Collaborative clustering novelty*) Let P be probability distribution over $\mathcal{X}$. The novelty of the algorithm $\mathcal{A}$ for the sample size m with respect to the probability distribution P is

$$nov(\mathcal{A}_{\langle A^1,\dots,A^J\rangle}, P, m) = \underset{X\sim P^m}{\mathbb{P}}\left[\mathcal{A}_{\langle A^1,\dots,A^J\rangle}(X) \neq \bigoplus_{j=1}^{J} A^j(X^j)\right] \tag{8}$$

where $\mathcal{A}(X)$ is the collaborative or multi-view clustering and $\bigoplus_{j=1}^{J} A^j(X^j)$ is the concatenation of all local clusterings.

Then, the novelty of algorithm $\mathcal{A}$ with respect to the probability distribution P is

$$nov(\mathcal{A}, P) = \limsup_{m\to\infty} nov(\mathcal{A}, P, m) \tag{9}$$

$\mathcal{A}$ satisfies the novelty property for distribution P if $nov(\mathcal{A}, P) > 0$

Yet, while novelty is often described as a desirable property, in collaborative clustering (and in unsupervised ensemble learning as we will see later), there is also a need that the results found a the global level after the collaborative step remain *consistent* with the local data when projected onto the local views. This leads us to the notion of consistency:

Definition 12 (*Collaborative clustering consistency*) Let P be probability distribution over $\mathcal{X}$. Let d be a clustering distance. Let $\mathcal{A}$ be a collaborative clustering algorithm. The consistency of the algorithm $\mathcal{A}$ for the sample size m with respect to the probability distribution P is

$$cons(\mathcal{A}_{\langle A^1,\dots,A^J\rangle}, P, m) = \underset{X\sim P^m}{\mathbb{E}}\left[d_P\left(\mathcal{A}_{\langle A^1,\dots,A^J\rangle}(X), \bigoplus_{j=1}^{J} A^j(X^j)\right)\right] \tag{10}$$

The consistency of algorithm $\mathcal{A}$ with respect to the probability distribution P is

$$cons(\mathcal{A}, P) = \limsup_{m\to\infty} cons(\mathcal{A}, P, m) \tag{11}$$

Please note that this definition of consistency for collaborative clustering has no link with the consistency of regular clustering algorithms as it was defined by Kleinberg [27]. Intuitively, our consistency measures the distance of the global clustering produced by the collaboration to the clustering produced by concatenation of local algorithms.

Two remarks can be made about novelty and consistency: The first one is that obviously these notions are very specific to the case of collaborative clustering and unsupervised ensemble learning (as we will see after), as it is obvious that without intermediary local clustering partitions, these notions simply do not exist. The second remark is that there exists a noticeable link between consistency and novelty, since novelty is actually a particular case of consistency based on the clustering distance defined as follows:

$$\forall \mathcal{C}, \mathcal{C}', \quad d_P^{\mathbb{I}}(\mathcal{C}, \mathcal{C}') = \mathbb{I}(\mathcal{C} \neq \mathcal{C}') \tag{12}$$

It can be verified easily that the function $d_P^{\mathbb{I}}$ is clustering distance.

However, although novelty is a specific case of consistency for a given distance, consistency and novelty are not equivalent in general, for an arbitrary clustering distance. This means that consistent algorithms are not necessarily concatenations. This is mainly due to the fact that clustering distances do not satisfy $d_P(\mathcal{C}, \mathcal{C}') = 0 \Rightarrow \mathcal{C} = \mathcal{C}'$. The converse is true however: $nov(\mathcal{A}, P) = 0$ implies $cons(\mathcal{A}, P) = 0$, whatever clustering distance is used to compute the consistency.

Example 7 Using Hamming distance as a local clustering distance and the canonical multi-view distance as d_P, we have seen previously that local partitions which would differ on a set of P-measure zero (in particular a finite set) would have a zero distance while being distinct. Consider then a trivial collaborative algorithm which changes the cluster of one single point in all the local partitions. Such an algorithm would be consistent, and yet the produced partitions would be distinct.

Finally, it is worth mentioning that while producing novel solutions is generally considered a desirable property for any multi-view or collaborative methods, this might not always be the case: We can for instance imagine a scenario where local solutions are already optimal but different and where novelty might mean sub-optimal solutions everywhere. Another more standard scenario would be a local view (or several local views) having too strong an influence and forcing other views to change otherwise fine but too different local solutions.

4.3 Stability of Collaborative Clustering

Now that we have defined the notion of stability for regular clustering, as well as key notions from collaborative and multi-view clustering, we have two goals: The first one is to derive a notion of stability for collaborative clustering algorithms. And second, we want to know how this notion of collaborative clustering stability can be linked to the notions of novelty (Definition 11) and consistency (Definition 12), and more importantly to the stability of the local algorithms since collaborative clustering has the particularity of using sets of regular clustering algorithms whose stability is already clearly defined (See Definition 8).

4.3.1 Stability of Risk-Minimizing Collaborative Clustering Algorithms

Theorem 1 below shows a direct adaptation of Ben-David's key theorem on clustering stability (Theorem 10 in [2]) to collaborative clustering.

Theorem 1 *If P has a unique minimizer C^* for risk $\mathcal{R}$, then any $\mathcal{R}$-minimizing collaborative clustering algorithm which is risk converging is stable on P.*

Proof Let $\mathcal{A}$ be a collaborative clustering algorithm on $\mathbb{X} = \mathbb{X}^1 \times \cdots \times \mathbb{X}^J$. Let us also consider a bijection $\nu : \mathbb{N}^J \to \mathbb{N}$. Then, based on collaborative algorithm $\mathcal{A}$, one can build a clustering algorithm $\tilde{\mathcal{A}}$ such that the clustering $\tilde{\mathcal{C}}$ induced by a sample S for $\tilde{\mathcal{A}}$ is such that $\tilde{\mathcal{C}} = \nu \circ (\mathcal{A}(S))$. For simplicity purposes, we will denote this algorithm $\tilde{\mathcal{A}} = \nu \circ \mathcal{A}$. We call d_P the global clustering distance and $\tilde{d}_P$ its associated local distance such that $\tilde{d}_P(\tilde{\mathcal{C}}_1, \tilde{\mathcal{C}}_2) = d_p(\nu^{-1} \circ \tilde{\mathcal{C}}_1, \nu^{-1} \circ \tilde{\mathcal{C}}_2)$.

Using these two clustering distances in the definition of stability, the following lemma is straightforward:

Lemma 1 *If $\tilde{\mathcal{A}} = \nu \circ \mathcal{A}$ is stable (for a distance $\tilde{d}_P$), then $\mathcal{A}$ is stable (for the distance d_P).*

If $\mathcal{A}$ is $\mathcal{R}$-minimizing, then $\tilde{\mathcal{A}}$ is $\tilde{\mathcal{R}}$-minimizing with $\tilde{\mathcal{R}}(P, \tilde{\mathcal{C}}) = \mathcal{R}(P, \nu^1 \circ \tilde{\mathcal{C}})$. It is direct that $Opt_{\mathcal{R}}(\tilde{P}) = Opt_{\mathcal{R}}(P)$ and that $\tilde{\mathcal{A}}$ is risk-converging. It is also direct that P has a unique minimizer $\tilde{\mathcal{C}}^*$ associated to $\tilde{\mathcal{R}}$.

Combining all the previous results together, it follows that $\tilde{\mathcal{A}}$ is $\tilde{\mathcal{R}}$-minimizing and risk converging. Since P has a unique minimizer for $\tilde{\mathcal{R}}$, then using [2] we have that $\tilde{\mathcal{A}}$ is stable. Lemma 1 guarantees the result.

Theorem 1 above implies that collaborative clustering algorithms can be treated exactly the same way as standard clustering algorithms when it comes to stability analysis. As we have seen previously, since $\mathbb{N}^J$ and $\mathbb{N}$ are equipotent, a collaborative clustering can be interpreted as a clustering of $X \subseteq \mathbb{X}$, and therefore there is a direct adaptation of stability from regular clustering to collaborative clustering when using the clustering distance from Proposition 4.

The result of Theorem 1 is extremely general and does *not* depend on the choice of a specific clustering distance: it shows the stability (*relative to a fixed distance d_P*) of risk-minimizing collaborative algorithms with a unique minimizer for distance d_P. The question remains open to know whether state-of-the-art collaborative clustering algorithms are stable with respect to a reasonable clustering distance (for instance the canonical distance).

4.3.2 Stability and Consistency

Stability is a notion introduced in regular clustering to describe how an algorithm is affected by slight changes in the data. We have introduced consistency as a measure of how strongly the collaboration affects the local decisions. This notion is inherent to multi-view and collaborative techniques. We will now show that, even though these two notions are intrinsically of different natures, they are strongly connected.

A first result on collaborative clustering stability can be shown about the concatenation of clustering algorithms. Proposition 6 below states that a concatenation of local algorithms is stable provided that the local algorithms are stable.

Proposition 6 *Suppose that the local algorithms A^j are stable for distance $d^j_{P_j}$. Then the concatenation of local algorithms $\mathcal{A} = \bigoplus_{j=1}^{J} A^j$ is stable for the canonical distance.*

Proof Let X_1 and X_2 be two samples drawn from distribution P. Then we have:

$$d_P(\mathcal{A}(X_1), \mathcal{A}(X_2)) = \frac{1}{J}\sum_{j=1}^{J} d^j_{P_j}\left((\mathcal{A}(X_1))^j, (\mathcal{A}(X_2))^j\right) \tag{13}$$

$$= \frac{1}{J}\sum_{j=1}^{J} d^j_{P_j}\left(A^j(X^j_1), A^j(X^j_2)\right) \tag{14}$$

Because of the linearity of the expected value, it comes that:

$$stab(\mathcal{A}, P, m) = \frac{1}{J}\sum_{j=1}^{J} stab(A^j, P^j, m) \tag{15}$$

Hence the stability of $\mathcal{A}$.

This result is rather intuitive, since the concatenation corresponds to a collaborative algorithm that does nothing. From this point of view, it is expected that the unmodified results of stable local algorithms will remain stable. This result is a consequence of the choice of the canonical distance and may be invalid for other distances. We note here that conserving the stability of concatenation is a desirable property for the choice of a clustering distance on the total space.

More interestingly, using the notion of consistency, the same result can be applied to get a more generic result (still valid only for the canonical distance):

Theorem 2 *Let $\mathcal{A}_{\langle A^1,\ldots,A^J\rangle}$ be a collaborative clustering algorithm. Then the stability of $\mathcal{A}$ relatively to the canonical distance is upper-bounded as follows:*

$$stab(\mathcal{A}, P) \leq cons(\mathcal{A}, P) + \frac{1}{J}\sum_{j=1}^{J} stab(A^j, P^j) \tag{16}$$

Proof Let X_1 and X_2 be two samples drawn from distribution P. Since the canonical distance satisfies the triangular inequality, we have:

$$d_P(\mathcal{A}(X_1), \mathcal{A}(X_2)) \leq d_P\left(\mathcal{A}(X_1), \left(\bigoplus_{j=1}^{J} A^j\right)(X_1)\right) \tag{17}$$

$$+ d_P\left(\left(\bigoplus_{j=1}^{J} A^j\right)(X_1), \left(\bigoplus_{j=1}^{J} A^j\right)(X_2)\right) \tag{18}$$

$$+ d_P\left(\left(\bigoplus_{j=1}^{J} A^j\right)(X_2), \mathcal{A}(X_2)\right) \tag{19}$$

Then, by taking the expected value of this expression, we obtain:

$$stab(\mathcal{A}, P, m) \leq 2 \times \mathbb{E}_{X\sim P^m}\left[d_P\left(\mathcal{A}(X), \left(\bigoplus_{j=1}^{J} A^j\right)(X)\right)\right] \tag{20}$$

$$+ \mathbb{E}_{X_1,X_2\sim P^m}\left[d_P\left(\left(\bigoplus_{j=1}^{J} A^j\right)(X_1), \left(\bigoplus_{j=1}^{J} A^j\right)(X_2)\right)\right] \tag{21}$$

which is the result we wanted.

This result has the advantage of being generic since it makes no assumption on the nature of the collaboration process. It also has the direct consequence that any consistent collaborative algorithm working from stable local results is stable for the canonical distance. However, this corollary is quite limited since the consistency assumption is extremely strong and does not apply to most practical cases where the collaborative process is expected to find results that differ from the simple concatenation of the local results from each views.

4.3.3 Stability of Contractive Collaborative Algorithms

We have seen that any multi-view algorithm can be, at least theoretically, decomposed into two steps: a local step where local algorithms compute a first partition, and a collaborative step in which the partitions are refined by exploiting the collaboration. Mathematically, we expressed this property in Proposition 3 by defining $\mathcal{A}^{ens,j}$ as an unsupervised ensemble learning algorithm producing a partition of $\mathbb{X}^j$, and setting $\mathcal{A}^{ens}(C^1, \ldots, C^J) = (\mathcal{A}^{ens,1}(C^1, \ldots, C^J), \ldots, \mathcal{A}^{ens,J}(C^1, \ldots, C^J))$. With these notations, any collaborative clustering algorithm $\mathcal{A}_{\langle A^1,\ldots,A^J\rangle}$ can be expressed as $\mathcal{A}_{\langle A^1,\ldots,A^J\rangle} = \mathcal{A}^{ens} \circ \left(\bigoplus_{j=1}^{J} A^j\right)$. We will show that a desirable property, for this generalized ensemble learning algorithm to guarantee the global stability, is to be

Lipschitz continuous with respect to clustering distance d_P (which implies, in this context, being contractive[2]).

Theorem 3 *Let $\mathcal{A}_{\langle A^1,\ldots,A^J\rangle}$ be a collaborative clustering algorithm, which decomposes into $\mathcal{A}^{ens} \circ \left(\bigoplus_{j=1}^{J} A^j\right)$. Suppose that $\mathcal{A}^{ens}$ is Lipschitz continuous for the canonical distance d_P, in the sense that there exists $K \in (0, 1]$ such that for all $\mathcal{C}, \mathcal{C}'$, $d_P(\mathcal{A}^{ens}(\mathcal{C}), \mathcal{A}^{ens}(\mathcal{C}')) \leq K d_P(\mathcal{C}, \mathcal{C}')$. Then if all $A^1, \ldots, A^J$ are stable, $\mathcal{A}_{\langle A^1,\ldots,A^J\rangle}$ is also stable.*

Proof Let us consider two samples X_1 and X_2 drawn from the distribution P. From there, we have:

$$d_P\big(\mathcal{A}(X_1) - \mathcal{A}(X_2)\big) = d_P\left(\mathcal{A}^{ens} \circ \left(\bigoplus_{j=1}^{J} A^j\right)(X_1) - \mathcal{A}^{ens} \circ \left(\bigoplus_{j=1}^{J} A^j\right)(X_2)\right) \tag{22}$$

Since $\mathcal{A}^{ens}$ is a Lipschitz contraction function, there exists a real constant $0 < K \leq 1$ such that:

$$d_P\big(\mathcal{A}(X_1) - \mathcal{A}(X_2)\big) \leq K\ d_P\left(\left(\bigoplus_{j=1}^{J} A^j\right)(X_1) - \left(\bigoplus_{j=1}^{J} A^j\right)(X_2)\right) \tag{23}$$

Since the A^j are stable for all j, from Eq. (23) and Proposition 6 we directly infer the stability of $\mathcal{A}$.

This proposition is interesting but raises the question of what would be required in practice and from an algorithm point of view for a collaborative algorithm to be a Lipschitz continuous function. On the other hand, when looking at Eq. (23) and considering what it means for a collaborative algorithm to be a contraction mapping ($K \leq 1$), we see that the partitioning of the two samples X_1 and X_2 drawn from the distribution P should be closer after the collaborative process. It turns out that this is exactly what is expected from a collaborative algorithm. Therefore, being a contraction mapping seems to be a necessary property that any collaborative algorithm should have. However, the mathematical analysis ends here, and it would be up to algorithms developers to demonstrate that their collaborative methods indeed have this property for all possible scenarii.

From there, once again we have a proposition that looks promising, as it may validate that all 'well-designed' collaborative algorithms are stable given that the local algorithms are stable too. However, we cannot be sure that it applies to any existing collaborative algorithm, as such a demonstration has never been done for any existing method. Furthermore, most of the existing implementations of collaborative

---

[2] The fact that the Lipschitz constant K must be lower than 1 is due to the convention that the clustering distances are defined between 0 and 1.

clustering algorithms rely on local algorithms that are known to be unstable: K-Means, FC-Means, EM for the GMM, SOM and GTM. Thus, these methods are already excluded from the scope of this proposition.

5 Open Questions

In the history of learning theory, clustering has always remained marginal compared to supervised learning, and in particular to classification. Within the broad domain of clustering, the question of a theoretical analysis of multi-view methods is even less represented. With the formal treatment we proposed in this chapter, we aimed to give good foundations for future theoretical works on multi-view and collaborative clustering, by clarifying the involved concepts and providing first fundamental results. However, it will not escape the reader's attention that there is still a long way to having solid theoretical results. In particular, multiple questions remain open and should be investigated in future works:

Choice of a multi-view distance. The multi-view distance is the core notion conditioning the definition of stability. Because of the unsupervised nature of clustering, there is no objective way to qualify the quality of a produced partition, and in particular stability can be defined only with regards to a chosen distance. Therefore, choosing which distance to use is essential for stability to reflect interesting properties of the algorithms.

We have introduced in this chapter the *canonical multi-view clustering distance*, a simple linear combination of local clustering distances. This choice is obviously the most straightforward way to define a multi-view clustering distance that exhibits some intuitive properties. Actually, we have shown that it leads to fundamental results, in particular Theorems 2 and 3. But this multi-view clustering distance may not be entirely satisfactory, since it ignores, in its definition, a core issue of multi-view problems: the interdependence between views. By taking marginal distributions in the local space, the canonical multi-view distance essentially ignores that views could be correlated. For instance, if two views are identical, an independent stability in each of these two views is not sufficient.

It is clear from this remark that other multi-view clustering distances should be investigated and that their theoretical properties should be analyzed. It is less clear though how to build such distances. We first notice that the currently used definition is just an extension from the definition of clustering distance for regular clustering. This definition does not constrain the distances d_P on the probability distribution P, which would be reasonable to have for multi-view distances. In addition, we observed that some intuitive results are established only for this specific clustering distance, for instance Proposition 5 (global stability if and only if stability on all views) and Proposition 6 (concatenation of stable algorithms is stable). We think that these properties should be added to the characterization of reasonable multi-view clustering distances.

Unicity of the minimizer for risk-minimizing multi-view algorithms. The stability theorem demonstrated by Ben-David et al. [2] is a fundamental theoretical result regarding clustering. By noticing that multi-view clustering can be seen as similar to regular clustering, we proposed with Theorem 1 a variant of this theorem for the multi-view and collaborative cases. This theorem is the most general we presented in this chapter, since it is not restricted by a specific choice of a clustering distance, however it is, at the same time, the least informative: indeed, we did not find evidences that standard collaborative clustering techniques satisfy the conditions of the theorem.

Multiple works in collaborative clustering have used an objective function of the form:

$$R(P, \mathcal{C}) = \sum_{j=1}^{J} \left(R^j(P^j, \mathcal{C}^j) + \sum_{i \neq j} \Delta(P, \mathcal{C}^i, \mathcal{C}^j) \right) \tag{24}$$

which corresponds to a trade-off between staying close to a local optimum and minimizing the differences between the views. It is not direct whether such a risk has a unique minimizer. Even when the local risk functions R^j have all a unique minimizer under the marginal probabilities, the divergence term $\Delta(P, \mathcal{C}^i, \mathcal{C}^j)$ brings in some perturbations and could affect the unicity of a minimizer. Intuitively here, it appears that when the Δ term is negligible, the existence of unique minimizers locally should guarantee the existence of a unique minimizer for the global risk. This result is very much in line with Theorem 2, since the multi-view stability is here relative to the local stability, but also to minimal perturbations introduced by the collaboration (i.e. consistency).

As a complement to the stability theorem, Ben-David et al. also proved that, in case there is no unique minimizers and some symmetry in the risk, the clustering algorithm is necessarily unstable. This result has not been presented in this chapter but could be of particular interest for the case of collaborative clustering. Indeed, in this context, it is not rare that two corrections of a partition are completely equivalent, which would lead to unstability. Characterizing this effect, in particular for risks of the form presented above, seems like a promising direction.

Stabilization of a collaboration. The results we presented in this chapter revolve around one main question: does the collaboration maintain the stability of the results? We could see that this is not clear, and that other factors can enter into account. Consistency has been introduced as a measure of the novel information contained into the collaboration, compared to a simple concatenation of the local partitions. If the consistency is maxed, nothing guarantees that some perturbing information has not been exchanged as well, which may cause unstability.

However, the converse question is still open: can a collaboration of unstable algorithms be stable? We have seen in Example 6 that a trivial choice of a multi-view clustering distance can lead to a stable collaboration if at least one of the local algorithms is stable, no matter if the other are or are not stable. Another trivial example would regard constant algorithms, which by definition will be stable for any input local algorithm. These two examples are trivial, either because of a degenerate choice of a multi-view clustering distance, or of a collaborative algorithm. They show however that the question has no simple answer and requires further investigation.

The question of the stabilization of a collaboration can be seen from two opposite angles: (1) If some local algorithms are unstable, is it possible to stabilize them with a reasonable collaboration? and (2) In a stable collaboration of (a potentially high number of) stable local algorithms, can changing only one local algorithm affect the stability of the collaboration? These two questions have strong practical implications.

Decomposition of the algorithms. At multiple points in the chapter, we have seen that a collaborative clustering algorithm can be decomposed into two steps: a concatenation of the local algorithms, followed by some unsupervised ensemble learning to make the collaboration. Although some algorithms directly implement this decomposition, many others do not and for them it becomes difficult to use results based on it, including Theorem 3. It may be then important to know whether some properties of the second phase (such as Lipschitz-continuity) can be inferred when the corresponding ensemble learning algorithm is not given explicitly.

Stability of consensus. In our analysis so far, we have considered only the most consensual definition of collaborative clustering where we have several algorithms working on multiple sites to first produce a local solution and then collaborate to improve each local solution without searching for a consensus partition of space. This corresponds to producing a multi-view partition of index J. However, as we have discussed in the state of the art section, clustering frameworks under the name collaborative clustering are a broad spectrum ranging from fully multi-view clustering to unsupervised ensemble learning (See Table 1). An adaptation of the theoretical notions presented in Sect. 4 to algorithms with consensus may present interesting peculiarities.

We note that notions such as stability, consistency or novelty do not make sense for unsupervised ensemble learning, which does not take data as input. We remind that these three measures are relative to various behaviours of the clustering algorithm when data are drawn from a specified distribution.

For other multi-view methods with consensus, apart from the inherent difficulty of defining the task (on which space is the produced partition defined?), we note that there is no natural "neutral" algorithm, i.e. an algorithm having no effect (such as the concatenation in Sect. 4). A possibility could be to rely on the majority vote operator.

6 Conclusion

In this chapter, we made an attempt to rethink collaborative clustering in comparison with the better known fields of multi-view clustering and ensemble learning. This formalization was needed to understand the interconnections between these various fields and to initiate a proper study of the theoretical properties of collaborative clustering. For that purpose, we extended key clustering notions such as clustering *stability* to the context of collaborative clustering, and we identified the additional key notions of *novelty* and *consistency* that are important for typical collaborative clustering applications.

Convinced of the importance to firstly defining a problem correctly before being able to solve it, we formalized the different branches of collaborative clustering, which have evolved during the last decade without being properly classified. This formal look into these algorithms made it clear that multi-view and collaborative clustering methods can all be seen as matching the definition proposed by Grozavu and Bennani [23]: collaborative methods should have an intermediary step with local results computed with local algorithms and should not aim for a consensus. Next, we demonstrated that collaborative algorithms matching this definition can be treated as multi-view clustering algorithms.

The theoretical study we proposed for collaborative and multi-view methods offers a clean basis for further investigations onto the theoretical properties of multi-view methods. Some challenges and open questions have been presented in Sect. 5 and we wish that our work may help to better consider the properties of existing and future collaborative clustering methods. And we also hope that our attempt at a formally defining the different branches of collaborative clustering will lead to a better integration of the somehow different family of collaborative clustering algorithms inside the multi-view and ensemble learning communities.

References

1. Arivazhagan, M.G., Aggarwal, V., Singh, A.K., Choudhary, S.: Federated learning with personalization layers (2019)
2. Ben-David, S., Von Luxburg, U., Pál, D.: A sober look at clustering stability. In: International Conference on Computational Learning Theory. pp. 5–19. Springer (2006)
3. Bickel, S., Scheffer, T.: Multi-view clustering. In: Proceedings of the 4th IEEE International Conference on Data Mining (ICDM 2004), 1-4 November 2004, Brighton, UK. pp. 19–26. IEEE Computer Society (2004). https://doi.org/10.1109/ICDM.2004.10095
4. Bishop, C.M., Svensén, M., Williams, C.K.I.: GTM: the generative topographic mapping. Neural Comput. **10**(1), 215–234 (1998)
5. Bonawitz, K., Eichner, H., Grieskamp, W., Huba, D., Ingerman, A., Ivanov, V., Kiddon, C., Konečný, J., Mazzocchi, S., McMahan, B., Overveldt, T.V., Petrou, D., Ramage, D., Roselander, J.: Towards federated learning at scale: system design. In: Talwalkar, A., Smith, V., Zaharia, M. (eds.) Proceedings of Machine Learning and Systems 2019, MLSys 2019, Stanford, CA, USA, March 31 – April 2. mlsys.org. https://proceedings.mlsys.org/book/271.pdf (2019)
6. Carlsson, G.E., Mémoli, F.: Characterization, stability and convergence of hierarchical clustering methods. J. Mach. Learn. Res. **11**, 1425–1470. http://portal.acm.org/citation.cfm?id=1859898 (2010)
7. Coletta, L.F.S., Vendramin, L., Hruschka, E.R., Campello, R.J.G.B., Pedrycz, W.: Collaborative fuzzy clustering algorithms: some refinements and design guidelines. IEEE Trans. Fuzzy Syst. **20**(3), 444–462 (2012)
8. Cornuéjols, A., Wemmert, C., Gançarski, P., Bennani, Y.: Collaborative clustering: why, when, what and how. Inf. Fusion **39**, 81–95 (2018)
9. Diao, E., Ding, J., Tarokh, V.: Heterofl: computation and communication efficient federated learning for heterogeneous clients. CoRR **abs/2010.01264**. https://arxiv.org/abs/2010.01264 (2020)

10. Dunn, J.C.: A fuzzy relative of the isodata process and its use in detecting compact well-separated clusters. J. Cybern. **3**(3), 32–57 (1973)
11. Falih, I., Grozavu, N., Kanawati, R., Bennani, Y., Matei, B.: Collaborative multi-view attributed networks mining. In: 2018 International Joint Conference on Neural Networks, IJCNN 2018, Rio de Janeiro, Brazil, July 8–13, 2018. pp. 1–8. IEEE (2018). https://doi.org/10.1109/IJCNN.2018.8489183
12. Filali, A., Jlassi, C., Arous, N.: SOM variants for topological horizontal collaboration. In: 2nd International Conference on Advanced Technologies for Signal and Image Processing, ATSIP 2016, Monastir, Tunisia, March 21–23, 2016. pp. 459–464. IEEE (2016). https://doi.org/10.1109/ATSIP.2016.7523117
13. Filali, A., Jlassi, C., Arous, N.: A hybrid collaborative clustering using self-organizing map. In: 14th IEEE/ACS International Conference on Computer Systems and Applications, AICCSA 2017, Hammamet, Tunisia, October 30 – Nov. 3, 2017. pp. 709–716. IEEE Computer Society (2017). https://doi.org/10.1109/AICCSA.2017.111
14. Forestier, G., Wemmert, C., Gançarski, P.: Multisource images analysis using collaborative clustering. EURASIP J. Adv. Signal Process. **2008** (2008). https://doi.org/10.1155/2008/374095
15. Forestier, G., Wemmert, C., Gançarski, P.: Semi-supervised collaborative clustering with partial background knowledge. In: Workshops Proceedings of the 8th IEEE International Conference on Data Mining (ICDM 2008), December 15–19, 2008, Pisa, Italy. pp. 211–217. IEEE Computer Society (2008). https://doi.org/10.1109/ICDMW.2008.116
16. Forestier, G., Wemmert, C., Gançarski, P.: Towards conflict resolution in collaborative clustering. In: 5th IEEE International Conference on Intelligent Systems, IS 2010, 7–9 July 2010, University of Westminster, London, UK. pp. 361–366. IEEE (2010). https://doi.org/10.1109/IS.2010.5548343
17. Forestier, G., Wemmert, C., Gançarski, P., Inglada, J.: Mining multiple satellite sensor data using collaborative clustering. In: Saygin, Y., Yu, J.X., Kargupta, H., Wang, W., Ranka, S., Yu, P.S., Wu, X. (eds.) ICDM Workshops 2009, IEEE International Conference on Data Mining Workshops, Miami, Florida, USA, 6 December 2009. pp. 501–506. IEEE Computer Society (2009). https://doi.org/10.1109/ICDMW.2009.42
18. Foucade, Y., Bennani, Y.: Unsupervised collaborative learning using privileged information. CoRR **abs/2103.13145**. https://arxiv.org/abs/2103.13145 (2021)
19. Gançarski, P., Salaou, A.: FODOMUST: une plateforme pour la fouille de données multi-stratégie multitemporelle. In: de Runz, C., Crémilleux, B. (eds.) 16ème Journées Francophones Extraction et Gestion des Connaissances, EGC 2016, 18-22 Janvier 2016, Reims, France. Revue des Nouvelles Technologies de l'Information, vol. E-30, pp. 481–486. Éditions RNTI. http://editions-rnti.fr/?inprocid=1002204 (2016)
20. Gançarski, P., Wemmert, C.: Collaborative multi-step mono-level multi-strategy classification. Multimed. Tools Appl. **35**(1), 1–27 (2007)
21. Ghassany, M., Grozavu, N., Bennani, Y.: Collaborative clustering using prototype-based techniques. Int. J. Comput. Intell. Appl. **11**(3) (2012). https://doi.org/10.1142/S1469026812500174
22. Ghassany, M., Grozavu, N., Bennani, Y.: Collaborative multi-view clustering. In: The 2013 International Joint Conference on Neural Networks, IJCNN 2013, Dallas, TX, USA, August 4–9, 2013. pp. 1–8. IEEE (2013). https://doi.org/10.1109/IJCNN.2013.6707037
23. Grozavu, N., Bennani, Y.: Topological collaborative clustering. Aust. J. Intell. Inf. Process. Syst. **12**(3). http://cs.anu.edu.au/ojs/index.php/ajiips/article/view/1216 (2010)
24. Hafdhellaoui, S., Boualleg, Y., Farah, M.: Collaborative clustering approach based on dempster-shafer theory for bag-of-visual-words codebook generation. In: Meurs, M., Rudzicz, F. (eds.) Advances in Artificial Intelligence - 32nd Canadian Conference on Artificial Intelligence, Canadian AI 2019, Kingston, ON, Canada, May 28-31, 2019, Proceedings. Lecture Notes in Computer Science, vol. 11489, pp. 263–273. Springer (2019). https://doi.org/10.1007/978-3-030-18305-9_21
25. Jiang, Y., Chung, F.L., Wang, S., Deng, Z., Wang, J., Qian, P.: Collaborative fuzzy clustering from multiple weighted views. IEEE Trans. Cybern. **45**(4), 688–701 (2015). https://doi.org/10.1109/TCYB.2014.2334595

26. Jiang, Z.L., Guo, N., Jin, Y., Lv, J., Wu, Y., Liu, Z., Fang, J., Yiu, S., Wang, X.: Efficient two-party privacy-preserving collaborative k-means clustering protocol supporting both storage and computation outsourcing. Inf. Sci. **518**, 168–180 (2020)
27. Kleinberg, J.M.: An impossibility theorem for clustering. In: Becker, S., Thrun, S., Obermayer, K. (eds.) Advances in Neural Information Processing Systems 15 [Neural Information Processing Systems, NIPS 2002, December 9–14, 2002, Vancouver, British Columbia, Canada]. pp. 446–453. MIT Press. https://proceedings.neurips.cc/paper/2002/hash/43e4e6a6f341e00671e123714de019a8-Abstract.html (2002)
28. Kohonen, T.: The self-organizing map. Neurocomputing **21**(1–3), 1–6 (1998)
29. von Luxburg, U.: Clustering stability: an overview. Found. Trends Mach. Learn. **2**(3), 235–274 (2009)
30. Mitra, S., Banka, H., Pedrycz, W.: Rough-fuzzy collaborative clustering. IEEE Trans. Syst. Man Cybern. Part B **36**(4), 795–805 (2006)
31. Murena, P., Sublime, J., Matei, B., Cornuéjols, A.: An information theory based approach to multisource clustering. In: Lang, J. (ed.) Proceedings of the Twenty-Seventh International Joint Conference on Artificial Intelligence, IJCAI 2018, July 13–19, 2018, Stockholm, Sweden. pp. 2581–2587. ijcai.org (2018). https://doi.org/10.24963/ijcai.2018/358
32. Ngo, L.T., Dang, T.H., Pedrycz, W.: Towards interval-valued fuzzy set-based collaborative fuzzy clustering algorithms. Pattern Recognit. **81**, 404–416 (2018)
33. Pedrycz, W.: Collaborative fuzzy clustering. Pattern Recognit. Lett. **23**(14), 1675–1686 (2002)
34. Pedrycz, W.: Knowledge-Based Clustering - From Data to Information Granules. Wiley (2005)
35. Pedrycz, W., Rai, P.: Collaborative clustering with the use of fuzzy c-means and its quantification. Fuzzy Sets Syst. **159**(18), 2399–2427 (2008)
36. Pokhrel, S.R.: Federated learning meets blockchain at 6g edge: A drone-assisted networking for disaster response. In: Proceedings of the 2nd ACM MobiCom Workshop on Drone Assisted Wireless Communications for 5G and Beyond. p. 49–54. DroneCom '20, Association for Computing Machinery, New York, NY, USA (2020). https://doi.org/10.1145/3414045.3415949
37. Shen, Y., Pedrycz, W.: Collaborative fuzzy clustering algorithm: some refinements. Int. J. Approx. Reason. **86**, 41–61 (2017)
38. Strehl, A., Ghosh, J., Cardie, C.: Cluster ensembles - a knowledge reuse framework for combining multiple partitions. J. Mach. Learn. Res. **3**, 583–617 (2002)
39. Sublemontier, J.: Unsupervised collaborative boosting of clustering: an unifying framework for multi-view clustering, multiple consensus clusterings and alternative clustering. In: The 2013 International Joint Conference on Neural Networks, IJCNN 2013, Dallas, TX, USA, August 4–9, 2013. pp. 1–8. IEEE (2013). https://doi.org/10.1109/IJCNN.2013.6706911
40. Sublime, J., Grozavu, N., Cabanes, G., Bennani, Y., Cornuéjols, A.: From horizontal to vertical collaborative clustering using generative topographic maps. Int. J. Hybrid Intell. Syst. **12**(4), 245–256 (2015)
41. Sublime, J., Lefebvre, S.: Collaborative clustering through constrained networks using bandit optimization. In: 2018 International Joint Conference on Neural Networks, IJCNN 2018, Rio de Janeiro, Brazil, July 8–13, 2018. pp. 1–8. IEEE (2018). https://doi.org/10.1109/IJCNN.2018.8489479
42. Sublime, J., Matei, B., Cabanes, G., Grozavu, N., Bennani, Y., Cornuéjols, A.: Entropy based probabilistic collaborative clustering. Pattern Recognit. **72**, 144–157 (2017)
43. Sublime, J., Matei, B., Murena, P.: Analysis of the influence of diversity in collaborative and multi-view clustering. In: 2017 International Joint Conference on Neural Networks, IJCNN 2017, Anchorage, AK, USA, May 14–19, 2017. pp. 4126–4133. IEEE (2017). https://doi.org/10.1109/IJCNN.2017.7966377
44. Sublime, J., Troya-Galvis, A., Puissant, A.: Multi-scale analysis of very high resolution satellite images using unsupervised techniques. Remote Sens. **9**(5), 495 (2017)
45. Vanhaesebrouck, P., Bellet, A., Tommasi, M.: Decentralized collaborative learning of personalized models over networks. In: Singh, A., Zhu, X.J. (eds.) Proceedings of the 20th International Conference on Artificial Intelligence and Statistics, AISTATS 2017, 20–22 April 2017, Fort Lauderdale, FL, USA. Proceedings of Machine Learning Research, vol. 54, pp. 509–517. PMLR. http://proceedings.mlr.press/v54/vanhaesebrouck17a.html (2017)

46. Wemmert, C., Gançarski, P., Korczak, J.J.: A collaborative approach to combine multiple learning methods. Int. J. Artif. Intell. Tools **9**(1), 59–78 (2000)
47. Yu, F., Tang, J., Cai, R.: Partially horizontal collaborative fuzzy c-means. Int. J. Fuzzy Syst. **9**, 198–204 (2007)
48. Zimek, A., Vreeken, J.: The blind men and the elephant: on meeting the problem of multiple truths in data from clustering and pattern mining perspectives. Mach. Learn. **98**(1–2), 121–155 (2015)
49. Zouinina, S., Grozavu, N., Bennani, Y., Lyhyaoui, A., Rogovschi, N.: Efficient k-anonymization through constrained collaborative clustering. In: IEEE Symposium Series on Computational Intelligence, SSCI 2018, Bangalore, India, November 18–21, 2018. pp. 405–411. IEEE (2018). https://doi.org/10.1109/SSCI.2018.8628635

An Optimal Transport Framework for Collaborative Multi-view Clustering

Fatima-Ezzahraa Ben-Bouazza, Younès Bennani, and Mourad El Hamri

Abstract Research on Multi-View Clustering (MVC) has become more and more attractive thanks to the richness of its application in several fields. Unlike traditional clustering, each view maintains different information to complete each other. On the other hand, the collaborative framework aims mainly to exchange knowledge between collaborators to improve their local quality. The purpose of this chapter is to present a new framework for Collaborative Multi-View Clustering (Co-MVC) based on Optimal Transport (OT) theory. The main idea of the proposed approaches is to perform a collaborative learning step to create a coordination map between multiple views that will ensure an optimal fusion for building a consensus solution. The intuition behind performing the collaboration step is to exploit the pre-existing structure learned in each view to improve the classical consensus clustering. In this chapter we propose two approaches: Collaborative Consensus Projection Approach (CoCP) that aims to perform the consensus in the global space of the data, and a Collaborative Consensus with New Representation (CoCNR) that seeks to encode a new data representation based on local ones. Both approaches are based on the entropy regularized Wasserstein distance and the Wasserstein barycenters between the data distribution in each view. Extensive experiments are conducted on multiple datasets to analyze and evaluate the proposed approaches.

Keywords Multi-view clustering (MVC) · Co-training · Subspace clustering · Collaborative learning · Multi-task learning · Optimal transport (OT) · Wasserstein distance

---

F.-E. Ben-Bouazza (✉)
High School of Biomedical Engineering, Mohammed VI University of Health Sciences, Casablanca, Morocco
e-mail: fbenbouazza@um6ss.ma

F.-E. Ben-Bouazza · Y. Bennani · M. El Hamri
LIPN–CNRS UMR 7030, Université Sorbonne Paris Nord, Villetaneuse, France
e-mail: Younes.Bennani@sorbonne-paris-nord.fr

M. El Hamri
e-mail: Mourad.ElHamri@sorbonne-paris-nord.fr

LaMSN, La Maison des Sciences Numériques, Saint-Denis, France

W. Pedrycz and S. Chen (eds.), *Recent Advancements in Multi-View Data Analytics*, Studies in Big Data 106, https://doi.org/10.1007/978-3-030-95239-6_5

1 Introduction

Unsupervised clustering methods have become recently more and more popular thanks to their ability to regroup unlabeled data which was very difficult to realize for the human being. In recent years many new clustering algorithms have been developed, while existing methods have also been modified and improved. The richness of these techniques is provided from the difficulty of proposing a universal approach that could be applied to all available data types. In fact, each method has a bias caused by the objective of creating the clusters. Therefore, two different methods can provide very different clustering results from the same data. Moreover, the same algorithm may provide different results depending on its initialization or parameters.

To overcome this problem, some methods suggest using different clustering results to benefit from the potential diversity. These methods use the information provided from different clusters computed in different circumstances. Multi-view clustering is one of these powerful methods where the objective is to form consistent clusters of similar topics by combining information from multi-view features, rather than the traditional clustering method using a single set of features. The importance of this method lies in the diversity of features which ensures not only consistent clustering but also a better interpretation.

On the other hand, collaborative learning became lately one of the attractive approaches in multi-models clustering framework [11]. Its main idea is to exchange information between distant models to improve local quality optimally while respecting the confidentiality of each model.

In this chapter, we aim to improve consensus clustering by creating a prior interaction between the views, through collaborative learning. Moreover, the proposed approaches are presented under the optimal transport theory. We propose two main algorithms of consensus clustering, to learn improved data structure from the distribution's data on each view through the collaboration step, while an aggregation step tends to get a unified model with better quality.

The rest of the chapter will be organized as follows: in Sect. 2, we present the related works of the multi-view clustering, in Sect. 3, we introduce the optimal transport theory. In Sect. 4, we present the proposed approach and motivation behind this work, while in Sect. 5, we detailed the conducted experiment with a careful comparison and interpretation. Finally, we conclude the proposed work and discuss future works and possible extensions.

2 Related Works

The idea of combining multiple models came to highlight the paradigm of distributed data, where the main objective is to get a unified view of the global dataset while ensuring higher quality and better visualization of the data. This fusion of information must be done respecting many properties to enhance the quality and to avoid losing the knowledge learned from the models. To do so, we recall the most important conditions to be checked:

- Robustness: the combination must perform better model than single clustering algorithms regarding a better complexity, and avoiding the over-fitting.
- Consistency: the learned model in ensemble clustering must avoid losing the real structure of the data.
- Novelty: the ensemble model must guarantee to find better results that are not attainable with a single clustering algorithm.
- Stability: we must obtain lower sensitivity to noise and outliers.

The challenge in ensemble clustering is how to construct an appropriate consensus function. Mainly the consensus function must contain two principal parts: the first one must be capable to detect the best clustering model for each set, and the second part, intends to improve these models by finding the right fusion, in order to learn a better model than a single view model. In the state of the art, there is two main consensus function approaches: the objective co-occurrence and the median partition.

The idea behind the first approach consists to find the best cluster label in the consensus partition that must be linked to each object. Among this approach, the consensus is obtained through a voting process among each object, where we analyze how many times the object belongs to the same cluster, so it can be associated with the consensus partition [2].

In the second approach, the consensus model is obtained based on an optimization problem regarding the local models. The main idea is to maximize the similarity between the local partition. The main challenge is how to choose an appropriate measure of similarity, where either it can count pairs that measures the agreement between two pairs objects [35], or set matching measures, which are based on cardinally set comparison [26].

On the other hand, learning from multi-view data to get a unified solution that captured all the information, was introduced in [42] by Yarowsky, where the multi-view learning is applied for word sense disambiguation. Mainly, the approach is based on a voting algorithm which is represented by two different classifiers: the first one is based on the local context as the first view, and the second one contains the other occurrences, designed as a second view.

In [7], the authors present a co-training approach based on two hypotheses trained on distinct views. The algorithm aims to improve the global learning quality of two classifiers with the highest confidence instances from unlabeled data. This approach requires that the views are independent.

In the same context, in [9], a Co-EM algorithm was presented as a multi-view consensus version of the Expectation-Maximization algorithm for semi-supervised learning. In 2004, Bickel and Scheffer [6] studied the problem where the set of attributes can be split randomly into two subsets. These approaches seek to optimize the agreement between the views. The authors described two different algorithms, an EM-based algorithm which gives very significant results, and an agglomerative multi-view algorithm which seems to be less efficient than single view approaches.

To summarize, multi-view clustering is considered as a basic task for several subsequent analyses in machine learning, in particular for ensemble clustering [37], also named aggregation clustering. The goal of this framework is to collect all the cluster information coming from different sources that represent the same dataset, or from different runs of the same clustering algorithm, to form a consensus clustering that includes all the information. This approach becomes a framework from multi-view clustering when it is applied to a clustering task with a multi-view description of the data [36, 41].

3 Optimal Transport

Well known for many centuries for its logistical and economic applications, optimal transport is currently undergoing a spectacular revival due to its unsuspected links with fluid mechanics, partial differential equations, and more recently with machine learning, as it gives a powerful way with several appealing properties to compare probability measures in a Lagrangian setting. Hence, numerous machine learning domains use it for modeling tasks, computing solutions and providing theoretical guarantees, like domain adaptation [12, 13], clustering [5, 8, 24], sub-space clustering [4], generative models [25] and semi-supervised learning [18, 19, 34].

3.1 Monge's Formulation

Optimal transport [33, 38] is a long-standing mathematical problem whose theory has matured over the years. Its birth is commonly dated back to 1781, with the problem introduced by the French mathematician Gaspard Monge in his thesis on the theory of cuttings and embankments [27] for military applications. The motivation for the problem was to find the optimal way of transporting sand piles from one place (déblais) to another place (remblais). This problem can be formulated as the following non-convex optimization problem: Given two probability spaces $(\mathcal{X}, \mu)$ and $(\mathcal{Y}, \nu)$ and a measurable cost function $c : \mathcal{X} \times \mathcal{Y} \rightarrow \mathbb{R}^{+}$, which represents the effort needed to move a unit of mass from $x \in \mathcal{X}$ forward to $y \in \mathcal{Y}$, the problem of Monge aims at finding the transport map $\mathcal{T} : \mathcal{X} \rightarrow \mathcal{Y}$, that transforms the measure μ to the measure ν and which minimizes the global transportation cost (1):

$$\inf_{\mathcal{T}}\{\int_{\mathcal{X}} c(x, \mathcal{T}(x)) d\mu(x) | \mathcal{T}\#\mu = \nu\}, \tag{1}$$

where $\mathcal{T}\#\mu$ denotes the push-forward operator of the measure μ through the map $\mathcal{T}$, defined by: for all measurable subset $B \subseteq \mathcal{Y}$, $\mathcal{T}\#\mu(\mathcal{B}) = \mu(\mathcal{T}^{-1}(\mathcal{B}))$.

The problem of Monge is quite difficult since it is not symmetric, it does not allow mass splitting, and the non-linearity in $\mathcal{T}$ makes it very difficult to analyze.

Moreover, the uniqueness and even the existence of a solution are not guaranteed, which is the situation if μ is a Dirac measure while the measure ν isn't.

3.2 *Kantorovich's Relaxation*

Gaspard Monge is the founding father of optimal transport according to several contemporary mathematicians. However, in his analysis, Monge did not even wonder about rigorous proof of the existence and the uniqueness of the optimal transport map, and his problem remained an open question for many years where results about the existence and uniqueness were limited to special cases. After a while, until the work of the Soviet mathematician Leonid Kantorovitch in 1942 [22], who faced optimization problems related to railway supply chain during World War II, proposed relaxation to the problem of Monge. In particular, his work allowed to give both existence and duality results and paved the way to a modern mathematical treatise of the problem.

The main underlying idea behind the formulation of Kantorovitch is to consider a probabilistic mapping instead of a deterministic map $\mathcal{T}$ to push the measure μ forward to the measure ν. In fact, in Kantorovich's relaxation, mass splitting into several fractions is allowed, which then permits the different fractions to be transported towards several destinations. This translates mathematically by replacing the push-forward of a measure by a probabilistic coupling γ in $\Pi(\mu, \nu)$, where $\Pi(\mu, \nu)$ is the set of probability measures over the product space $\mathcal{X} \times \mathcal{Y}$ such that both marginals of γ are μ and ν, more formally, $\Pi(\mu, \nu) = \{\gamma \in \mathcal{P}(\mathcal{X} \times \mathcal{Y}) | proj_{\mathcal{X}}\#\gamma = \mu, proj_{\mathcal{Y}}\#\gamma = \nu\}$. The joint probability γ is a new way to describe the displacement of mass from μ towards ν: for each pair $(x_i, y_j) \in \mathcal{X} \times \mathcal{Y}$, γ specify how much mass goes from x_i to y_j.

The relaxed formulation of Kantorovich, known as the Monge-Kantorovich problem, can then be formulated as follows:

$$\inf_{\gamma}\{ \int_{\mathcal{X}\times\mathcal{Y}} c(x, y)\, d\gamma(x, y) \,|\, \gamma \in \Pi(\mu, \nu) \}, \tag{2}$$

The minimizers for this problem are named optimal transport plans between μ and ν. The good news about this formulation is that the minimizers of this problem are always well existent under very general assumptions. It is the case when $\mathcal{X}$ and $\mathcal{Y}$ are Polish spaces where the cost function c is lower semi-continuous.

3.3 *Discrete Settings*

In the discrete settings, where the access to the measures μ and ν is only available via finite collections $X = (x_1, \ldots, x_n) \subset \mathcal{X}$ and $Y = (y_1, \ldots, y_m) \subset \mathcal{Y}$, the measures

μ and ν can be formulated as $\mu = \sum_{i=1}^{n} a_i \delta_{x_i}$ and $\nu = \sum_{j=1}^{m} b_j \delta_{y_j}$, where a and b are stochastic vectors in $\sum_n$ and $\sum_m$ respectively.

The values of the cost function c can be compactly represented by a rectangular cost matrix C_{XY} of size $n \times m$, the relaxation of Kantorovich becomes then the following linear program:

$$\min_{T \in U(a,b)} \langle T, C_{XY} \rangle_F \tag{3}$$

where $U(a, b) = \{T \in \mathcal{M}_{n\times m}(\mathbb{R}^+) \mid T 1_m = a, \quad T^{\mathbf{T}} 1_n = b\}$ is the transportation polytope that operates as a realizable set, and $\langle T, C_{XY} \rangle_F = trace(T^{\mathrm{T}} C_{XY})$ is the Frobenius dot-product of matrices.

In the special case where $n = m$ and μ and ν are uniform measures, the transportation polytope is then the perfect matching polytope (the assignment polytope) of order n composed of doubly stochastic matrices, and the solutions of the linear program lying in the corners of this polytope are permutation matrices. It is worth noting that Kantorovich's relaxed formulation is symmetric, in the sense that a coupling $T \in U(a, b)$ if and only if $T^{\mathbf{T}} \in U(b, a)$.

3.4 Wasserstein Distance

In the case where $\mathcal{X} = \mathcal{Y}$ is a space endowed with some metric d, then obviously it may be used as the cost function, i.e., $c = d^p$ for $p \in [1, +\infty[$. Then, the total transportation cost in the problem of Monge-Kantorovich defines the so-called p-Wasserstein distance:

$$\mathrm{W}_p(\mu, \nu) = \inf_{\gamma \in \Pi(\mu,\nu)} \left(\int_{\mathcal{X}^2} d^p(x, y)\, d\gamma(x, y) \right)^{1/p}, \quad \forall \mu, \nu \in \mathcal{P}_p(\mathcal{X}) \tag{4}$$

which is a metric over the space of probability measures with finite moments of order p, $\mathcal{P}_p(\mathcal{X})$. This space is called the Wasserstein space of order p.

The Wasserstein distance provides a powerful means to compare probability measures in a geometrically faithful way even when their supports do not overlap. This metric has many other benefits as its intuitive formulation, which can capture the whole underlying geometry of the measures by relying on the cost function $c = d^p$ encoding the geometry of the input space $\mathcal{X}$. Secondly, the fact that the Wasserstein distance mastery weak convergence is another major advantage which means that it is an excellent choice for machine learning tasks.

3.5 Wasserstein Barycenter

Since the p-Wasserstein distance metrizes the space $\mathcal{P}_p(\mathcal{X})$ it permits the extension of geometric concepts such as the notion of barycenter [1].

Let $\{\nu_1, \ldots, \nu_N\}$ be a set on N probability measures in $\mathcal{P}_p(\mathcal{X})$ and the vector $\lambda = (\lambda_1, \ldots, \lambda_N) \in \sum_N$ of N positive weights summing to 1. The Wasserstein barycenter of $\{\nu_1, \ldots, \nu_N\}$ with respect to $\lambda = (\lambda_1, \ldots, \lambda_N)$ is any probability measure $\tilde{\mu}$ in $\mathcal{P}_p(\mathcal{X})$ which satisfies:

$$\sum_{i=1}^{N} \lambda_i \mathrm{W}_p^p(\tilde{\mu}, \nu_i) = \inf_{\mu \in \mathcal{P}_p(\mathcal{X})} \sum_{i=1}^{N} \lambda_i \mathrm{W}_p^p(\mu, \nu_i), \tag{5}$$

In the case where $\{\nu_1, \ldots, \nu_N\}$ are discrete measures in $\mathcal{P}(\mathcal{X})$, each supported on a finite set of s_i elements, then there exists a Wasserstein barycenter $\tilde{\mu}$ in $\mathcal{P}^S(\mathcal{X})$, composed of probability measures with at most S support point in $\mathcal{X}$, where $S = \sum_{i=1}^{N} s_i - N + 1$.

There are fast and efficient algorithms that find local solutions of the Wasserstein barycenter problem over $\mathcal{P}^S(\mathcal{X})$ for $S \geq 1$ [15].

3.6 Entropy-Regularization

Optimal transport is a powerful paradigm to compare probability measures. However, its influence in machine learning has long been limited and neglected in favor of simpler φ-divergences (Kullback-Leibler, Jensen-Shannon, Hellinger, Total Variation) or integral probability metrics (Maximum Mean Discrepancy) because of:

- its computational complexity (optimal transport has a heavy computational price tag): As stated above, discrete optimal transport is a linear program. Therefore it is possible to solve it with the simplex algorithm or interior-point methods using network flow solvers. However, the optimal transport problem scales cubically on the sample size when we compare two discrete measures supported on n atoms. in general metric space, the computation complexity is $\mathcal{O}(n^3 log(n))$ [29], which is often too costly in practice.
- its statistical burden (optimal transport suffers from the curse of dimensionality): Let μ be a probability measure over $\mathbb{R}^d$, and $\hat{\mu}_n$ its empirical estimation, then we have $\mathbb{E}[\mathrm{W}_p(\mu, \hat{\mu}_n)] = \mathcal{O}(n^{\frac{-1}{d}})$. The empirical distribution $\hat{\mu}_n$ becomes less and less representative as the dimension d of the space $\mathbb{R}^d$ becomes large, so that in the convergence of $\hat{\mu}_n$ to $\hat{\mu}$ in Wasserstein distance is slow.

Entropy-regularization has emerged as a solution to the computational burden of optimal transport [14], and to the sample complexity properties of its solutions [21].

The entropy-regularized optimal transport problem reads:

$$\inf_{\gamma \in \Pi(\mu, \nu)} \int_{\mathcal{X} \times \mathcal{Y}} c(x, y)\, d\gamma(x, y) + \varepsilon \mathcal{H}(\gamma), \tag{6}$$

where $\mathcal{H}(\gamma) = \int_{\mathcal{X} \times \mathcal{Y}} \log(\frac{d\gamma(x,y)}{dxdy})\, d\gamma(x, y)$ is the entropy of the transport plan γ.

The entropy-regularized discrete problem is defined as follows:

$$\min_{T \in U(a,b)} \langle T, C_{XY} \rangle_F - \varepsilon \mathcal{H}(T), \tag{7}$$

where $\mathcal{H}(T) = -\sum_{i=1}^{n} \sum_{j=1}^{m} t_{ij}(\log(t_{ij}) - 1)$ is the entropy of T

We focus here on the discrete version of the regularized problem. The main underlying idea behind this regularization is to use $-\mathcal{H}$ as a function to obtain approximate solutions to the Monge-Kantorovich problem by increasing the entropy of the original optimal transport plan. Hence, it encourages the regularized plan to become less sparse than the original transport plan has at most $m + n - 1$ nonzero elements [10].

The regularized problem has a unique solution, since its objective function is an ε-strongly convex function, as the function $\mathcal{H}$ is 1-strongly concave because its Hessian $\partial^2 \mathcal{H}(T) = -diag(\frac{1}{t_{ij}})$ and $t_{ij} \leq 1$. The regularized plan can be characterized as follows: let the vectors u and v be the exponential scaling of the dual variables α and β respectively, $u = \exp(\frac{\alpha}{\varepsilon})$ and $v = \exp(\frac{\beta}{\varepsilon})$, and $K = \exp(\frac{-C_{XY}}{\varepsilon})$, then the solution of the regularized problem has the following form $T_{\varepsilon}{}^* = diag(u)K diag(v)$. The vectors u, v should satisfy: $u \odot (Kv) = a \ and \ v \odot (K^{\mathbf{T}} u) = b$, where $\odot$ is Hadamard product. This problem is called the matrix scaling problem and it may be efficiently addressed by using an iterative process: Sinkhorn-Knopp algorithm [23], that update iteratively $u^{(l+1)} = \frac{a}{K v^{(l)}}, \ and \ v^{(l+1)} = \frac{b}{K^{\mathbf{T}} u^{(l+1)}}$, initialized using a vector $v^{(0)} = 1_m$.

Given a low parameter ε, the solution of the regularized problem T_{ε}^* will converge, w.r.t. the weak topology, to the optimal solution with maximal entropy within all optimal transport plans of the Monge-Kantorovich problem [30]. Sinkhorn's algorithm [14], is formally summarized in Algorithm 1:

Algorithm 1: Sinkhorn's algorithm

Input : $(x_i)_{i=1,\dots,n}, (y_j)_{j=1,\dots,m}, a, b$
Parameters: ε
1 $c_{i,j} = \|x_i - y_j\|^2, \ \forall i, j \in \{1, \dots, n\} \times \{1, \dots, m\}$
2 $K = exp(-C_{XY}/\varepsilon)$
3 Initialize $v \leftarrow 1_m$
4 **while** *not converged* **do**
5 $\quad u \leftarrow \frac{a}{Kv}$
6 $\quad v \leftarrow \frac{b}{K^T u}$
7 **return** $\mathbf{T}_{\varepsilon}^* = diag(u)K diag(v)$

4 Proposed Approaches

In this section, we highlight how multi-view clustering can be seen as an optimal transport problem. Moreover, we will present a new framework of multi-view clustering based on collaborative learning.

4.1 Motivation

To prove that our method is reasonable from a theoretical point of view, we will explain the basic principles of multi-view clustering and how to transform it into an optimal transport problem. Multi-view clustering can be divided into two steps: the local step which aims to cluster the data in each view, and the global step which tends to aggregate this information *-centroids of clusters-* to reach a consensus and represent all views at the same time.

We consider $X = \left\{X^1, \ldots, X^r\right\}$ with r multiple views, where $X^v = \left\{x_1^v, \ldots, x_n^v\right\}$ is the vth view composed of n instances.In general, the existing methods of from a unified view or a consensus, seek to maximize some objective function that combines the basic partitioning $H = \{h_1, h_2, .., h_r\}$ given by some algorithm to find a consensus partitioning h, where the choice of the utility function U is very important.

$$\Gamma(h, H) = \sum_{i=1}^{r} w_i U(h, h_i) \tag{8}$$

where $\Gamma : \mathbb{Z}_+^n \times \mathbb{Z}_+^{nr} \mapsto \mathbb{R}$ is the consensus function, and $U : \mathbb{Z}_+^n \times \mathbb{Z}_+^r \mapsto \mathbb{R}$ the utility function with $\sum_i^r w_i = 1$. This problem can be transformed to a minimization problem without changing its nature by using different distances like Mirkin distance [26]. Moreover, [39] proved that the consensus problem is equivalent to K-means problem under some assumption defined as follow.

Definition 1 A [39] utility function U is a K-means consensus clustering utility function, if $\forall \quad H = \{h_1, \ldots, h_r\}$ and $K \geq 2$, there exists a distance f such that

$$\max_{h \in H} \sum_{i=1}^{r} w_i U(h, h_i) \Leftrightarrow \min_{h \in H} \sum_{k=1}^{K} \sum_{x_l \in C_k} f(x_l^{(b)}, c_k) \tag{9}$$

holds for any feasible region H. Where $X^b = \{x_l^b \mid 1 \leq l \leq n\}$ be a binary dataset derived from the set of r basic partitioning H as follow:
$x_l^{(b)} =< x_{l,1}^{(b)}, \ldots, x_{l,i}^{(b)}, \ldots, x_{l,r}^{(b)} >,$ with $x_{l,i}^{(b)} =< x_{l,i1}^{(b)}, \ldots, x_{l,ij}^{(b)}, \ldots, x_{l,iK}^{(b)} >$ and
$x_{l,ij} = \begin{cases} 1 & L_{h_i}(x_l) = j \\ 0 & otherwise \end{cases}$ and c_k is the centroid of the cluster.

Table 1 Notations

Notations	
X	The global data such that $x_i \in \mathbb{R}^d$
μ	The distribution $\frac{1}{n}\sum_{i=1}^{n}\delta_{x_i}$
X^v	The view v such that $x_i \in \mathbb{R}^{d_v}$ with $d_v < d$
μ^v	The distribution of the view v $\mu^v = \frac{1}{n}\sum_{i=1}^{n}\delta_{x_i^v}$
d_v	The dimension of the view v
c_j^v	The centroid j in the view v between the data and the centroids c_j^v
C^v	The centroid matrix of the vth view
ν^v	The distributions of the centroids $\frac{1}{k_v}\sum_{j=1}^{k_v}\delta_{c_j^v}$
$L^v = \left\{l_{ij}^v\right\}$	The optimal transport matrix of the view v
c_k	The centroids of the consensus cluster
$L = \{l_{ik}\}$	The optimal transport matrix between the centroids of each view and the consensus centroids
ν	The distribution of the consensus centroids $\frac{1}{K}\sum_{k=1}^{K}\delta_{c_k}$
Π	The optimal transport matrix between x_i and c_k

Based on the idea that consensus clustering can be seen as the K-means problem, we transform it into an optimal transport problem based on the $Wasserstein$-distance. Specifically, the views are seen as distrusted sets, which we combine to form the best consensus represented as a unified distribution.

Besides its link with K-means clustering, with the intuitive optimal transport formulation., it captures the underlying geometry of each view by relying on the cost function which encodes the local structure, which makes it a reasonable choice for the local step. Moreover, it is a powerful tool to make meaningful comparisons between distributions even when their supports do not overlap, which makes it a good way to choose the best views for collaboration, to name a few.

In the next section, we detail the proposed approach, and how we improve the consensus clustering based on the optimal transport theory. Also, we propose two main consensus clustering methods which are validated by several experiments on different datasets.

4.2 Collaborative Multi-view Clustering

Let $X = \{x_1, x_2, \ldots, x_n\}, x_i \in \Omega \subset \mathbb{R}^d, 1 \leq i \leq n$ be the dataset made of d numerical attributes. Let $X^v = \left\{x_1^v, x_2^v, \ldots, x_n^v\right\}, x_i^v \in \mathbb{R}^{d_v}, d_v < d$, the subset of attributes processed by the view v.

4.2.1 Local Step

We consider the empirical measure of the data in the vth view: $\mu^v = \frac{1}{n}\sum_{i=1}^{n}\delta_{x_i^v}$. We seek to find a discrete probability measure ν^v including k finite number of support points which minimizes the following equation :

$$\min_{\nu^v \in \mathcal{P}_k(\mathcal{X})} W_{2,\lambda}^2(\mu^v, \nu^v) \tag{10}$$

This problem is considered as a special case where $N = 1$ of Wasserstein barycenter detailed in Sect. 3.5. Furthermore, we should notice that in [15], when $d = 1$ and $p = 2$ and without constraints on the weight over Σ_{k_v}, this problem is equivalent to Lloyd's algorithm. Thus to solve the optimization problem (10), we consider In what follows $p = 2$ and we proceed similarly to k-means clustering. The local step for the view v iteratively alternates between the assignment of each data to the nearest centroid and the centroids optimization $C^v = \{c_1^v, c_2^v, \ldots, c_{k_v}^v\}$.

Algorithm 2: Local view algorithm

Input : for the data of the vth view v, $X^v = \{x_i^v\}_{i=1}^n \in \mathbb{R}^{d_v}$ such that $d_v < d$ and k_v the number of clusters
the entropic constant λ

Output: The OT matrix $L^{v*} = \{l_{ij}^v\}$ and the centroids c_j^v

1 Initialize k_v, random centroids $c^v(0)$ with the distribution $\nu^v = \frac{1}{k_v}\sum_{j=1}^{k_v}\delta_{c_j}$;
2 **repeat**
3 Compute the OT matrix $L^v = \{l_{ij}^v\}$ $1 \le i \le n, 1 \le j \le k_v$;
4
$$L^{v*} = \underset{L^v \in \Pi(\mu^v, \nu^v)}{\operatorname{argmin}} W_{2,\lambda}^2(\mu^v, \nu^v);$$
5 Update the distribution centroids c_j^v:
6
$$c_j^v = \sum_i l_{ij} x_i^v \quad 1 \le j \le k_v;$$
7 **until** *until convergence*;
8 **return** $\{L^{v*}\}$ *and* $\{c_j^v\}_{j=1}^{k_v}$

The algorithm 2 describes how we cluster the data locally. It is an alternation between the Sinkhorn algorithm 1 to assign each data point to its nearest centroid, and the update of the centroids distribution to be the average weighted of the data point assigned to it. It should be noted that algorithm 2 is equivalent to K-means, but it allows a soft assignment instead of the hard one which means that $l_{ij}^v \in [0, \frac{1}{n}]$, also the term of regularization $-\frac{1}{\lambda}E(\gamma)$ will guarantee a solution with higher entropy, which means that the point will be more uniformly assigned to the clusters.

4.2.2 Collaborative Learning

The collaborative learning step aims to transfer the information between the models computed in each view. The main idea is to increase the local quality based on the received knowledge from distant models. This process is computed until the stabilization of clusters with higher quality. It's must be mentioned that collaborative learning is done in two simultaneous phases.

The first phase is built on the optimal transport matrix, which can be seen in this context as a similarity matrix between the local data distribution of each view. This phase aims to select the best distant collaborator to exchange information with. Thus, an optimized order of the collaboration will be learned intuitively. This phase also creates privileged knowledge that will be used at the consensus step to make the fusion more smooth and optimized. It must be mentioned that the idea of choosing the best collaborator at each iteration is based on comparing the diversity between local data structures and then choosing the median one. This approach was concluded from a heuristic work in [31], proving that the best collaborator to choose is the model with median diversity.

On the other hand, the second phase consists to transfer the knowledge between the models to increase the local quality of each collaborator. Mainly, this process is done by transporting the prototypes of the chosen model to influence the location of the local centroids, to get the better model clustering. Bearing in mind the same notation above (Table 1), we attempt to minimize the following objective function:

$$\underset{L^v, C^v}{\operatorname{argmin}}\{< L^v, \mathcal{C}(X^v, C^v) >_F - \frac{1}{\lambda}E(L^v) + \sum_{v'=1.v'\neq v}^{r} \alpha_{v',v}(< L^{v,v'}, \mathcal{C}(C^v, C^{v'}) >_F - \frac{1}{\lambda}E(L^{v,v'}))\} \tag{11}$$

The first term of the objective function details the local clustering, while the second term presents how the interaction between the collaborators is done through the optimal transport matrix. $\alpha_{v',v}$ are non-negative coefficients are based on a proportional sum of the diversity between the distribution of collaborator's data sets and gain between local quality. While $L^{v,v'}$ is the similarity matrix plan between the centroids of the vth and v'th collaborators and $\mathcal{C}$ is the transport cost.

It must be mentioned that in this work we proceed with a horizontal collaboration where the views share the same instances, with different features in each view. For more details about the collaborative learning algorithm, we suggest our recent work [3].

4.2.3 Collaborative Consensus Projection Approach (CoCP)

The collaborative Consensus Projection approach (CoCP) is based on the structure projections on the global space of the clusters obtained after the collaboration step in each view. The idea behind this projection is to transfer the structures to the real space of the data. Hence, the consensus model is computed using the real partition of the data.

Moreover, the aggregation of the prototypes to the global space makes the consensus model more similar to the real partition of the data, but with better quality compared to the classical clustering approaches.

Algorithm 3: Collaborative Consensus with projection (CoCP)

Input : $\{X^v\}_{v=1}^r$: the r views' data with distributions $\{\mu^v\}_{v=1}^r$
$\{k^v\}_{v=1}^r$: the numbers of clusters
λ : the entropic constant
$\{\alpha_{v,v'}\}_{v,v'=1}^{v,v'=r}$: the confidence coefficient matrix

Output: The partition matrix Π^* and the centroids $C = \{c_k\}_{k=1}^K$

1 Initialize the centroids $C^v = \left\{c_j^v\right\}_{j=1}^{k^v}$ randomly
2 Compute the associated distribution $\nu^v = \frac{1}{k^v}\sum_{j=1}^{k^v}\delta_{c_j^v}, \forall v \in \{1, \ldots, r\}$ using algorithm 2
3 Update the centroids of each view using the collaboration Eq. 11 [3]
4 Project the centroids on the global space based on the privileged optimal transport map between the views.
5 **repeat**
6 Compute the matrix centroids $C = \{c_k\}_{k=1}^{k=K}$ in the global space with K centroids :
7 Transfer the projected centroids to the consensus centroids $L = \left\{l_{jk}\right\} 1 \leq j \leq k_v, 1 \leq k \leq K, 1 \leq v \leq r;$
8
$$L^* = \underset{L \in \Pi(\nu^v, \nu)}{\text{argmin}} \; W_{2,\lambda}^2(\nu^v, \nu);$$
Update the consensus centroids c_k;
9
$$c_k = \sum_{i=1}^{n} l_{ik}.x_i \quad 1 \leq k \leq K;$$
10 **until** *Convergence*;
11 Compute the optimal transport plan between the data and the consensus centroids c_k;
12
$$\Pi^* = \underset{\pi_{ik} \in \Pi(\mu, \nu)}{\text{argmin}} \; W_{2,\lambda}^2(\mu, \nu)$$
return c_k *and* $\Pi^* = \{\pi_{ik}\}$

In algorithm 3, we detail the mechanism of the CoCP approach. After the collaboration phase, the first step consists to project the centroids of each view in the global space of the data. Then, re-cluster the projected centroids in the global space using the *Sinkhorn* algorithm to obtain a new prototype c_k. The last step aims to transport the new prototypes to the instances based on the Wasserstein distance.

4.2.4 Collaborative Consensus with New Representation (CoCNR)

The collaborative consensus with new representation (CoCNR) aims to ensemble the structure obtained in each view in order to rebuild a new representation of the data based on the partition matrices. This representation gives the posterior probability of each point belonging to each cluster in each view. In other words, it is a superposition of the multiple structures learned in each view that allows forming a consensus clustering that contains all the information already gotten from each view.

Algorithm 4: Collaborative Consensus with new representation (CoCNR)

Input : $\{X^v\}_{v=1}^r$: the r views' data with distributions $\{\mu^v\}_{v=1}^r$
$\{k^v\}_{v=1}^r$: the numbers of clusters
λ : the entropic constant
$\{\alpha_{v,v'}\}_{v,v'=1}^{v,v'=r}$: the confidence coefficient matrix
Output: The partition matrix Π^* and the centroids $C = \{c_k\}_{k=1}^K$

1 Initialize the centroids $C^v = \left\{c_j^v\right\}_{j=1}^{k^v}$ randomly
2 Compute the associated distribution $\nu^v = \frac{1}{k^v}\sum_{j=1}^{k^v} \delta_{c_j^v}, \forall v \in \{1, \ldots, r\}$ using algorithm 2
3 Update the centroids of each view using the collaboration Eq. 11 [3]
4 Initialize K centroids $c_k(0)$ with the distribution $\nu = \frac{1}{K}\sum_{k=1}^{K} \delta_{c_k}$;
5 Compute the new representation matix of the data;
6

$$X = concat(L_v) \quad for \quad 1 \leq v \leq r$$

repeat
7 | Compute the optimal transport plan $\Pi = \{\pi_{ik}\}$;
8 |

$$\Pi = \text{argmin } W_{2,\lambda}^2(\mu, \nu);$$

| Update the consensus centroids c_k;
9 |

$$c_k = \sum_{i=1}^{n} \pi_{ik}.x_i \quad 1 \leq k \leq K;$$

10 **until** *Convergence*;
11 **return** c_k *and* $\Pi = \{\pi_{ik}\}$

Algorithm 4, details the process of the CoCNP method. After the collaboration phase, the first step is to concatenate all the partition matrices and re-cluster them using *Sinkhorn* algorithm to obtain a better partition of the data and centroids containing information that emerged from each view.

5 Experimental Validation

In this section, we evaluate the proposed approaches on several datasets, we compare the proposed methods to classical consensus K-means clustering algorithm and the clustering of a single view.

5.1 Setting

We mentioned that the used datasets (Table 2) are available on UCI Machine Learning Repository [17].

5.1.1 Datasets

See Table 2

- Breast data: is Breast Cancer data set where the features are provided from a digitized image of a Fine Needle Aspirate (FNA) of a breast mass. The instances are described by 9 attributes based mainly on the characteristics of the cell nuclei, While the class attribute contains the prognosis (benign or malignant).
- Dermatology dataset: the differential diagnosis of erythemato-squamous diseases is a real problem in dermatology. They all share the clinical features of erythema and scaling, with very few differences. The diseases in this group are psoriasis, seborrhoeic dermatitis, lichen planus, pityriasis rosea, chronic dermatitis, and pityriasis rubra pilaris which gives 6 classes. Usually, a biopsy is necessary for the diagnosis but unfortunately, these diseases share many histopathological features as well. Another difficulty for the differential diagnosis is that a disease may show the features of another disease at the beginning stage and may have the characteristic features at the following stages.

Table 2 Some characteristics of the experimental real-world datasets

Datasets	#Instances	#Attributes	#Classes
Breast	699	9	2
Dermatology	358	33	6
Ecoli	332	7	6
Iris	150	4	3
PenDigits	10992	16	10
Satimage	4435	36	6
Wine	178	13	3

- Escherichia coli dataset (EColi): this dataset contains 336 instances describing cells measures. The original dataset contains 7 numerical attributes (we removed the first attribute containing the sequence name). The goal of this dataset is to predict the localization site of proteins by employing some measures about the cells. 4 main site locations can be divided into 8 hierarchical classes.
- Iris dataset (Iris): this dataset has 150 instances of iris flowers described by 4 integer attributes. The flowers can be classified into 3 categories: Iris Setosa, Iris Versicolour, and Iris Virginica. Class structures are "well behaved" and the class instances are balanced (50/50/50).
- Pen digits dataset: collected from 250 samples of 44 writers. The samples written by 30 writers are used for training, cross-validation, and writer-dependent testing, and the digits written by the other 14 are used for writer independent testing. This database is also available in the UNIPEN format.
- Sateimage dataset: consists of the multi-spectral values of pixels in 3x3 neighborhoods in a satellite image and the classification associated with the central pixel in each neighborhood. The aim is to predict this classification, given the multi-spectral values. In the sample database, the class of a pixel is coded as a number, and there are 6 classes. There are 36 predictive attributes (=4 spectral bands x 9 pixels in the neighborhood). In each line of data, the four spectral values for the top-left pixel are given first followed by the four spectral values for the top-middle pixel and then those for the top-right pixel, and so on with the pixels read out in sequence left-to-right and top-to-bottom. Thus, the four spectral values for the central pixel are given by attributes 17,18,19, and 20. If you like you can use only these four attributes while ignoring the others. This avoids the problem which arises when a 3x3 neighborhood straddles a boundary.
- Wine dataset: provided from the analysis of wines grown in the same area in Italy. The analysis was done on three cultivars based on a 13 chemical analysis to determine the best wine. The data contain no missing values and contains only numeric data, with a three-class target variable (Type) for classification.

5.1.2 Data Sets Pre-processing

To validate the proposed algorithms experimentally, we first start with a data pre-processing to create the local subset defined as the collaborator's sets(views). Mainly, we split each dataset into five views where each view captures the instances of the data in different feature spaces as explained in the Fig. 1. It must be mentioned that the construction of the views is done randomly while taking into consideration a threshold of similarity between the views.

5.1.3 Quality Measures

The proposed approaches were evaluated by internal and external indexes quality. We chose the Davies-Bouldin (DB) index as an internal index, we also added the

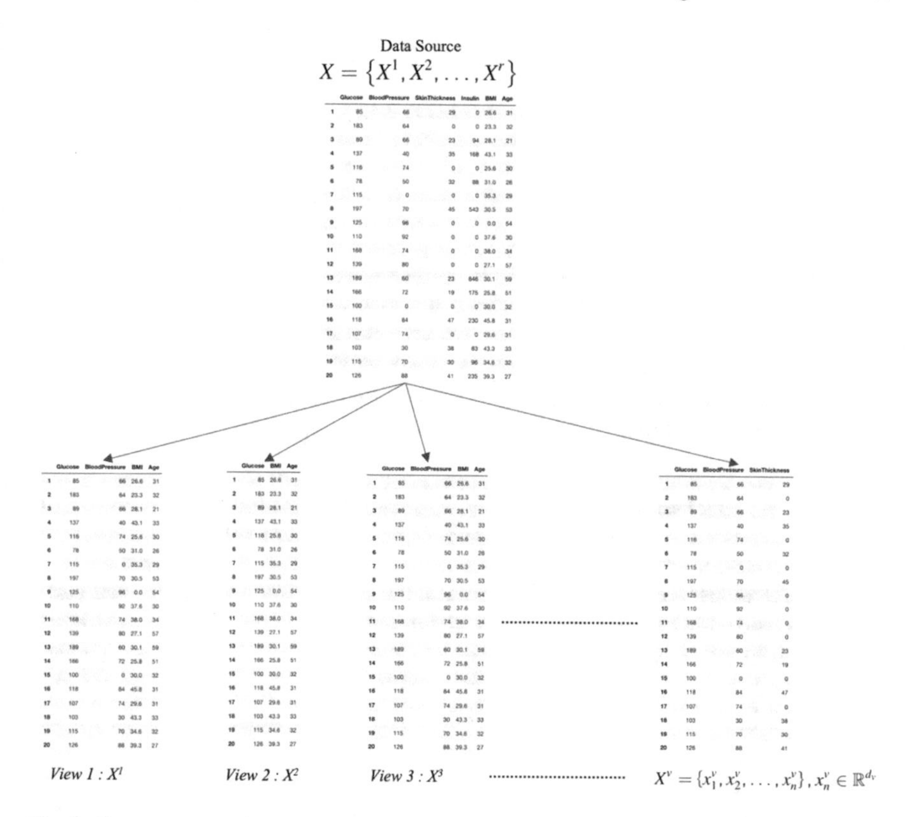

Fig. 1 Data pre-processing

Silhouette index since the DB metric is deployed as stopping criteria during the collaboration step [3]. Besides, since the datasets used in the experiments dispose of real labels, we add an external criterion: Adjusted Rand Index (ARI). The proposed indexes are defined in what follows.

Davies-Bouldin index (DB) [16] is defined as follow:

$$DB = \frac{1}{K} \sum_{k=1}^{K} \max_{k \neq k'} \frac{\Delta_n(c_k) + \Delta_n(c_{k'})}{\Delta(c_k, c_{k'})} \tag{12}$$

where K is the number of clusters, $\Delta(c_k, c_{k'})$ is the similarity between centroid's cluster c_k and $c_{k'}$, and Δ_n is the average similarity of all elements belonging to the cluster C_k and its centroid c_k.

The BD index aims to evaluate the clustering quality by measuring the compactness of the clusters, it is based on the ratio of the sum of within-clusters scatter to between-clusters separation. The lower the value of the DB index, the better the quality of the cluster.

The second index is the silhouette index S [32] defined as follow:

$$S = \frac{1}{K} \sum \frac{b(i) - a(i)}{\max(a(i), b(i))} \tag{13}$$

The *Silhouette* index attempts to measure the similarity of an object to its own cluster (cohesion) compared to other clusters (separation). It is based on the mean intra-cluster distance and the mean nearest-cluster distance for each sample. Where a_i is the average distance between the ith instance and instances belonging to the same cluster, and b_i is the average distance between the ith instance and instances belonging to other clusters. The silhouette index value is better when it is closer to one, which means that the instances are assigned to the right cluster.

The adjusted Rand index (ARI) [35, 40] defined as follows:

$$ARI = \frac{\sum_{ij} \binom{n_{ij}}{2} - \sum_i \binom{a_i}{2} \sum_j \binom{b_j}{2} / \binom{n}{2}}{\frac{1}{2}(\sum_i \binom{a_i}{2} + \sum_j \binom{b_j}{2}) - \sum_i \binom{a_i}{2} \sum_j \binom{b_j}{2} / \binom{n}{2}} \tag{14}$$

The ARI measures the similarity between the real partition and predicted partition, by considering all pairs of samples and counting pairs that are assigned in the same or different clusters. Where $n_{ij} = \mid -C_i \cap Y_j \mid$ and C_i is the ith cluster, and Y_j is jth real class provided from the real partition of the dataset. a_i is the number of instances included in the same cluster described by the same label, and b_j is the number of instances belonging to a different cluster with a different label. Thus, the values of the ARI index are close to 0 for random labeling independently of the number of clusters and samples, and exactly 1 when the partitions are identical (up to a permutation).

In Sect. 5.2, we detailed the comparison of the proposed algorithms based on the quality indexes above.

5.1.4 Tools and Instruments Used for the Experiments

One of the best proprieties of the Wasserstein distance is the vectored computation. More precisely, the computation of a n distances between histograms either from one to many, or form many to many can be achieved simultaneously using elementary linear algebra operations. On that account, we use the PyTorch version computation of Sinkhorn similarity, on GPGPU's. Moreover, We have parallelized the data collaborators to compute local algorithms at the same time. For the experiment results, we used Alienware area-51 m with GeForce RTX 2080/PCIe/SSE2 /NVIDIA Corporation graphic card.

5.2 Results and Validation

In this section, we evaluate the proposed approaches CoCNR and CoCP compared to the classical consensus clustering based on optimal transport approaches CNR

Table 3 Clustering performance on seven real-world datasets by $DavisBouldin$ index DB

Datasets	CoCNR	CoCP	CNR	CP	SVA
Breast	1.183	0.625	1.735	1.727	1.742
Dermatology	0.942	1.085	1.926	1.194	1.310
Ecoli	0.755	1.150	1.145	1.405	1.236
Iris	0.504	0.646	0.893	0.908	0.915
PenDigits	0.615	1.030	1.136	1.334	1.257
Satimage	1.005	1.078	1.011	1.221	1.274
Wine	0.755	0.551	1.308	0.556	0.556

and CP, we compared the proposed approach to a single view clustering based on $Sinkhorn$-means algorithm. Furthermore, we compared with a classical approach of the consensus clustering based on k-means. The comparison is done based on different quality indexes as mentioned in the previous section.

For this purpose, we compute the algorithm 2 on different views of each dataset to produce local models. Then, the local models collaborate to improve their local quality and to create the global map that will be used in the consensus step as privilege information.

5.2.1 Results over Different Quality Measures

The first experiment set was restricted to evaluate the compactness of the produced consensus models. Table 3 shows the results of different consensus approaches. As one can see, the collaboration step does improve the quality of the clusters for different datasets. Moreover, the results based on the $Davis - Bouldin$ index confirm that the best consensus approach is the CoCNR approach, where we compute a completely new representation based on the distribution of the local clusters in each view. Furthermore, the collaboration phase has improved the quality of the learned representation, thanks to the map resulting from the collaboration process that includes the similarity information between the local model's structure.

Table 4 detailed the results of the $Silhouette$ index, which compares the proposed consensus approaches to a single view approach, where each dataset is described with all the features. Similar to Davis-Bouldin's results, the silhouette values were improved by adding the collaboration step to the classical consensus either for the CP or CNR approach, which means that the collaboration has increased the quality of consensus models. Moreover, even with a collaboration step, the best consensus model is the CoCNR compared to the CoCP.

Furthermore, the Fig. 2 compares the $silhouette$ index results, as one can see, the radar CoCNR is over the radars of other approaches, which means that CoCNR is the best method following the $Silhouette$ index.

To summarize, according to $Davis - Bouldin$ and $Silhouette$ values, the map between the different views does improve the consensus model to improve the qual-

Table 4 Clustering performance on seven real-world datasets by *Silhouette* index

Datasets	CoCNR	CoCP	CNR	CP	SVA
Breast	0.695	0.543	0.508	0.533	0.531
Dermatology	0.612	0.431	0.537	0.331	0.371
Ecoli	0.736	0.586	0.669	0.381	0.319
Iris	0.979	0.684	0.697	0.472	0.549
PenDigits	0.526	0.313	408	0.331	0.314
Satimage	0.531	0.455	0.500	0.436	0.372
Wine	0.604	0.582	0.552	0.563	0.501

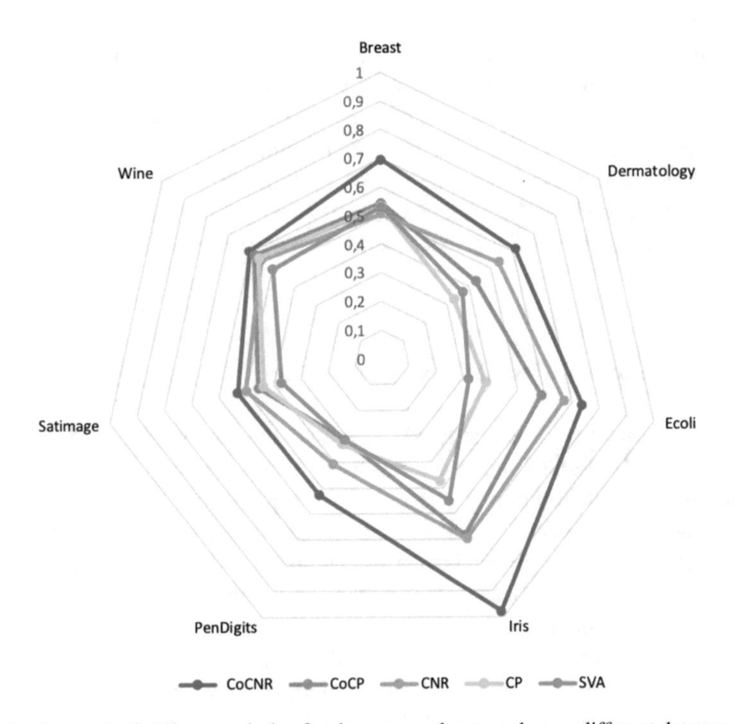

Fig. 2 Radar graph of *silhouette* index for the proposed approach over different datasets

ity of the clusters. However, even with collaborative learning, the best consensus approach in terms of compactness of the CoCNR.

On the other hand, we compared the proposed approach to a classical consensus approach detailed in [39], using the R_n index. In the paper [39] the authors compared the proposed approach based on multiple distances, we chose the most appropriate distance to compare with the proposed approaches. It must be mentioned that in the comparison we followed the data pre-processing as mentioned by the authors in [39].

Table 5 presents the R_n values over multiple datasets. To further evaluate the performance, we compute a measurement score as defined in [39]:

Table 5 Clustering performance on seven real-world datasets by normalized Rand index R_n

datasets	CoCNR	CoCP	CNR	CP	SVA	KCC
Breast	0.574	0.835	0.331	0.737	0.689	0.055
Dermatology	0.461	0.131	0.460	0.120	0.147	0.035
Ecoli	0.533	0.524	0.282	0.344	0.338	0.506
Iris	0.873	0.881	0.442	0.460	0.449	0.735
PenDigits	0.551	0.663	0.506	0.603	0.535	0.534
Satimage	0.369	0.538	0.357	0.470	0.467	0.450
Wine	0.251	0.351	0.226	0.214	0.2149	0.1448
Score	5.910	6,267	4.500	4.708	4.577	3.982

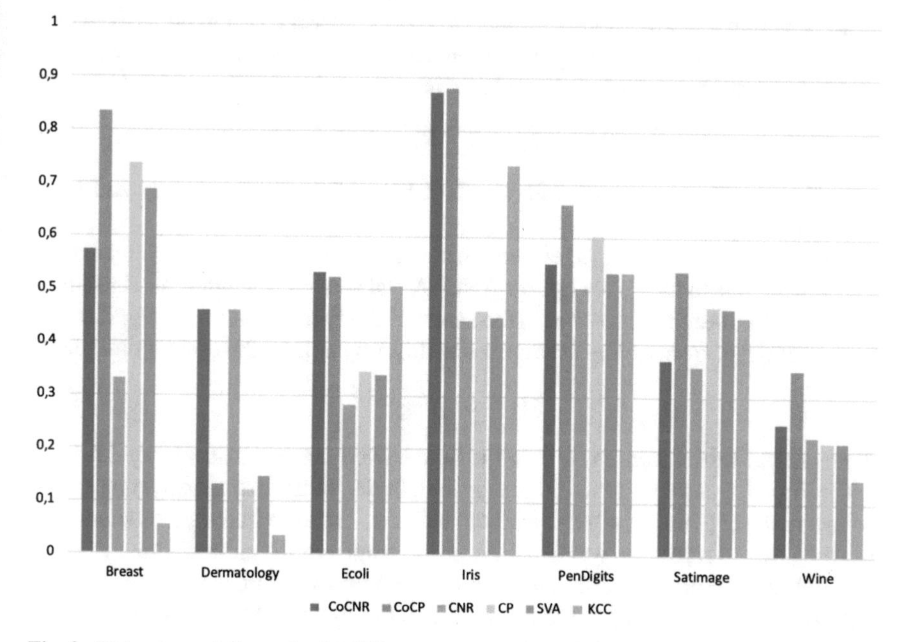

Fig. 3 Histograms of R_n results for different approaches over different datasets

$$Score(M_i) = \sum_j \frac{R_n(M_i, D_j)}{\max_i R_n(M_i, D_j)} \tag{15}$$

where $R_n(M_i, D_j)$ indicates the R_n value of M_i method on the D_j datasets. This score presents an overall evaluation of all the datasets. The Table 5 shows that the approaches based on optimal transport outperform the classical consensus based on k-means. Moreover, following the results of this measurement, and the Fig. 3, the collaboration phase has improved the consensus model for both approaches.

5.2.2 Discussion

To analyze the results presented in the previous section we used the Friedman test [20] which is a non-parametric alternative to the one-way ANOVA with repeated measures. It is designed to test for differences between groups when the necessary assumptions to conduct the one-way ANOVA with repeated measures have been violated (e.g., normality and sphericity). Which makes it an appropriate test for machine learning studies where such assumptions are most likely violated due to the nature of learning algorithms and datasets.

The Friedman test ranks the algorithms for each dataset independently, the best performing algorithm receiving the rank of 1, the second-best rank 2, etc. Let r_i^j be the rank of the jth of k algorithms on the ith of N datasets, the Friedman test compares the average ranks of algorithms, $R_j = \frac{1}{N}\sum_i^N r_i^j$. Under the null hypothesis, which assumes that all the compared algorithms are equivalent and consequently their ranks R_j should be the same, the Friedman statistic

$$\chi_F^2 = \frac{12N}{k(k+1)}\left[\sum_j^k R_j^2 - \frac{k(k+1)^2}{4}\right]$$

is distributed according to χ_F^2 with $k-1$ degrees of freedom, when N and k are sufficiently large.

If the null hypothesis is rejected, we can conduct a post-hoc test. The Nemenyi test [28] is analogous to the Tukey test for ANOVA and is employed when all algorithms are compared to each other. The performance of the two algorithms is significantly different if the corresponding average ranks differ by at least the critical difference

$$CD = q_\alpha\sqrt{\frac{k(k+1)}{6N}}$$

where critical values q_α are based on the Studentized range statistic divided by $\sqrt{2}$.

The Figs. 4, 5 and 6 present a critical difference diagram illustrating a projection of average ranks of algorithms on an enumerated axis. The algorithms are ordered from left (the best) to right (the worst). A thick line connects algorithms whose average ranks are not significantly different from the level of 5% significance.

As shown in Figs. 4 and 5, the collaboration has improved the consensus model, the CoCNR achieves significant improvement over the other proposed techniques (CoCP, CNR, CP, SVA). Furthermore, the test presented in the Fig. 6, compares the results of the R_n index to the KCC approach. As we can see, the best approach is CoCP followed by CoCNR, this can be explained by the fact that CoCP learns the consensus model on the real space of the data, consequently, the structure of the consensus model will be more similar to the real partition of data. While CoCNR encodes a new data structure that can be more qualified in terms of cluster compactness but not similar to real data partition as much as CoCP which explains to values of Table 5.

Compared to the nearest approach of the state of the art, the positive impact introduced by the optimal transport and the collaborative learning to multi-view consensus framework is:

- The proposed approaches are based on a strong and well-defended theory that becomes increasingly popular in the field of machine learning.

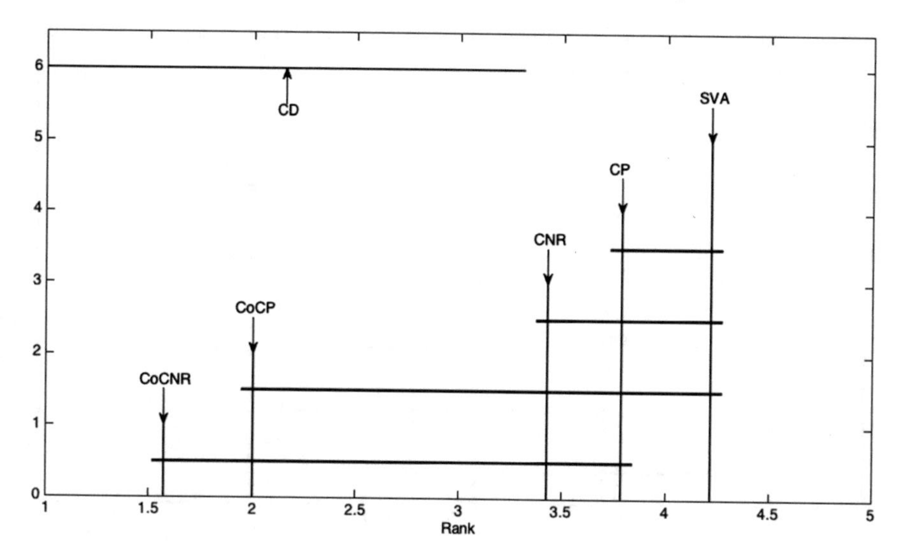

Fig. 4 Critical difference (CD) diagram using Nemenyi post hoc test performed after a Friedman test, comparing benchmark algorithms according to DB index. Approaches are ordered from left (the best) to right (the worst)

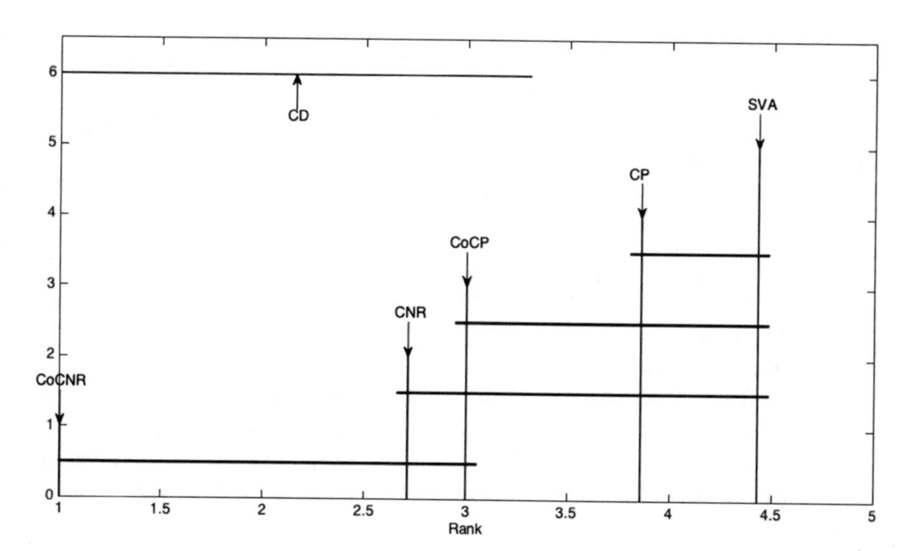

Fig. 5 Critical difference (CD) diagram using Nemenyi post hoc test performed after a Friedman test, comparing benchmark algorithms according to $Silhouette$ index. Approaches are ordered from left (the best) to right (the worst)

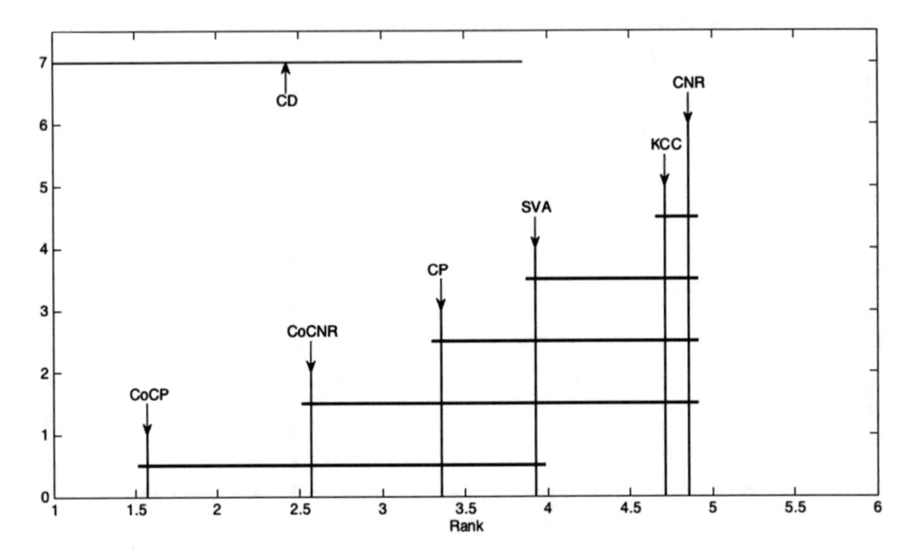

Fig. 6 Critical difference (CD) diagram using Nemenyi post hoc test performed after a Friedman test, comparing benchmark algorithms according to R_n index. Approaches are ordered from left (the best) to right (the worst)

- The experimental results highlighted the efficiency on both artificial and real datasets.
- The exchange of the information during the collaboration step improved in the local model of each view.
- The privileged information given by the collaborative step had increased the quality of the consensus model for both approaches.

Finally, the proposed approaches ensure a trade-off between the quality of the consensus model and the structure of the real data, which gives the ability to choose a qualified consensus model with a compact cluster or a model that takes into consideration the real structure of the data during the consensus process.

Although the presented approaches had given a significant result in terms of quality, they still suffer from some limitations, notably the curse of high dimensionality. This limitation is essentially linked to the optimal transport similarity computation, and to overcome it, we had increased the regularization penalty coefficient to avoid any possible over-fitting.

6 Conclusion

We have reviewed the multi-view clustering by exploring the optimal transport framework. The idea was to take advantage of this theory to build an optimal consensus model that unified the information learned in each view.

Furthermore, we optimized the consensus model computation by including a collaborative step that builds not only a qualified local model but also a global map containing prior information about the local structure of the clusters and measures the similarity between the views. This map guarantees a smooth computation and an optimized consensus model, stable and optimal. This was validated by the experimental results.

In this chapter, we proposed two main algorithms, the CoCNR which is based on encoding a global qualified space of the data where the consensus will be learned. The second approach CoCP is based on projecting the partitions into the global space to build a unified consensus model.

The results obtained show that optimal transport can provide for this machine learning task, a formal framework and algorithmic flexibility that marks an improvement in performance over the existing system. On the other hand, optimal transport is becoming increasingly popular in the field of machine learning, with several applications in data science under different learning paradigms. However, in a large dimension, we are often confronted with the intrinsic instability of optimal transport concerning input dimensions.

Indeed, the complexity of approximating Wasserstein's distance can grow exponentially in size, making it difficult to estimate this distance accurately. A multi-view approach can help to mitigate this phenomenon by breaking down the overall problem into sub-problems, each representing a view. Future works include the investigation of the learning capabilities of new data representations through our proposed approaches. Furthermore, we aim to include the feature selection tools to create an accurate sub-space that captured qualified information. Thus the consensus will get à higher quality.

References

1. Agueh, M., Carlier, G.: Barycenters in the Wasserstein space. SIAM J. Math. Anal. **43**(2), 904–924 (2011)
2. Ayad, H.G., Kamel, M.S.: On voting-based consensus of cluster ensembles. Pattern Recognit. **43**(5), 1943–1953 (2010)
3. Ben-Bouazza, F.-E., Bennani, Y., Cabanes, G., Touzani, A.: Unsupervised collaborative learning based on optimal transport theory. J. Intell. Syst. **30**(1), 698–719 (2021)
4. Ben-Bouazza, F.-E., Bennani, Y., Touzani, A., Cabanes, G.: Subspace guided collaborative clustering based on optimal transport. In: International Conference on Soft Computing and Pattern Recognition, pp. 113–124. Springer (2020)
5. Ben Bouazza, F.-E., Cabanes, G., Bennani, Y., Touzani, A.: Collaborative clustering through optimal transport. In: International Conference on Artificial Neural Networks. Springer (in press)
6. Bickel, S., Scheffer, T.: Multi-view clustering. In: ICDM, vol. 4, pp. 19–26 (2004)
7. Blum, A., Mitchell, T.: Combining labeled and unlabeled data with co-training. In: COLT, pp. 92–100. ACM (1998)
8. Ben Bouazza, F.E., Bennani, Y., El Hamri, M., Cabanes, G., Matei, B., Touzani, A.: Multi-view clustering through optimal transport. Aust. J. Intell. Inf. Process. Syst. **15**(3), 1–9 (2019)
9. Brefeld, U., Scheffer, T.: Co-em support vector learning. In: ICML, pp. 16–24. ACM (2004)

10. Brualdi, R.A.: Combinatorial Matrix Classes, vol. 13. Cambridge University Press (2006)
11. Cornuejols, A., Wemmert, C., Gancarski, P., Bennani, Y.: Collaborative clustering: why, when, what and how. Inf. Fusion **39**, 81–95 (2018)
12. Courty, N., Flamary, R., Habrard, A., Rakotomamonjy, A.: Joint distribution optimal transportation for domain adaptation. In: Guyon, I., Luxburg, U.V., Bengio, S., Wallach, H., Fergus, R., Vishwanathan, S., Garnett, R. (eds.) Advances in Neural Information Processing Systems, vol. 30, pp. 3730–3739. Curran Associates, Inc. (2017)
13. Courty, N., Flamary, R., Tuia, D., Rakotomamonjy, A.: Optimal transport for domain adaptation. IEEE Trans. Pattern Anal. Mach. Intell. **39**(9), 1853–1865 (2016)
14. Cuturi, M.: Sinkhorn distances: lightspeed computation of optimal transport. In: Advances in Neural Information Processing Systems, pp. 2292–2300 (2013)
15. Cuturi, M., Doucet, A.: Fast computation of Wasserstein barycenters. In: International Conference on Machine Learning, pp. 685–693. PMLR (2014)
16. Davies, D.L., Bouldin, D.W.: A cluster separation measure. IEEE Trans. Pattern Anal. Mach. Intell. **1**(2), 224–227 (1979)
17. Dua, D., Graff, C.: UCI machine learning repository (2017)
18. El Hamri, M., Bennani, Y., Falih, I.: Inductive semi-supervised learning through optimal transport (in press). In: International Conference on Neural Information Processing. Springer (2021)
19. El Hamri, M., Bennani, Y., Falih, I.: Label propagation through optimal transport. In: 2021 International Joint Conference on Neural Networks (IJCNN), pp. 1–8. IEEE (2021)
20. Friedman, M.: The use of ranks to avoid the assumption of normality implicit in the analysis of variance. J. Am. Stat. Assoc. **32**(200), 675–701 (1937)
21. Genevay, A., Chizat, L., Bach, F., Cuturi, M., Peyré, G.: Sample complexity of Sinkhorn divergences. In: The 22nd International Conference on Artificial Intelligence and Statistics, pp. 1574–1583. PMLR (2019)
22. Kantorovich, L.V.: On the translocation of masses. In: Dokl. Akad. Nauk. USSR (NS), vol. 37, pp. 199–201 (1942)
23. Knight, P.A.: The Sinkhorn–Knopp algorithm: convergence and applications. SIAM J. Matrix Anal. Appl. **30**(1), 261–275 (2008)
24. Laclau, C., Redko, I., Matei, B., Bennani, Y., Brault, V.: Co-clustering through optimal transport (2017). arXiv preprint arXiv:1705.06189
25. Martin Arjovsky, S.C., Bottou, L.: Wasserstein generative adversarial networks. In: Proceedings of the 34th International Conference on Machine Learning, Sydney, Australia (2017)
26. Mirkin, B.G.: Additive clustering and qualitative factor analysis methods for similarity matrices. J. Classif. **4**(1), 7–31 (1987)
27. Monge, G.: Mémoire sur la théorie des déblais et des remblais. De l'Imprimerie Royale (1781)
28. Nemenyi, P.B.: Distribution-Free Multiple Comparisons. Princeton University (1963)
29. Pele, O., Werman, M.: Fast and robust earth mover's distances. In: 2009 IEEE 12th International Conference on Computer Vision, pp. 460–467. IEEE (2009)
30. Peyré, G., Cuturi, M., et al.: Computational optimal transport: with applications to data science. Foundations and Trends® in Machine Learning, 11(5-6), pp. 355–607 (2019)
31. Rastin, P., Cabanes, G., Grozavu, N., Bennani, Y.: Collaborative clustering: how to select the optimal collaborators? In: 2015 IEEE Symposium Series on Computational Intelligence, pp. 787–794. IEEE (2015)
32. Rousseeuw, P.J.: Silhouettes: a graphical aid to the interpretation and validation of cluster analysis. J. Comput. Appl. Math. **20**, 53–65 (1987)
33. Santambrogio, F.: Optimal Transport for Applied Mathematicians, vol. 55, (58–63), p. 94. Birkäuser, New York (2015)
34. Solomon, J., Rustamov, R., Guibas, L., Butscher, A.: Wasserstein propagation for semi-supervised learning. In: International Conference on Machine Learning, pp. 306–314 (2014)
35. Steinley, D.: Properties of the Hubert-Arable adjusted rand index. Psychol. Methods **9**(3), 386 (2004)
36. Tao, Z., Liu, H., Li, S., Ding, Z., Fu, Y.: From ensemble clustering to multi-view clustering. In: IJCAI (2017)

37. Vega-Pons, S., Ruiz-Shulcloper, J.: A survey of clustering ensemble algorithms. Int. J. Pattern Recognit. Artif. Intell. **25**(03), 337–372 (2011)
38. Villani, C.: Optimal Transport: Old and New, vol. 338. Springer Science & Business Media (2008)
39. Wu, J., Liu, H., Xiong, H., Cao, J., Chen, J.: K-means-based consensus clustering: a unified view. IEEE Trans. Knowl. Data Eng. **27**(1), 155–169 (2014)
40. Wu, J., Xiong, H., Chen, J.: Adapting the right measures for k-means clustering. In: SIGKDD, pp. 877–886. ACM (2009)
41. Xie, X., Sun, S.: Multi-view clustering ensembles. In: International Conference on Machine Learning and Cybernetics, vol. 1, pp. 51–56 (2013)
42. Yarowsky, D.: Unsupervised word sense disambiguation rivaling supervised methods. In: 33rd Annual Meeting of the Association for Computational Linguistics (1995)

Data Anonymization Through Multi-modular Clustering

Nicoleta Rogovschi, Younès Bennani, and Sarah Zouinina

Abstract A growing interest in data privacy protection is mostly motivated by countries and organizations' desire to demonstrate democracy and practice guidelines by opening their data. Anonymization of data refers to the process of removing sensitive data's identifiers while keeping their structure and also the information type [35]. A fundamental challenge with data anonymization is achieving a balance between the data's worth and the degree of disclosure. The k-anonymity framework is a well-known framework for data anonymization. It states that a dataset is anonymous if and only if it contains at least $k - 1$ similar items. In this chapter, we introduce two ways for attaining k-anonymity: k-CMVM and Constrained-CMVM. All methods rely on collaborative topological unsupervised learning [27, 32] to generate k-anonymous content. The k-CMVM method selects the k layers automatically, but the contrainted version does it through exploration. Additionally, we boosted the performance of these two strategies by including the pLVQ2 weighted vector quantization method. These techniques were evaluated on the basis of two metrics: separability and structural utility. The obtained experimental results indicate very good performance.

Keywords Multi-modular clustering · Collaborative topological clustering · Data privacy protection · Data anonymization · k-anonymization

---

N. Rogovschi (✉)
LIPADE, Université de Paris, Paris, France
e-mail: nicoleta.rogovschi@u-paris.fr

Y. Bennani
LIPN UMR 7030 CNRS, Université Sorbonne Paris Nord, Villetaneuse, France
e-mail: youenes.bennani@sorbonne-paris-nord.fr

LaMSN, La Maison des Sciences Numériques, Saint-Denis, France

S. Zouinina
Inventiv IT, Levallois-Perret, France
e-mail: sarah.zouinina@linventiv-it.fr

W. Pedrycz and S. Chen (eds.), *Recent Advancements in Multi-View Data Analytics*, Studies in Big Data 106, https://doi.org/10.1007/978-3-030-95239-6_6

1 Introduction

Theses days, data is deeply engrained in almost every area of human life.

Sensors, social networks, mobile applications, and the internet of things collect data for the purpose of processing, analyzing, changing, and learning from all of this. To exploit obtained data without affecting its privacy, several criteria must be met, most notably those relating to the safety of the humans contained in the dataset. Data anonymization is a technique for maintaining data privacy that has been used for quite some time for statistical reasons.

Data anonymization represents the practice of removing sensitive information's identity by keeping its structure and without loosing the knowledge.

This is usually done by hiding one or more variables to keep the data's distinctive properties. The motivation of governments and institutions to share their data is driving the rising interest in data anonymization. While open data is an attractive research subject, it is also difficult since the data must be permanently anonymised with a minimal chance of re-identification and of acceptable quality for analytic purposes. Sharing the closest to truth data enables service providers to experiment with novel mining algorithms, extracting insights from large datasets without exposing the identities of persons contained in the dataset.

Aware of the vital significance of establishing a fit between privacy and utility, numerous algorithms to resolving this issue have been presented. The first strategies mainly relied on the randomization technique, which includes introducing noise to data. This approach has been shown to be ineffective due to the feasibility of data reconstruction.

The introduction of k-anonymization technique reduced the danger of data privacy violation employing randomization. Anonymization is accomplished by first removing critical identifiers such as a person's name and address, followed by generalizing and/or suppressing pseudo-identifiers such as a person's date of birth, ZIP code, gender, and age. The k value must be selected in such a way that the database's information is retained. The method has been extensively investigated as a result of its popularity [1, 20, 23]. Due to the fact that k-anonymity is a group-based technique, clustering was identified as a primary objective [2, 4, 14, 15].

Data loss is minimized by microaggregating k items and substituting them with group representative data [6]. However, the clustering algorithms shown here utilize the k-means algorithm, which is prone to local optima and bias. k-anonymity is achieved through microaggregation, and we propose two ways for achieving this goal in this chapter: k-CMVM and Constrained-CMVM. For the purpose of generating k-anonymous data, both systems rely on topological collaborative clustering. The first one calculates the k levels, whereas the second one discovers them by research and experimentation. To accomplish this, we leverage the Self Organizing Maps (SOM) topological structure [17] and their reduced aptitude towards local optimas. We will employ SOM as a clustering method because it has been shown to perform well in actual applications including visualization and pattern recognition. Clustering outcomes are increased by the collaborative learning technique. After the topographical learning process is complete, the "similar" samples will be clustered into clusters of similar patterns. These clusters may be described more succinctly by their barycenters or other statistical moments, because we consider this information is more manipulable than the original.

In the second section of the chapter, we will discuss discriminative information and how it may impact the anonymization process in scenarios when labels are applied. The Learning Vector Quantization Method (LVQ) is used to investigate the supervision [17]. We will choose a variant that assigns weights to each attribute, which results in a greater preservation of the anonymized dataset's utility; this method is called pLVQ2 and was discussed in [34].

The rest of this chapter is divided into the following sections: Sect. 2 describes the theoretical foundations and state of the art, Sect. 3 examines the various techniques presented for anonymization, Sect. 4 illustrates the various obtained results, and the last section discusses the conclusions and perspectives.

2 Background Theory

This part will lay the theoretical background for the approaches that will be discussed in the remainder of the articlechapter. We present the framework of k-anonymity via microaggregation in the Sect. 2.1, an overview of multi-view topographic unsupervised learning in Sect. 2.2, and finally, in the Sect. 2.2, we review the expressions and descriptions used throughout this work.

2.1 *State of the Art: Existed Approaches for Achieving k-Anonymity via Microaggregation*

Venkatasubramanian [33] divides anonymization techniques into three categories:

- To begin, there are statistical approaches that offered privacy metrics in terms of variance; the bigger the variance, the more private the disturbed data.
- Second, there are probabilistic techniques, which try to quantify the concept of background knowledge that an entity may hold. The researchers drew on information theory and Statistical modeling, as well as more explicit concepts of information flow.
- Secure multi-party computations are the third class of procedures; these techniques were inspired by the field of cryptography, and the volume of information released is quantified in terms of the quantity of information available to the attacker. One of the most intriguing applications of these ideas is Yao's The Millionaire Problem, in which two billionaires compete to identify who is wealthier without providing any information about their respective riches.

Microaggregation is a technique for restricting disclosure that tries to protect the confidentiality of data subjects in microdata deployments. It could be used in place of generalization and suppression to generate k-anonymous data sets in which the identity of each individual is hidden within a collection of k things. Despite its generalization, microaggregation alters data in a variety of ways to maximize its value, including increasing the granularity of information, minimizing the impact of anomalies, and eliminating quantitative data discretization.

Microaggregation clusters records into tiny aggregates or groupings of at minimum k size. Instead of sharing the attribute Vi's original value for a particular entry, the average of the group's values is reported. To reduce risk of data loss, it is best if the groups are as homogenous as feasible.

The well researched privacy-preserving technique is k-anonymity. The approach implies that all data relating to an individual is maintained in a table of characteristics and records.

Sweeney described a strategy for anonymizing a database that involves suppressing or generalizing the quasi-identifiers to the point where each record is unrecognizable from at least k−1 data. Quasi-identifiers are characteristics that, when used alone, do not reveal anything about their holders. However, when aggregated, quasi-identifiers may reveal the holder's identification. This strategy emphasized the concept of combining comparable components in order to anonymize them.

The objective of traditional k-anonymity is to minimize data loss, as data can be disguised in a variety of ways depending on the approach used. It is preferable to avoid broad generalizations and to make fewer suppressions. Indeed, strategies for addressing k anonymity are driven by some user preferences or rules [5, 10, 29]. When conducting data analysis, the k anonymous data should have enough information about the respondents to aid in pattern discovery activities.

As with probabilistic methods, partitionning consists of classification in supervised learning and clustering in unsupervised learning. Li et al. presented the first approach that combines data partitionning and anonymization in [21]. The technique generates representations from the dataset by identifying equivalence classes with less than k records. It calculates the distance between the detected equivalence class and the others and merges it with the nearest equivalence class to construct a cluster of at least k items with the least amount of information distortion. While this approach produces accurate computational results, it is quite tedious.

The k-member clustering technique was described in [6] and creates interrelated clusters containing at least k entries. An item is added to a cluster with minimum information loss. This process is repeated until a cluster has at least k members. Greedy algorithm based clustering introduced by Loukides et al. [22], captures the data's utility while maintaining its privacy by giving quality metrics that consider the variable, tuples' variety, and a clustering method. This approach is comparable to prior efficient k-member clustering algorithms [6], but now with the constraint of maximising sensitive data divergences (confidentiality) and minimizing quasi-identifier similarities (usefulness). Further research on anonymization using clustering were introduced in literature [3, 14, 16].

Due to the ineffectiveness of removing important identifiers, microdata were revealed utilizing the microaggregation approach [31].

The next section describes how we anonymize microdata via multi-view topological collaborative microaggregation. To anonymize the data, we first select the number of views to explore, then randomly divide it vertically and generate a SOM for each view to retrieve the related prototypes.

2.2 Multi-view Unsupervised Collaborative Clustering

Data mining's ultimate purpose is to discover and classify patterns in data. Assuming we have a group of datasets represented by many sets of characteristics, acquiring knowledge about these data equates to acquiring knowledge about each family of characteristics separately. This is referred to as multi-view data minning, in which each view of the database enables the extraction of distinct patterns within the examined data.

From the other side, collaborative learning seeks to create statistical techniques for recovering topological invariants from the analyzed dataset [9, 12, 24, 26]. In this work, we are interested in models that simultaneously reduce dimension and accomplish clustering. SOM models are frequently used for visualization and unsupervised topographical grouping because they enable projection in tiny areas that are typically two dimensional.

The collaborative system is employed to increase the clustering results of the SOM, and the outputs of multiple self-organizing maps are compared. The SOM method is used to cluster each dataset.

The basic principle of the collaboration among different SOM maps is that if an object from the iith subset is mapped on the jth cell in the iith SOM map, the same data from the jjth dataset will be reflected on the same j cell in the jjth map or one of its neighboring cells. In these other terms, cells associated with distinct maps must be able to record comparable inputs.

As a result, the classic SOM optimization algorithm was changed by include a collaboration component.

The cost function is made of two components in the following formal expression:

$$R^{[ii]}(\chi, w) = R^{[ii]}_{SOM}(\chi, w) + (\lambda^{[jj]}_{[ii]})^2 R^{[ii]}_{Col}(\chi, w) \tag{1}$$

with

$$R^{[ii]}_{SOM}(\chi, w) = \sum_{i=1}^{N} \sum_{j=1}^{|w|} K^{[ii]}_{\sigma(j,\chi(x_i))} \|x^{[ii]}_i - w^{[ii]}_j\|^2 \tag{2}$$

$$R^{[ii]}_{Col}(\chi, w) = \sum_{jj=1, jj \neq ii}^{P} \sum_{i=1}^{N} \sum_{j=1}^{|w|} \left(K^{[ii]}_{\sigma(j,\chi(x_i))} - K^{[jj]}_{\sigma(j,\chi(x_i))} \right)^2 * D_{ij} \tag{3}$$

Algorithm 1: The Topographic Collaborative Multi-view Method

Input: Dataset $V[ii]$ containing P views
Output: P topographic maps $\{w[ii]\}_{ii=1}^{P}$
Local Phase:
for $ii = 1$ to P **do**
Construct a SOM map for the view $V[ii]$
$w^{[ii]} \leftarrow \text{argmin}_w \left[R_{SOM}^{[ii]}(\chi, w)\right]$
Calculate Davies-Bouldin index for the obtained $SOM[ii]$ using $w^{[ii]}$
$DB_{Beforecollab}^{[ii]} \leftarrow DB^{[ii]}$
end forCollaborative learning phase:
for $ii = 1$ to P **do**
for $jj = 1, jj \neq ii$ to P **do**

$$\lambda_{[ii]}^{[jj]}(t+1) \leftarrow \lambda_{[ii]}^{[jj]}(t) + \frac{\sum_{i=1}^{N} \sum_{j=1}^{|w|} K_{\sigma(j,\chi(x_i))}^{[ii]}}{2\sum_{i=1}^{N} \sum_{j=1}^{|w|} \left(K_{\sigma(j,\chi(x_i))}^{[ii]} - K_{\sigma(j,\chi(x_i))}^{[jj]}\right)^2}$$

$$w_{jk}^{[ii]}(t+1) \leftarrow w_{jk}^{[ii]}(t) + \frac{\sum\limits_{i=1}^{N} K_{\sigma(j,\chi(x_i))}^{[ii]} x_{ik}^{[ii]} + \sum\limits_{jj=1, jj \neq ii}^{P} \sum\limits_{i=1}^{N} \lambda_{[ii]}^{[jj]} L_{ij} x_{ik}^{[ii]}}{\sum\limits_{i=1}^{N} K_{\sigma(j,\chi(x_i))}^{[ii]} + \sum\limits_{jj=1, jj \neq ii}^{P} \sum\limits_{i=1}^{N} \lambda_{[ii]}^{[jj]} L_{ij}}$$

$DB_{AfterCollab}^{[ii]} \leftarrow DB^{[ii]}$
if $DB_{AfterCollab}^{[ii]} \geq DB_{BeforeCollab}^{[ii]}$ **then**
$w_{jk}^{[ii]}(t+1) \leftarrow w_{jk}^{[ii]}(t)$
end if
end for
end for

$$with \ \ D_{ij} = \|x_i^{[ii]} - w_j^{[ii]}\|^2 \tag{4}$$

The total number of views is denoted by P, N states for the total number of data, and $|w|$—for the total number of neurons which constitue the ii SOM map.

$\chi\ (xi)$ is the assignment function that enables for the determination of the Best Matching Unit (BMU); it does this by applying the Euclidean distance to identify the cell with the nearest prototype first from data x_i.

The collaboration connection λ is estimated, which indicates the significance of collaboration between two SOMs, i.e. establishing a link between all views and maps. It has a value in the range [1–10], with 1 being a neutral connection and 10 denoting the maximum collaboration within a map. Its value changes with each collaborative iteration. As seen in Algorithm 1, this value is dependent on the topographical relationship of the two cooperation maps in the case of collaborative learning.

The following is the definition of this cost function.:

$$K_{\sigma(i,j)}^{[cc]} = exp\left(-\frac{\sigma^2\,(i,j)}{T^2}\right) \tag{5}$$

$\sigma(i, j)$ is the distance measured between neurons on the map, expressed as the length of their shortest path.

$K^{[cc]}_{\sigma(i,j)}$ is the $SOM[cc]$ neighborhood function between i and j.

The temperature parameter T determines the size of a cell's neighborhood influence.

T can be reduced between $Tmax$ and $Tmin$.

The value of this function fluctuates between maps because it is based on the closest prototype to the data, which may not be the same for all SOM maps.

In fact, while working with a SOM map, the method compares the map's prototypes and the map mstructure (the neighborhood information).

3 Proposed Methods of Anonymization

The data is arranged in a table of rows/objects and variables, with each row described as a tuple; tuples do not have a unique identifier. Each row has an sorted m-tuple with the elements $< a_1, a_2, .., a_j, .., a_m >$.

Notation Consider the table $T\{A_1, A_2, .., A_m\}$, which has a finite number of tuples equivalent to the variables $T\{A_1, A_2, .., A_m\}$. Assuming the value of $T = \{A_1, A_2, .., A_m\}$, $\{A_l, .., A_k\} \subseteq \{A_1, A_2, .., A_m\}$

For $t \in T$, $t[A_l, .., A_k]$ corresponds to the tuple of components $x_l, .., x_k$ of $A_l, .., A_k$ in T.

Suppose a table T with a size of $n \times m$, where m represents the size of features and n represents the size of items. The table is indicated by the notation $T = \{A_1, A_2, .., A_m\}$. □

Definition 1 k-anonymity

$AT\{A_1, A_2, .., A_m\}$ represents a table, then OT is shown to be k-anonymous if and only if every tuple in AT must have at minimum k occurrences, otherwise OT is seen to be k-anonymous.

Definition 2 Internal validation index: The Davies Bouldin

The DB measure [7] is built on a cluster similarity measure Rij that is a proportion of the dispersion measure si and the cluster disparity dij [19]. Rij must meet the following requirements:

$$R_{ij} \geq 0$$
$$R_{ij} = R_{ji}$$
$$R_{ij} = 0 \quad if \quad s_i = s_j = 0, \quad s_i = \frac{1}{\|c_i\|} \sum_{x \in c_i} d(x, c_i)$$
$$R_{ij} > R_{ik} \quad if s_j > s_k, \quad d_{ij} = d_{ik}$$
$$R_{ij} > R_{ik} \quad if s_j = s_k, \quad d_{ij} < d_{ik}$$
$$R_{ij} = \frac{s_i + s_j}{d_{ij}} \quad where \quad d_{ij} = dist(w_i, w_j)$$
$$DB = \frac{1}{n_c} \sum_{i=1}^{n_c} R_i \quad where \quad R_i = \max_{j=1,..,n_c} R_{ij}, \quad i = 1..n_c$$

Under which wi denotes the cell prototypes, nc denotes the number of neurons, and ci denotes the ith neuron.

Davies-Bouldin is a cluster validity indicator that quantifies the clustering's internal quality.

The use of this validity measure is motivated by our ability to analyze how components of the same cluster are comparable. So, while searching for the optimal number of clusters, the DB index is minimized [30].

3.1 *Pre-Anonymization Step*

3.1.1 k-CMVM

In this study, we propose to incorporate a pre-anonymization phase into the method, allowing for the option of two distinct levels of anonymity. The first employs BMUs (k-CMVM) prototypes, while the second employs a linear mixture of models (Constrained CMVM).

When introduced by Kohonen, Self Organizing Maps appeared to be a straightforward yet powerful algorithm for generating "order out of disorder" [13] by constructing a one- or two-dimensional layer of neurons to capture the important variables contained in an input space. SOMs are built on competitive learning. The neurons are appropriately adjusted to the different input patterns throughout this competition. Their places become meaningfully organized in relation to one another.

The winning neuron or Best Matching Unit is the best tuned neuron; in this situation, we opted to represent the input data using its associated prototypes, i.e. Best Matching Unit. Due to the fact that the SOM generates a map of cells, or clusters, and each cluster is specified via its prototype, the object's prototype is the prototype of the cluster to which it relates.

3.1.2 Constrained CMVM

Kohonen expanded the usage of the SOM in [18] by demonstrating that rather than describing inputs through using "Best Matching Unit," i.e., the "Winning neuron," they are described using a linear mixture of the reference vectors. This innovative technique evaluates incoming data and takes into consideration it using a collection of models that more precisely characterize the object. In comparison to the conventional SOM learning method, which employs only the BMUs. The linear combination of models maintains the information more effectively.

Assume that each input is a Euclidean vector x with dimension n. The prototype SOM matrix is indicated by the symbol M of dimension (pxn), where p denotes the number of cells in the SOM. We minimize the following formulae to obtain the parameters of the models:

$$\min \left\| M'\alpha - x \right\|, \tag{6}$$

here $alpha$ is a vector of non-negative scalars denoted by the symbol $alphai$. The restriction of non-negativeness is critical while handling using evaluation metrics as inputs, as their negative values are meaningless.

There are numerous approaches to solving the aforementioned goal function. The most often used and easy optimization technique is gradient-descent optimization. It is an iterative method that may take the non-negativity requirement into consideration.

The current fitting issue falls within the category of quadratic optimization, for which various techniques have been built over time. A solution in a single pass is founded on the Kuhn Tucker theorem [11].

3.2 Fine Tuning

One of the difficulties in machine learning is mining multi-view data that is dispersed across several sites. To address this issue, we suggest to employ collaborative clustering. This technique is capable of dealing with data from many sources, i.e. multiple sets containing the same persons in distinct feature dimensions, wherein the data type may be distinct.

In other words, each dataset is a subset of a broader global dataset that contains data on the same population. This approach naturally addresses the curse of dimensionality, as the algorithm considers each component independently, resulting in a more accurate conclusion.

The k-CMVM and Constrained-CMVM algorithms (2 and 3, respectively) construct traditional SOM for every view of the dataset and utilize the collaborative model to communicate topographic information across collaborators, as detailed in the methodology 1. It uses the Davies Bouldin score [7] as a stopping criterion, which is a clustering assessment index that represents the effectiveness of the clustering. If DB lowers, the collaboration is beneficial; if it grows, the collaboration is stopped and the initial map is used. A positive collaboration is one where collaborators increase one another's clustering quality; a negative collaboration, on the other side, is one in which a collaborator negatively impacts another by lowering the quality of its clustering. By using the database index as a stopping condition, we can control how the views interact and ensure that collaboration helps the clustering exclusively. As a result of the collaboration, we are able to acquire more homogenous clusters by including topographical information from all views.

We discovered that using a linear mixture of models produced superior results than anonymizing the data using BMUs since it retains the majority of the information. The dataset's de-anonymized segments are then reorganized in the same manner as the original dataset. The second portion of each algorithm creates a significant difference between the two. On the one hand, k-CMVM provides a pre-anonymized dataset that will be fine-tuned using a SOM model whose map size is determined using the Kohonen algorithm [17]. The resulting dataset is recoded using prototypes

Algorithm 2: k-CMVM.

Input: $OT(A_1, A_2, .., A_m)$ a table to anonymize
P number of views $V^{[ii]}$
Output: $AT(A_1, A_2, .., A_m)$ Anonymized table
k anonymity level **Collaboration process:**

- Create P views $V^{[ii]}$

$$V^{[ii]} \leftarrow OT(A_j, .., A_l); \quad Where\ (j,l) \in 1, .., m \tag{7}$$

- Calculate $w^{[ii]}$ with the collaboration approach 1 with all $V^{[ii]}$

Pre-Anonymization:
For each $V^{[ii]}$, $ii = 1$ to P:

- Compute the BMU of each item j in $V^{[ii]}$ using corresponding $w_{jc}^{[ii]}$ with c represanting the matching cell:
$$\text{argmin}(X_i^{[ii]} - w_{jc}^{[ii]})$$
- Code each data j of OT with its corresponding vector: $X'_j \leftarrow [w_{jc_{(1)}}^{[1]}, w_{jc_{(2)}}^{[2]}, ..., w_{jc_{(q)}}^{[P]}]$, where $c_{(q)}$ is the index of the cell associated with data j.

Fine-tuning and anonymization:

- Learn a global SOM based on the anonymized dataset. OT'
- For each c in neurons 1 to n_c
  - Substitute the jth data of OT' by w'_{jc}: $X''_j \leftarrow [w'_{jc_{(1)}}, w'_{jc_{(2)}}, ..., w'_{jc_{(q)}}]$
  - Output results in $AT(A_1, A_2, .., A_m)$
- Output level of anonymity:

$$k \leftarrow count(occurences AT(j)) \tag{8}$$

of the BMU's closest neighbors, and the dataset's level of anonymity is calculated. The k levels are not predefined; rather, the model determines them automatically.

On the other side, the fine tuning stage for the Constrained-CMVM technique is as follows: we apply a constrained SOM on the pre-anonymized dataset. To obtain a restricted map, we first build a SOM based on the outputs of the pre-anonymization phase. A preset number of k layers of anonymity is used, and items from neurons that violate the constraint of k cardinality are reassigned to the nearest neurons.

This procedure alters the map's topology, but aids in the creation of groups of at least k components in each neuron. We use the best matching unit to code the objects of each neuron, resulting in a k-anonymized dataset.

Then, we review the many k levels to see which one best matches our specifications.

To summarize, the suggested anonymization approaches employ a multi-view strategy in order to deal with complicated data and data from many sources. Additionally, this approach is utilized to retain the dataset's quality while recoding and to avoid the dimensionality curse. The user specifies the number of subsets to be utilized

Algorithm 3: Constrained-CMVM

Input: $OT(A_1, A_2, .., A_m)$ a dataset to anonymize
P number of views $V^{[ii]}$
k anonymity level
Output: $AT(A_1, A_2, .., A_m)$ Anonymized data
Multi-view Clustering process:

- Create P views $V^{[ii]}$

$$V^{[ii]} \leftarrow OT(A_j, .., A_l); \quad Where\ (j, l) \in 1, .., m \tag{9}$$

- Calculate $w^{[ii]}$ using the collaboration approach 1 with all $V^{[ii]}$.

Pre-Anonymization:
For each $V^{[ii]}$, $ii = 1$ to P:

- Find the linear mixture of SOM models for every object j in $V^{[ii]}$.

$$c(j)^{[ii]} \leftarrow \sum_{l=1}^{l=q} \delta_l w_{jk}^{[ii]} \tag{10}$$

where $c(j)^{[ii]}$ is the coding of the jth data of the $[ii]$th view.
δ_l are the coefficients of linear mixture of models.

- Code each data j of OT its corresponding vector. $X'_j \leftarrow [c^{[1]}(j), c^{[2]}(j), .., c^{[P]}(j)]$.

Constrained Clustering and Anonymization:

- Learn a global SOM from the anonymized dataset OT'.
- Let $|.|$ represents the number of elements in a cluster: find the clusters where $|E_c| \leq k$.
- Redistribute these elements among the remaining cells in such a way that each remaining cell contains at least k elements.
- Using the matching BMUs, recode the table OT' and output the results in AT.

for collaboration, which is proportional to the amount of the data. The algorithm 1 constructs the maps by transferring topological information across cooperated maps. After pre-anonymizing the data, it is fine-tuned using BMUs or grouped by cell under the restriction of k items.

3.3 Integrating Discriminatory Power

By comparing the outcomes of data anonymization using the k-CMVM (method mentioned in 2) and the Constrained-CMVM (method referred in 3), we intended to determine when data is tagged and how supervision may affect the performance of the anonymized results.

We employed the Learning Vector Quantization (LVQ) approach to examine this problem. This choice was taken because it has the ability to improve clustering results by considering the class of each item. The proposed approach is trained on a selection of data that most closely matches the training set.

Because it is based on hebbian learning, the LVQ technique is best recognized for its simplicity and quick convergence. Indeed, it is a prototype-based technique that develops a collection of prototypes inside the area of the examined input data and uses then to classify previously unknown cases.

Algorithm 4: Weighting Pattern Features Adaptively During Learning

---

Initialization :
Initialize the weights P based on :

$$p_j^i = \begin{cases} 0, & when\ \text{i}\neq\text{j} \\ 1, & when\ \text{i=j} \end{cases}$$

The k-means technique is used to choose the codewords for each class.
Learning Process:

1. Present a learning object x.
2. Let $w_i \in C_i$ be the closest prototype vector to x.

- **if** $x \in C_i$, then go to 1
- **else then**
  - let $w_j \in C_j$ be the second nearest prototype vector
  - **if** $x \in C_j$ then
    - a symmetrical window win is set around the mid-point of w_i and w_j.
    - **if** x falls within win, then
      Prototypes updating :
    - w_i is moved away from x based on :

$$w_i(t+1) = w_i(t) + \alpha(t)[Px(t) - w_j(t)] \tag{11}$$

    - w_j is moved closer x based on :

$$w_j(t+1) = w_j(t) - \alpha(t)[Px(t) - w_j(t)] \tag{12}$$

    - for the rest of the prototypes

$$w_k(t+1) = w_k(t) \tag{13}$$

      Weighting Patterns features:
    - adapt p_k^k based on:

$$p_k^k(t+1) = p_k^k(t) - \beta(t)x^k(t)(w_i^k(t) - w_j^k(t)) \tag{14}$$

    - go to 1.

Where $\alpha(t)$ and $\beta(t)$ are the learning rates

LVQ was created for classification issues with pre-existing data sets that may be used to monitor the system's learning. LVQ is a non-parametric function, which implies that no assumptions regarding the content of the function being approximated are made.

The learning rate decays linearly from an initial value to near zero over the course of training. The more complicated the distribution of classes, the more prototypes will be required; in some cases, thousands may be required. Multiple rounds of the LVQ training method are recommended for more rigorous usage, with the first pass learning rapidly to generate prototypes and the second pass learning slowly and running for an extended period of time (perhaps 10-times more epochs).

Each class in the LVQ model is defined by a fixed prototype with the same dimension as the data to be classified. LVQ iteratively modifies the prototypes. The learning technique begins by grouping the data using a clustering approach and then doing classification on the clusters' prototypes using LVQ. We chose to monitor the clustering outcomes by relocating the center clusters' utilizing the pLVQ2 method provided in algorithm 4 for each approach. We utilize the pLVQ2 [34] because this enhanced

We use the pLVQ2 [34] because it is an enhanced version of the LVQ that considers the attributes of every variable and adjusts the weights of every variable according to its contribution to discrimination. The algorithm is divided into two layers: the first layer generates weight values that are then passed to the LVQ2 algorithm.

The cost function for this approach is as follows:

$$R_{pLVQ2}(x, w, P) = \begin{cases} \|Px - w_j\|^2 - \|Px - w_i\|^2, \; If \; C_k = C_j \\ 0, \quad otherwise \end{cases} \tag{15}$$

Here, Ck signifies the class k, $xinCk$ denotes a data point, P denotes the weighting coefficient matrix, wi denotes the codeword vector closest to Px, and wj denotes the codeword vector second closest to Px. The k-CMVM and Constrained-CMVM models in the pLVQ2 combined with the Collaborative Paradigm enhance the utility of anonymized data.

The use of pLVQ2 occurs during cooperation across cluster centers in order to enhance the Collaboration's outcomes during the pre- and post-anonymization stages.

4 Validation of the Approaches

4.1 Description of the Datasets

Our techniques previously described, k-CMVM, k-CMVM++, Constrained-CMVM, and Constrained-CMVM++, were evaluated on different datasets given by the University of California, Irvine Machine Learning Repository [8] (Table 1):

Table 1 Description of the datasets

Datasets	#Instances	#Attributes	#Class
DrivFace	606	6400	3
Ecoli	336	8	8
Glass	214	10	7
Spam base	4601	57	2
Waveform	5000	21	3
Wine	178	13	3
Yeast	1484	8	10

- The DrivFace database contains sequences of images of human driving in real-world situations. It is made of 606 samples with a resolution of $6400x480$ pixels each, taken over many days from four drivers (2 women and two men) with a variety of face characteristics such as spectacles and beard.
- Protein localization sites are included in the Ecoli and Yeast datasets. Each characteristic used to classify a protein's localization location is a score (between 0 and 1) assigned to a particular aspect of the protein sequence. The higher the score, the more likely it is that the protein sequence contains this characteristic.
- The glass dataset contains information about the oxide content of the glass, which is used to define its kind. The classification of glass types was prompted by a criminological investigation.
- The Spam base dataset has 57 characteristics that provide information on the frequency with which certain phrases are used, the frequency with which capital letters are used, and other indicators for determining whether an e-mail is spam or not.
- Waveforms are used to describe three distinct types of waves with the addition of noise. Each class is created by combining two of three "base" waves, and each instance is created by adding noise to each property (mean 0, variance 1).
- Wine data represent the outcome of a chemical study of three distinct varieties of wine cultivated in the same region of Italy. The research identified the concentrations of thirteen components contained in every of the three wine varieties.

4.2 Utility Measures and Statistical Analysis

Microaggregation's effect on the usefulness of anonymized data is quantified by the performance of the resulting machine learning model [25, 28]. We constructed a decision tree model and used it to classify the anonymized data in order to assess its value. We next evaluated the separability usefulness of the outputs from both techniques before and after incorporating the discriminant information to have a better understanding of how much data quality was sacrificed for anonymization.

Table 2 The separability usefulness of the base methods MDAV, k-CMVM, and constrained CMVM is compared before and after the discriminant information is included

	DrivFace	Ecoli	Glass	Spam base	Waveform	Wine	Yeast
Original	92.2	82.4	69.6	91.9	76.9	88.8	83.6
MDAV	89.1	75.6	61.2	70.1	69.8	68.4	83.4
k-CMVM	90.3	84.5	82.4	86.4	83.0	69.7	86.3
k-CMVM++	92.4	98.8	94.4	87.1	88.4	70.5	100
Constrained-CMVM	93.2	85.1	75.2	90.6	81.5	74.2	87.4
Constrained-CMVM++	94.1	86.3	85.9	91.5	88.4	77.8	88.7

The pre-anonymization step was critical in constructing anonymous objects by views; rather than coding the entire example with a single model, we coded each section with the BMU in the case of k-CMVM or with a linear mixture of neighboring models in the case of Constrained-CMVM, and then with fine tuning to add an additional component of anonymization. We present the outcomes of the four methods after anonymization in Table 2. We'd want to refer to the correctness of a dataset as the separability utility, as a high utility indicates a high degree of separability across clusters. The titles in Table 2 relate to the ones that follow:

- **Original**: The first utility of dataset separability applying a decision tree model with tenfold cross-validation.
- k-**CMVM**: The separability utility of the dataset was calculated using multi-view clustering with collaboration between the views using the Kohonen Heuristic to establish the optimal map size. Pre-anonymization samples were created using BMUs.
- **Constrained-CMVM**: After using the Kohonen Heuristic to find the appropriate map size, the pre-anonymization samples were coded using the Linear Mixture of Models.
- The $^{++}$ in the name of the algorithms relates to discriminant version.

5 Conclusion

We discussed four microaggregation strategies for data anonymization in detail in this chapter: CMVM and Constrained CMVM, which use the Collaborative Multi-View paradigm, and CMVM++ and Constrained CMVM++, which we introduced to improve the performance of the anonymized dataset when using the ground truth labels. The data above indicate the efficacy and significance of the approaches.

We began by experimenting with multi-view clustering, which we believe is an excellent strategy for dealing with data originating from multiple sources and containing high-dimensional components. Second, we proved that collaborative topographical clustering improves clustering quality, hence boosting the accuracy of the model. Thirdly, pre-anonymization using a Linear Mixture of SOM results in a higher separability utility than pre-anonymization with BMUs. Fourth, we uncovered a beneficial trade-off between the utility of separability and the level of anonymity. Finally, we evaluated the constraints and potential of incorporating discriminant information in the absence of known ground truth labels, comparing its performance to that of the state-of-the-art and CMVM & Constrained CMVM.

An expansion of this study will attempt to discover an aggregate of all the framework presented with the goal of optimizing the anonymized sets' privacy and utility.

We are investigating further methods of data anonymization and have discovered that 1D clustering may be used to anonymize data without erasing the information contained inside. Additionally, we would want to investigate novel approaches for anonymizing imbalanced datasets.

References

1. Agrawal, D., Aggarwal, C.C.: On the design and quantification of privacy preserving data mining algorithms. In: Proceedings of the Twentieth ACM SIGMOD-SIGACT-SIGART Symposium on Principles of Database Systems, pp. 247–255. ACM (2001)
2. Alabdulatif, A., Khalil, I., Reynolds, M.C., Kumarage, H., Yi, X.: Privacy-preserving data clustering in cloud computing based on fully homomorphic encryption. In: Alias, R.A., Ling, P.S., Bahri, S., Finnegan, P., Sia, C.L. (eds.) 21st Pacific Asia Conference on Information Systems, PACIS 2017, Langkawi, Malaysia, July 16–20, p. 289 (2017)
3. Bhaladhare, P.R., Jinwala, D.C.: A clustering approach using fractional calculus-bacterial foraging optimization algorithm for k-anonymization in privacy preserving data mining. Int. J. Inf. Secur. Priv. **10**(1), 45–65 (2016)
4. Biswas, C., Ganguly, D., Roy, D., Bhattacharya, U.: Privacy preserving approximate k-means clustering. In: Zhu, W., Tao, D., Cheng, X., Cui, P., Rundensteiner, E.A., Carmel, D., He, Q., Yu, J.X. (eds.) Proceedings of the 28th ACM International Conference on Information and Knowledge Management, CIKM 2019, Beijing, China, November 3–7, pp. 1321–1330. ACM (2019)
5. Ciriani, V., De Capitani di Vimercati, S., Foresti, S., Samarati, P.: *k*-anonymity. In: Yu, T., Jajodia, S. (eds.) Secure Data Management in Decentralized Systems, volume 33 of Advances in Information Security, pp. 323–353. Springer (2007)
6. Dai, C., Ghinita, G., Bertino, E., Byun, J.-W., Li, N.: TIAMAT: a tool for interactive analysis of microdata anonymization techniques. Proc. VLDB Endow. **2**(2), 1618–1621 (2009)
7. Davies, D., Bouldin, D.: A cluster separation measure. IEEE Trans. Pattern Anal. Mach. Intell. **2**, 224–227 (1979)
8. Dheeru, D., Karra Taniskidou, E.: UCI machine learning repository (2017)
9. Falih, I., Grozavu, N., Kanawati, R., Bennani, Y.: Topological multi-view clustering for collaborative filtering. In: Ozawa, S., Tan, A.-H., Angelov, P.P., Roy, A., Pratama, M. (eds.) INNS Conference on Big Data and Deep Learning 2018, Sanur, Bali, Indonesia, 17–19 April 2018, volume 144 of Procedia Computer Science, pp. 306–312. Elsevier (2018)

10. Fiore, M., Katsikouli, P., Zavou, E., Cunche, M., Fessant, F., Le Hello, D., Aïvodji, U.M., Olivier, B., Quertier, T., Stanica, R.: Privacy in trajectory micro-data publishing: a survey. Trans. Data Priv. **13**(2), 91–149 (2020)
11. Gentleman, W.: Solving least squares problems. SIAM Rev. **18**(3), 518–520 (1976)
12. Grozavu, N., Cabanes, G., Bennani, Y.: Diversity analysis in collaborative clustering. In: 2014 International Joint Conference on Neural Networks, IJCNN 2014, Beijing, China, July 6–11, 2014, pp. 1754–1761. IEEE (2014)
13. Haykin, S.S.: Neural Networks and Learning Machines, vol. 3. Pearson Upper Saddle River (2009)
14. Hegde, A., Möllering, H., Schneider, T., Yalame, H.: SoK: efficient privacy-preserving clustering. Proc. Priv. Enhancing Technol. **2021**(4), 225–248 (2021)
15. Islam, M.Z., Brankovic, L.: Privacy preserving data mining: a noise addition framework using a novel clustering technique. Knowl. Based Syst. **24**(8), 1214–1223 (2011)
16. Kabir, M.E., Mahmood, A.N., Wang, H., Mustafa, A.K.: Microaggregation sorting framework for k-anonymity statistical disclosure control in cloud computing. IEEE Trans. Cloud Comput. **8**(2), 408–417 (2020)
17. Kohonen, T.: Self-Organizing Maps. Springer, Berlin (1995)
18. Kohonen, T.: Description of input patterns by linear mixtures of SOM models. In: Proceedings of WSOM, vol. 7 (2007)
19. Kovács, F., Legány, C., Babos, A.: Cluster validity measurement techniques. In: 6th International Symposium of Hungarian Researchers on Computational Intelligence. Citeseer (2005)
20. LeFevre, K., DeWitt, D.J., Ramakrishnan, R.: Mondrian multidimensional k-anonymity. In: Proceedings of the 22nd International Conference on Data Engineering, ICDE'06, pp. 25–25. IEEE (2006)
21. Li, J., Chi-Wing Wong, R., Fu, A.W., Pei, J.: Achieving k-anonymity by clustering in attribute hierarchical structures. In: International Conference on Data Warehousing and Knowledge Discovery, pp. 405–416. Springer (2006)
22. Loukides, G., Shao, J.: Capturing data usefulness and privacy protection in k-anonymisation. In: Proceedings of the 2007 ACM Symposium on Applied Computing, pp. 370–374. ACM (2007)
23. Mehta, B., Rao, U.P., Gupta, R., Conti, M.: Towards privacy preserving unstructured big data publishing. J. Intell. Fuzzy Syst. **36**(4), 3471–3482 (2019)
24. Mitra, S., Banka, H., Pedrycz, W.: Collaborative Rough Clustering. In: PReMI, pp. 768–773 (2005)
25. Pallarès, E., Rebollo-Monedero, D., Rodríguez-Hoyos, A., Estrada-Jiménez, J., Mezher, A.M., Forné, J.: Mathematically optimized, recursive prepartitioning strategies for k-anonymous microaggregation of large-scale datasets. Expert Syst. Appl. **144**, 113086 (2020)
26. Pedrycz, W.: Collaborative fuzzy clustering. Pattern Recognit. Lett. **23**(14), 1675–1686 (2002)
27. Rastin, P., Cabanes, G., Grozavu, N., Bennani, Y.: Collaborative clustering: how to select the optimal collaborators? In: IEEE Symposium Series on Computational Intelligence, SSCI 2015, Cape Town, South Africa, December 7–10, 2015, pp. 787–794. IEEE (2015)
28. Rodríguez-Hoyos, A., Estrada-Jiménez, J., Rebollo-Monedero, D., Mezher, A.M., Parra-Arnau, J., Forné, J.: The fast maximum distance to average vector (F-MDAV): an algorithm for k-anonymous microaggregation in big data. Eng. Appl. Artif. Intell. **90**, 103531 (2020)
29. Ros-Martín, M., Salas, J., Casas-Roma, J.: Scalable non-deterministic clustering-based k-anonymization for rich networks. Int. J. Inf. Secur. **18**(2), 219–238 (2019)
30. Saitta, S., Raphael, B., Smith, I.F.: A bounded index for cluster validity. In: International Workshop on Machine Learning and Data Mining in Pattern Recognition, pp. 174–187. Springer (2007)
31. Soria-Comas, J., Domingo-Ferrer, J., Sánchez, D., Martinez, S.: t-closeness through microaggregation: strict privacy with enhanced utility preservation. IEEE Trans. Knowl. Data Eng. **27**(11), 3098–3110 (2015)
32. Sublime, J., Grozavu, N., Cabanes, G., Bennani, Y., Cornuéjols, A.: From horizontal to vertical collaborative clustering using generative topographic maps. Int. J. Hybrid Intell. Syst. **12**(4), 245–256 (2015)

33. Venkatasubramanian, S.: Measures of anonymity. In: Privacy-Preserving Data Mining, pp. 81–103. Springer (2008)
34. Yacoub, M., Bennani, Y.: Features selection and architecture optimization in connectionist systems. Int. J. Neural Syst. **10**(5), 379–395 (2000)
35. Zouinina, S., Bennani, Y., Rogovschi, N., Lyhyaoui, A.: A two-levels data anonymization approach. In: Maglogiannis, I., Iliadis, L., Pimenidis, E. (eds.) Artificial Intelligence Applications and Innovations - 16th IFIP WG 12.5 International Conference, AIAI 2020, Neos Marmaras, Greece, June 5–7, 2020, Proceedings, Part I, volume 583 of IFIP Advances in Information and Communication Technology, pp. 85–95. Springer (2020)

Multi-view Clustering Based on Non-negative Matrix Factorization

Nistor Grozavu, Basarab Matei, Younès Bennani, and Kaoutar Benlamine

Abstract Clustering is a machine learning technique that seeks to uncover the intrinsic patterns from a dataset by grouping related objects together. While clustering has gained significant attention as a critical challenge over the years, multi-view clustering have only a few publications in the literature.

Non-negative matrix factorization (NMF) is a data mining technique which decompose huge data matrices by placing constraints on the elements' non-negativity. This technique has garnered considerable interest as a serious problem with numerous applications in a variety of fields, including language modeling, text mining, clustering, music transcribing, and neuroscience (gene separation). This lack of negative simplifies the interpretation of the generated matrices. Two different matrices are used: one is used to represent the cluster prototypes of a dataset, and the other is used to represent data partitions. The investigated datasets are frequently non-negative, and they occasionally contain sparse representations. Ding et al. [5] demonstrated that the orthogonal NMF and K-means hard clustering are equivalent. Kim and Park addressed this similarity and suggested a sparse non-negative matrix factorization (NMF) technique for data clustering. Their technique outperforms both K-means and NMF in terms of result consistency. We propose a research of multi-view clustering algorithms in this chapter and provide a novel methodology called multi-view non-negative matrix factorization, in which we investigate the collaboration between

N. Grozavu (✉)
ETIS–CNRS UMR 8051, CY Cergy Paris University, 33 Bd. du Port, 95000 Cergy, France
e-mail: nistor.grozavu@cyu.fr

B. Matei · Y. Bennani · K. Benlamine
LIPN–CNRS UMR 7030, Sorbonne Paris Nord University, 99 av. J-B Clément, 93430 Villetaneuse, France
e-mail: matei@lipn.univ-paris13.fr

Y. Bennani
e-mail: bennani@lipn.univ-paris13.fr

K. Benlamine
e-mail: benlamine@lipn.univ-paris13.fr

LaMSN, La Maison des Sciences Numériques, Saint-Denis, France

W. Pedrycz and S. Chen (eds.), *Recent Advancements in Multi-View Data Analytics*, Studies in Big Data 106, https://doi.org/10.1007/978-3-030-95239-6_7

several NMF models. This strategy was evaluated using a variety of datasets, and the experimental findings demonstrate the suggested approach's usefulness.

Keywords Multi-view clustering · Collaborative clustering · Non-negative matrix factorization · Projection

1 Introduction

Data clustering is a critical step in the information extraction process from databases. It seeks to find the intrinsic structures of a set of items by grouping objects with similar characteristics. This task is more challenging than supervised classification since the number of clusters to be discovered is typically unknown, making it difficult to assess the quality of a clustering division. This endeavor has gotten even more difficult over the last two decades as available data sets have become more complex due to the arrival of multi-view data sets, dispersed data, and data sets with varying scales of structures of interest (e.g. hierarchical clusters).

Due to the increasing complexity of an already tough task, lone clustering algorithms struggle to produce competitive results with a high degree of certainty. However, much as in the real world, similar problems can be solved more readily by combining several methods to improve both the quality and reliability of the outputs.

In recent years, NMF has gotten a lot of attention for its ability to cluster non-negative data [1, 14], and it has been applied in a variety of domains such as feature selection, dimensionality reduction, and clustering, as well as text mining.

Paatero (1994) [19] introduced the NMF method, which is an unsupervised clustering methodology in which a data matrix is factored into (usually) two matrices, where each matrix has the property that none of the matrices has any negative components. The absence of negatives makes it easier to interpret the matrices that have been constructed. Two different matrices are used: one is used to represent the prototypes of clusters, and the other is used to represent data partitions.

The NMF method is a powerful technique that is utilized in a variety of critical applications. Frequently, the datasets being studied are non-negative and, on occasion, have a sparse representation. Non-negativity is associated with probability distributions in machine learning, but sparseness is associated with feature selection.

Ding et al. [5] demonstrated that the orthogonal NMF and k-means method are equivalent. Kim and Park [14] highlighted this equivalence when they introduced a sparse NMF technique for data clustering. This approach outperforms both the k-means and the NMF in terms of result consistency. Collaborative Clustering [6, 20, 22]is a new challenge in data mining, with only a few publications on the subject [3, 4, 7, 20, 22].

In this work, we suppose that we have a collection of sparse data sets scattered across multiple places; the data could pertain to consumers of banks, stores, or medical organizations, for example. The data sets may contain information about multiple individuals whose characteristics are characterized by the same variables; in this situation, a vertical collaboration technique is offered. The horizontal technique is used to collaborate on data sets that describe the same objects but with distinct variables [13]. This approach is analogous to multi-view clustering in that it is applied to data that is multi-represented, i.e., the same collection of objects characterized by many representations (features) [18].

In this chapter, we introduce a new method called Multi-view Non-negative Matrix Factorization. We investigate the collaboration between several NMF methods and then use Karush-Kuhn-Tucker (KKT) conditions to find a consensus of the various collaborations in order to solve the problem of NMF non-uniqueness and obtain a more stable solution.

The remainder of this chapter is divided into the following sections: In Sect. 2, we discuss similar research and review the notion of Non-negative Matrix Factorization. In Sect. 3, we demonstrate the proposed technique, which is developed from the NMF algorithm, in a multi-view environment. Section 4 proposes and analyzes certain numerical experiences. The conclusion makes some suggestions for future investigation.

2 Related Works

Unlike the vast majority of distributed data clustering (DDC) algorithms [24], which form a consensus having accest to all data, the main challenge of collaboration is that the clustering method are applied locally, but collaborate by exchanging the extracted knowledge [20].

We differentiate two types of collaborative approaches in general [9, 21]: *Vertical collaboration* refers to any instances in which multiple algorithms work on disparate data sets that share comparable clusters or distributions. And *Horizontal collaboration* addresses situations in which many algorithms collaborate on the same objects, which are subsequently described from multiple perspectives.

Collaborative techniques are typically divided into two stages [20]:

1. *Local clustering*: Each method will perform clustering on the local view and will obtain a local clustering result.
2. *Collaborative clustering*: The method exchange their results and attempt to confirm or modify their models in order to achieve more accurate clustering results.

These two processes are occasionally followed by an aggregate step, which seeks to achieve consensus on the collaboratively produced final findings. We will not cover the aggregation step in this study since it is a distinct field and because, depending on the application, aggregation may not always be desirable, for example, when the divisions between the various views or locations are in conflict.

Rather than that, we will concentrate on the collaborative step, in which the algorithms exchange information with the purpose of mutual development.

The quality of categorization following collaboration is contingent upon the clustering strategy employed during the local phase. Earlier work on collaborative clus-

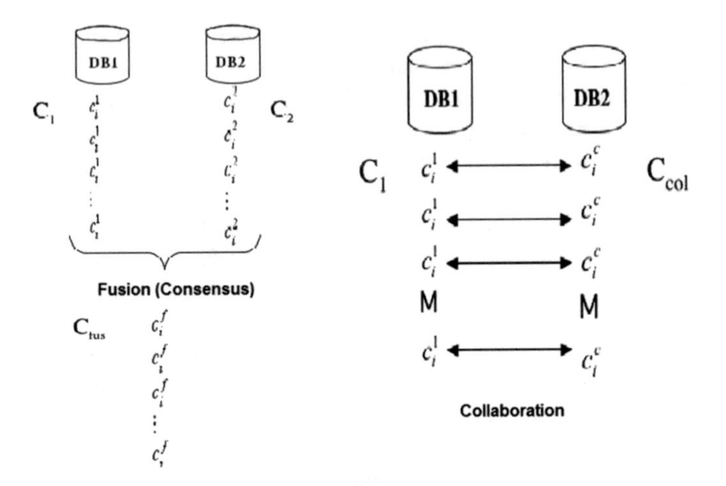

Fig. 1 Building a consensus (left) versus Collaborative Clustering (right)

tering relied on deterministic models, such as fuzzy C-means (FCM) [20] and self-organizing maps (SOM) citation [11].

Similarly to the comparison of deterministic and probabilistic models, we propose a collaborative clustering approach in this study that is based on a non-negative matrix factorization (NMF) in its local phase, a technique that is an alternative to classical topological methods adapted for sparse data.

Collaboration is crucial in that clustering algorithms function locally (that is, on separate data sets), but collaborate by exchanging information about their conclusions [20]. As a result, data sites share information granules during the learning process, while maintaining data confidentiality.

Collaborative clustering consists of two phases: a local phase and a collaborative phase. On a database-by-database basis, the local phase would apply a clustering method based on prototypes. The collaboration phase tries to integrate all categories related with other databases obtained during the local phase.

The Fig. 1 illustrates the difference between consensus clustering and collaborative clustering.

2.1 Topological Collaborative Clustering

The local data is clustered using a the Kohonen map (SOM: Self Organizing Maps).

To facilitate the formulation, we assume that the constructed maps from several datasets have the same sizes (number of cells).

An observation from one dataset is projected on the ii]th neuron in another dataset, and the same observation from another dataset is projected on the jj]th neuron in another map, and so on.

So the traditional SOM objective function gets a new term representing collaboration. A collaborative parameter is added to this function to indicate the confidence and collaboration between the $[ii]$ and $[jj]$ classifications. During the collaborative learning process, a new collaboration step is included to assess the value of cooperation. Two parameters are added to compute the collaboration's relevance: one to adapt distant clustering information, and the other to weight the collaborative clustering connection.

Analyses of the SOM algorithm's objective function reveal some parallels to the k-means issue.

Indeed, SOM prototypes can be understood as local centers subject to additional topological constraints.

As a result, it is very logical to apply the same collaborative model to SOM maps. That is, modifying the SOM cost function appropriately by adding a collaborative term to transmit the topological information. Grozavu and Bennani [12] used this strategy, emphasizing the map's ability to improve its overall clustering performance.

The proposed cost function is made of two expressions in formal terminology:

$$R_H^{[ii]}(W) = R_Q^{[ii]}(W) + R_{Col-H}^{[ii]}(W). \tag{1}$$

The first part in the Eq. (1) indicates the conventional SOM cost function:

$$R_Q^{[ii]} = \sum_{\substack{jj=1 \\ jj \neq ii}}^{P} \sum_{i=1}^{N} \sum_{j=1}^{|w|} K_{\sigma(j,\chi(x_i))}^{[ii]} \|x_i^{[ii]} \text{-} w_j^{[ii]}\|^2. \tag{2}$$

At the conclusion of this paragraph, the classical SOM principle is recalled. The collaborative method is denoted by the second term in the right hand side of the Eq. (1):

$$R_{Col-H}^{[ii]} = \sum_{\substack{jj=1 \\ jj \neq ii}}^{P} \sum_{i=1}^{N} \sum_{j=1}^{|w|} L_{ij} \|x_i^{[ii]} \text{-} w_j^{[ii]}\|^2. \tag{3}$$

We have used the notation as follows:

$$L_{ij} = \left(K_{\sigma(j,\chi(x_i))}^{[ii]} - K_{\sigma(j,\chi(x_i))}^{[jj]} \right)^2. \tag{4}$$

This factor simulates the best matched unit distance between two distinct maps when comparing two different maps. A numerical estimate of the difference in representation between the two maps is obtained by comparing the neighborhood distances between a data point i and the neuron j in both maps, as shown in Fig. 1. If the distance between each data point is comparable in the two SOMs for each data point, it indicates that the bmus of data points is the same (or at the very least that they are close neighbors) in both maps. Because it is based purely on the topology of the two maps and the construction of a neighborhood distance, the factor provides an excellent means of avoiding the problem of needing to be in the same feature space as the other map. Nevertheless, as described in Eq. (5), the neighborhood function $Ksigma$ is reliant on the feature space of the data in order to compute a distance metric $sigma(i, j)$. As a result, in order to do any kind of comparison between the neighborhood, we must first equalize the neighborhood functions across the different feature areas in the community.

In the Eqs. (2) and (3), the parameter P indicates the number of datasets (or classifications), N indicates the number of observations, and $|w|$ indicates the number of prototype vectors from the ii SOM map (the number of neurons). In [?], [12], the authors employ a gradient descent technique to tackle this optimization problem.

During the first phase of the collaboration step, the value of the collaboration connection β is determined.

This parameter provides for the determination of the significance of collaboration between two datasets; the value is determined by the topological similarity of the two collaboration maps.

In the Eqs. (2) and (3), $\chi(x_i)$ is the assignment function that allows for the determination of the Best Matching Unit (BMU); it does so by selecting the neuron with the closest prototype from the input x_i using the Euclidean distance:

$$K_{\sigma(i,j)}^{[cc]} = \exp\left(-\frac{\sigma^2(i,j)}{T^2}\right),$$

where T is the temperature manage the cell's neighboring effect on the map, T can take values in the interval: $Tmax$ and $Tmin$.

The neighborhood metric $K\sigma(i,j)^{[cc]}$ has the same nature across all maps, but its value changes across them: it is determined by the nearest neuron to the data, which is not generally the same across all SOM maps.

To compute the matrix of collaborative prototypes, the following gradient optimization technique is used:

$$w^{*[ii]} = \text{argmin}_W \left[R_H^{[ii]}(W)\right]$$

The updated rule obtained is as follows:

$$w_{jk}^{*[ii]}(t+1) = w_{jk}^{*[ii]}(t) + \frac{\sum_{i=1}^{N} K_{\sigma(j,\chi(x_i))}^{[ii]} x_{ik}^{[ii]} + \sum_{\substack{jj=1 \\ jj\neq ii}}^{P} \sum_{i=1}^{N} L_{ij} x_{ik}^{[ii]}}{\sum_{i=1}^{N} K_{\sigma(j,\chi(x_i))}^{[ii]} + \sum_{\substack{jj=1 \\ jj\neq ii}}^{P} \sum_{i=1}^{N} L_{ij}} \tag{5}$$

where L_{ij} is defined by the Eq. (4). Indeed, while collaborating with a SOM map, the method considers the map's prototypes and topology through the neighborhood function. The horizontal collaboration approach is defined as follows:

Vertical cooperation occurs when all databases share the same features (same description space), but include distinct sets of observations. The number and size of prototype vectors for all SOM mappings will be equal to || in this scenario. The fundamental concept of collaboration in this circumstance is as follows: Using the Euclidean distance, a neuron k in the $[ii]$th SOM map and the identical neuron k in the $[jj]$th SOM map should be extremely similar. In other terms, cells associated with distinct maps should reflect clusters of related observations. That is why, during the cooperation stage, the goal function of the traditional SOM has been changed to include this restriction on the different maps.

Algorithme 1: The horizontal collaboration algorithm

for $t = 1$ to N_{iter} **do**

1. Local step:
for each map $[ii]$, $ii = 1$ to P **do**
Find the prototypes minimizing the SOM cost function:

$$w^* = \text{argmin}_w \left[\sum_{i=1}^{N} \sum_{j=1}^{|w|} K^{[ii]}_{\sigma(j,\chi(x_i))} \|x_i^{[ii]} - w_j^{[ii]}\|^2 \right]$$

end for
2. Collaboration phase:
for each map $[ii]$, $ii = 1$ to P **do**
Update the neurons of the $[ii]$ map by using (5) that minimize the cost function $R_H^{[ii]}$ of the horizontal collaboration.
end for
end for

The concept of enabling collaboration by introducing a cooperative matching term to restrict the similarity of clustering components from various databases has shown its potential to generate enhanced clustering solutions. However, this technique has inherent constraints that restrict its use. Indeed, this method necessitates that all maps contain the same amount of neurons. This is a stringent condition that is unlikely to be met in situations where each agent specifies the size of its learning map independently. Collaboration is impossible in this circumstance. As a result, new methods of transmitting clustering information must be considered.

Topological Vertical Collaborative Clustering
By definition, collaboration requires that all local solutions have similar structures in order to communicate clustering knowledge. For instance, collaboration between several K-means results implies that all segmentation solutions have exactly the same amount c of prototypes. This may be understood by examining at the shape

of the K-means loss function for the site $[ii]$ in the case of vertical collaboration expression [20]:

$$Q[ii] = L[ii] + \sum_{[jj]=1,[jj]\neq[ii]}^{P} \beta_{[ii][jj]} C[ii][jj] \tag{6}$$

with

$$L[ii] = \sum_{n=1}^{N[ii]} \sum_{k=1}^{K[ii]} (g_{nk}^{[ii]})^2 \|x_n^{[ii]} - w_k^{[ii]}\|^2$$

and

$$C[ii][jj] = \sum_{n=1}^{N[ii]} \sum_{k=1}^{K[ii]} (g_{nk}^{[ii]})^2 \|w_k^{[ii]} - w_k^{[jj]}\|^2 \tag{7}$$

where the matrix $G[ii] = (g_{nk}^{[ii]})$ and the scalar $K[ii]$ indicate the partition matrix and the number of clusters obtained for the site $[ii]$, respectively.

By an examination of Eq. (7), collaboration framework demands that the number of components cl be constant across all locations during the local map learning process.

A comparison of the problem (7) and the formulation of vertical collaboration for SOM maps demonstrates that the limitation of collaboration application occurs for the same reason. Indeed, the collaboration term is created by comparing prototypes term by term. As a result, each clustering must have an equal number of components.

To extend the collaboration of K-means with a variable number of clusters, Coletta and al introduce a different method, defining the collaboration by adding virtual points to the database [2].

The critical concept is to introduce prototypes directly into other databases in order to change the K-means solution.

This enriched database incorporates clustering information, which results in some changes in the positions of the centers.

The collaborative impact is created by the manner in which the observations are introduced. Indeed, each virtual point is assigned a weight to regulate the effect of collaboration. This technique has the advantage of being extremely versatile and allowing for collaboration across many K-means clustering algorithms with varying cluster sizes.

One disadvantage of this technique is that collaboration is very dependent on the number of clusters $\overline{K}[ii]$ and the weighting parameter, i.e. if the $\overline{K}[ii]$ is small, the collaboration's effect will be minimal. A solution of this problem is the use of topological clustering, i.e. the SOM, to convey a greater number of prototypes than K-means.

To add prototypes to other datasets, all records must include components from the same features space, i.e. the approach is limited to vertical collaboration.

In the following, we present a multi-view non-negative matrix factorization method suited for sparse data in this work, as well as a study of the suggested approach's stability, based on this topological strategy.

2.2 Classical NMF

We will briefly discuss the traditional Non Negative Matrix Factorisation and its analogous to K-means method in this section. Let $X = (\mathbf{x}_1, \mathbf{x}_2, \ldots, \mathbf{x}_N) \in \mathbb{R}_+^{M\times N}$, be a non-negative data matrix with M objects and N attributes, $\mathbf{x}_n \in \mathbb{R}_+^{M\times 1}$ denotes the nth attribute of X. Assume that K is a constant input parameter.

The NMF approximates X to a low rank using the product of two non-negative matrices. FG in which case the factors represent $F = (\mathbf{f}_1, \mathbf{f}_2, \ldots, \mathbf{f}_K) \in \mathbb{R}_+^{M\times K}$, $G = (\mathbf{g}_1, \mathbf{g}_2, \ldots, \mathbf{g}_K) \in \mathbb{R}_+^{K\times N}$ and T represents the operator that has been transposed. A constrained optimization problem might be used to model the NMF, with the error function being minimized in the following ways:

$$\mathcal{L}(X, F, G) = \|X - FG\|^2 \tag{8}$$

Using the matrix form, the constraints problem has the following values:$F, G \geq 0$. It is essential to note that the function $\mathcal{L}$ reflects the Frobenius norm of the matrix difference $X - FG$.

Researchers Golub and Paatero [19, 25] first investigated the breakdown of NMFs in the early 1990s. In order to address this matrix approximation problem, several families of algorithms have been suggested, including the algorithms [19, 25]. For example, the initial method introduced [16] makes use of a multiplicative update formula and is built on a gradient descent strategy that typically develops in two stages: initially, the equation is optimized in accordance to G, and then the equation is minimized in accordance to F. Another option is to employ the alternative of least squares approach [8].

Additionally, Lee and Seung's work [16] demonstrates that NMF has an inherent segmentation characteristic for the attributes of X, because in Eq. (9) the factor may be assumed cluster prot$(\mathbf{f}_1, \mathbf{f}_2, \ldots, \mathbf{f}_K)$ prototypes, whereas the factor $(\mathbf{g}_1, \mathbf{g}_2, \ldots, \mathbf{g}_K)$ could be considered as partition, each $\mathbf{g}_j \in \mathbb{R}+^{N\times 1}$. More precisely $\mathbf{g}_{jk} > 0$ denotes that input data $\mathbf{x}_j \in \mathbb{R}+^{M\times 1}$ belongs to the kth cluster which has the prototype $\mathbf{f}_k \in \mathbb{R}+^{M\times 1}$ the kth column of F.

As a result of G, the collection of columns $J = \{1, \ldots, N\}$ is partitioned into K clusters. $C = \{C_1, \ldots, C_K\}$ and $F \in \mathbb{R}_+^{M\times K}$ indicates the simplified cluster representation of X. Furthermore, the formula in Eq. (9) is identical to the formula associated with the K-means clustering method.

Algorithme 2: Alternate Least Square NMF

Input: data X, parameter K
Output: factors F, G
Initialization: randomly define F
for $t = 1$ **to** $Iter$ **do**
  update the matrix G using the expression $F^T F G^T = F^T X$
  assign 0 to for $G < 0$
  update the matrix F using the formula $G^T G F^T = G^T X^T$
  assign 0 for $F < 0$
end
Return factors F, G.

Algorithme 3: Clustering Step

Input: data X, partition matrix G, number of clusters K
Output: cluster labels C
for $j = 1$ **to** N **do**
  each $\mathbf{x}_j$ is affected to the k_0th group, obtained by:

$$k_0 = \text{argmax}_k\{\mathbf{g}_{jk}; k = 1, \ldots, K\}.$$

end
Return labels C.

2.3 Hard K-Means and NMF

The K-means approach was pioneered by Lloyd [17]. By solving the optimization problem specified by the respective sum of squared errors, K-means approach allows for the division of the set of columns $\{1, \ldots, N\}$ into K clusters $C = \{C_1, \ldots, C_K\}$.

$$\mathcal{L}_K = \sum_{k=1}^{K} \sum_{n \in C_k} \|\mathbf{x}_n - \mathbf{f}_k\|^2 \tag{9}$$

The notation $G \in \mathbb{R}+^{K \times N}$ denotes the binary classification matrix provided by $gkn = 1$, if the column $\mathbf{x}n \in Ck$ exists, and 0 if it does not. As a result, the functional in Eq. (9) has the following form:

$$\mathcal{L}_K = \sum_{k=1}^{K} \sum_{n=1}^{N} g_{kn} \|\mathbf{x}_n - \mathbf{f}_k\|^2 \tag{10}$$

For a fixed partition $C = \{C_1, \ldots, C_K\}$ we consider $n_k = |C_k|$ the number of objects corresponding to the group k. Afterwards, we have the cluster prototypes $F = (\mathbf{f}_1, \mathbf{f}_2, \ldots, \mathbf{f}_k)$ are:

$$\mathbf{f}_k = \frac{1}{n_k} X \cdot \mathbf{g}_k, \quad k = 1, \ldots, K \tag{11}$$

and the indicator clustering matrix $(\mathbf{g}_1, \mathbf{g}_2, \ldots, \mathbf{g}_K)^T \in \mathbb{R}_+^{K \times N}$ are given by:

$$\mathbf{g}_k = \frac{1}{\sqrt{n_k}}(0, \ldots, 0, \underbrace{1, \ldots, 1}_{n_k}, 0, \ldots, 0)^T \in \mathbb{R}_+^{1 \times N}$$

In 2005, Ding et al. proved that the K-means clustering problem can be described as a matrix approximation problem, with the goal of clustering being to minimize the approximation error between the original data X and the reconstructed data [5].

The following equality is established for the goal function of hard K-means clustering by Ding et al. [5] , which is as follows:

$$\sum_{k=1}^{K} \sum_{n=1}^{N} g_{kn} \left\| \mathbf{x}_n - \mathbf{f}_k \right\|^2 = \| X - FG \|_F^2 , \tag{12}$$

where

$$X \in \mathbb{R}^{M \times N} \text{ is a matrix of vectors } \mathbf{x}_n \in \mathbb{R}^{M \times 1} \tag{13}$$

$$F \in \mathbb{R}^{M \times K} \text{ is a matrix of cluster centroids } \mathbf{f}_k \in \mathbb{R}^{M \times 1} \tag{14}$$

$$G \in \mathbb{R}^{K \times N} \text{is a matrix of binary indicator variables such that :} \tag{15}$$

$$g_{kn} = \begin{cases} 1 \ , & \text{if } \mathbf{x}_n \in C_k \\ 0 \ , & \text{otherwise.} \end{cases} \tag{16}$$

For the functional of NMF decomposition defined in Eq. (8) by definition we have:

$$\begin{aligned} \| X - FG \|_F^2 &= \sum_{k=1}^{K} \left(\sum_{n=1}^{N} g_{nk} \left\| \mathbf{x}_n - \mathbf{f}_k \right\| \right)^2 \\ &= \sum_{k=1}^{K} \left(\sum_{n=1}^{N} g_{kn}^2 \left\| \mathbf{x}_n - \mathbf{f}_k \right\|^2 + \sum_{n' \neq n} H_{nn'} \right) \\ &= \sum_{k=1}^{K} \left(\sum_{n=1}^{N} g_{kn}^2 \left\| \mathbf{x}_n - \mathbf{f}_k \right\|^2 \right), \end{aligned} \tag{17}$$

where we have used:

$$H_{nn'} = g_{kn} g_{kn'} (\mathbf{x}_n - \mathbf{f}_k)^T (\mathbf{x}_{n'} - \mathbf{f}_k) = 0, \tag{18}$$

since G is an orthogonal matrix $GG^T = I$. Ding et al. [5] showed that (18) are close to zero and therefore $\mathcal{L}(X, F, G)$ is well approximated by $\mathcal{L}_K$ in the following sense:

$$\left| \mathcal{L}(X, F, G) - \sum_{n=1}^{N} \sum_{k=1}^{K} \mathbf{g}_{nk}^2 \, \|\mathbf{x}_n - \mathbf{f}_k\|^2 \right| \leq \mathcal{L}_K. \tag{19}$$

According to [10], this formulation of the K-means objective function reveals some interesting characteristics of the matrices F and G.

These intriguing features present a wide variety of options; in fact, taking these limitations on G into account allows us to build a variety of NMF algorithm variations.

Remember that K-means method is one of the most commonly used clustering approaches, having been pioneered by Lloyd [17] in the early eighties.

According to [16], the K-means method can be considered of as an approximation problem, with the clustering goal being to minimize the approximation error between the original data X and the newly generated matrix based on the cluster structures.

3 Multi-view Setting

The NMF enables the projection of enormous datasets into small regions; while it is typically two-dimensional, it is increasingly being utilized for unsupervised grouping and display of multidimensional datasets. In this part, we examine the information flow between finite clustering results achieved using an NMF model and those obtained in a multi-view context. Each dataset is grouped using a method based on NMFs. We are interested in the multi-view clustering technique because it allows for the comparison of data that are same but characterized differently by distinct factors. Here, all the distributed views share the same units but are described differently. In this situation, the number and size of centroid vectors will be same for all NMF factorizations.

As previously stated, let $X = (\mathbf{x}1, \mathbf{x}2, \ldots, \mathbf{x}N) \in \mathbb{R}+^{M \times N}$, be a dataset containing only non-negative values. In the case of a multi-view framework, we make the assumption that there is a finite number of V views. $\mathcal{A}^{(1)}, \ldots, \mathcal{A}^{(v)}$ that each focus on a distinct attribute of a data set consisting of $Mv \leq M$ numerical variables.

We notice that$X^{(v)} = \{\mathbf{x}1^{(v)}, \ldots, \mathbf{x}N^{(v)}\}, \mathbf{x}n^{(v)} \subseteq \mathbf{x}n$ represents the subset of attributes treated by a particular view $\mathcal{A}^{(v)}$. We are dealing with a discrete situation, in which each observation $\mathbf{x}n^{(v)}$ in $\mathbb{R}^{Mv \times 1}$. Locally, each view $\mathcal{A}^{(v)}$ employs a traditional NMF. As a result, the appropriate expression for the $\mathcal{A}^{(v)}$ approach is defined as follows:

$$\mathcal{L}_v(X^{(v)}, F^{(v)}, G^{(v)}) = \|X^{(v)} - F^{(v)} G^{(v)}\|^2 \tag{20}$$

Notice that the underscript v denotes the functional's view dependency. Here $F^{(v)} = (\mathbf{f}_1^{(v)}, \mathbf{f}_2^{(v)}, \ldots, \mathbf{f}_K^{(v)}) \in \mathbb{R}_+^{M_v \times K}$ indicates the matrix of cluster centroids and $G^{(v)} = (\mathbf{g}_1^{(v)}, \mathbf{g}_2^{(v)}, \ldots, \mathbf{g}_K^{(v)}) \in \mathbb{R}_+^{K \times N}$ indicates the matrix clustering indicators produced using $\mathcal{A}^{(v)}$.

The concept behind the multi-view technique is to include a matching term that constrains the similarity of clustering elements across solutions. Adding this information to a fixed view v from multiple sources is a well-known collaboration technique (see Pedrycz02). This exchange of clustering information has clearly demonstrated its potential to provide more accurate clustering results. To transmit the obtained local knowledge, all local clustering results must have similar structures, much as collaborative process requires [20].

From then, when a data is supplied from view v, the multi-view NMF algorithm optimizes the distance from that data and the corresponding prototypes of local NMF views $v' \neq v$ in order to include the knowledge from view v'.

This is a technique that requires collaboration. If an observation from the $X^{(v)}$th data set is mapped into the $k()$th centroid of the vth NMF view, it is supposed that after the collaboration, the same data utilized as an observation from the $X^{(v')}$th data set is mapped onto the same unit k centroid of the vth NMF view.

To do this, we will introduce the matrix $D(v)D^{(v)}$ of the euclidean distances between each data point X and the set of centroids $F^{(v)}$ in the lth local model, more precisely, we will get $Dkn^{(v)} = \|\mathbf{x}n^{(v)} - k^{(v)}\|$.

As a result, we establish the pairwise collaborative $\mathcal{C}(v, v')$ term between the lth and l'th NMF's as follows:

$$\mathcal{C}(v, v')(F^{(v)}, G^{(v)}) = \|(G^{(v)} - G^{(v')}) \circ D^{(v)}\|_F^2. \tag{21}$$

In other terms, the $\mathcal{C}(v, v')$ element is the weighted sum of the Euclidean distances between the observation $xn^{(v)}$ and all centroids in $F^{(v)}$, where $G^{(v)} - G^{(v')}$ represents the weight. Recall that when $G(v)G^{(v)}$ and $G(v')G^{(v')}$ agree, the collaborative factor is equal to zero and the optimization considers only the $lthl$th local NMF.

Additionally, because $\mathcal{C}(v, v') \geq 0$ by construction and $D^{(v)} \geq 0$ by definition, $\mathcal{C}(v, v')(F^{(v)}, G^{(v)}) = 0$ if and only if $G^{(v)} - G^{(v')} = 0$, i.e. the data $\mathbf{x}_n^{(v)}$ in the lth data set is affected to the same cluster.

Thus, the set of centroids $F^{(v)}$ and the set of matrix partitions $G^{(v)}$ are estimated iteratively and alternatively by minimizing the global objective functional:

$$\mathcal{J}(F, G) = \sum_{v=1}^{V} \left(\mathcal{L}_v(F^{(v)}, G^{(v)}) + \sum_{v' \neq v} \beta_{v,v'} \cdot \mathcal{C}(v, v')(F^{(v)}, G^{(v)}) \right) \tag{22}$$

We use the notations $F = (F^{(v)})_v$, $G = (G^{(v)})_v$, $v = 1, \ldots, V$.Here $\mathcal{L}_v$ is the vth local expression introduced in Eq. (20), $\beta = (\beta_{v,v'})_{v,v'}$ is indeed the degree of the collaboration with the respect to the constraint $\sum_{v' \neq l} \beta_{v,v'} = 1$. The traditional NMF's goal function has been changed to allow for the interchange of information across different factorizations. This fact is stated by the constant $Cv, v'(F)$, which ensures that the structure of previous datasets corresponds to the structure of the dataset.

Recall that the cost function described in Eq. (22) is first differentiable in all variables continuously.

As a result, a minimum exists at all times and may be discovered using nonlinear programming. The global functional minimization should adhere to the following constraints: $G^{(v)}G^{(v)} = I$ for all $v \in \{1, \dots, V\}$ as $G^{(v)}$ represents the obtained partition result.

We utilize the gradient descent technique to minimize the cost function in Eq. (22) Lee and Seung suggested using data-adaptive learning rates in 2001 [16].

For the parameter $\Theta \in \{F, G\}$ and the global functional $\mathcal{J}(F, G)$ the update formula in gradient descent algorithm writes:

$$\Theta = \Theta - \eta_\Theta \circ \nabla_\Theta(\mathcal{J}(F, G)) \tag{23}$$

Notice that both parameters $\Theta \in \{F, G\}$ must meet the condition $\Theta \geq 0$. For all matrices $F = (F^{(v)})_v$, $G = (G^{(v)})_v$, $v = 1, \dots, V$, these restrictions should be evaluated component-wise.

All elements in the formulation of the learning rate $\eta\Theta$ for $\Theta \in \{F, G\}$ and all matrices $F = (F^{(v)})v$, $G = (G^{(v)})v$, $v = 1, \dots, V$ are positive in the gradient descent strategy defined in Eq. (23).We notice the presence of the subtraction Eq. (23), since this may add some negative components. In this situation, the requirement on non-negativity is violated. To overcome this, we follow Lee and Seung's approach [16] and design data-adaptive learning rates for the parameter $\Theta \in \{F, G\}$ and the functional $\mathcal{J}$ as follows:

$$\eta_\Theta = \frac{\Theta}{[\nabla_\Theta \mathcal{J}]_+}. \tag{24}$$

The multiplicative update rule for each partition matrix G writes:

$$G = G \circ \frac{[\nabla_G \mathcal{J}]_-}{[\nabla_G \mathcal{J}]_+} \tag{25}$$

The multiplicative update rule for each F, G writes:

$$F = F \circ \frac{[\nabla_F \mathcal{J}]_-}{[\nabla_F \mathcal{J}]_+} \tag{26}$$

The developed form of these equations are given in the Appendix in Eqs. (45) and (44). In these update rules, the numerator contains all of the gradient's negative terms, whilst the denominator contains all of the gradient's positive terms. Here, $\circ$ and the fraction line are used to represent matrix multiplication and division on an element-by-element basis for all matrices $F = (F^{(v)})_v$, $G = (G^{(v)})_v$, $v = 1, \dots, V$.

3.1 Optimized Weights for the Collaborative NMF to Improve Stability

We demonstrated in this part that collaborating with several NMF improves the conventional reconstruction error. In other words, by working on many NMF and then achieving consensus on these collaborations, we may minimize the standard error associated with NMF solutions.

We propose a novel collaborative approach in which we modify the objective function via a weighting strategy in order to mitigate the risk of negative collaboration: we investigate how optimizing the weights of a combination of functions of the type (22) can result in an optimal value for the global function and mitigate the risk of negative collaboration by further optimizing the weight fac.

Since $\beta_{v,v'} \geq 0$, let us consider the weight of the collaboration $\beta_{v,v'} = \tau^2_{v,v'}$ in what follows. We are mainly interested in finding the positive weights $\tau^2_{v,v'}$ that will determine the strength of the collaborative term. The minimization of the collaborative functional is based on the dual form of the problem. By using the Karush-Kuhn-Tucker conditions (KKT) [15] and assuming that the weights $\tau_{v,v'}$ respect also the condition $\forall v \quad \sum_{v' \neq l}^{V} \tau^2_{v,v'} = 1$. The results of the optimization under the Karush-Kuhn-Tucker conditions are shown below in Eq. (27). For all $v' \neq l$ we have:

$$\beta_{v,v'} = \frac{|C_{v,v'}|^2}{\left(\sum_{v''=1}^{V} |C_{v'',v'}|^2\right)} \tag{27}$$

We conclude that the weight factorization factor $\beta v,\, v'$ is proportional to the value of the collaboration term $|Cv,\, v'|$ normalized by the other collaboration terms by construction.

The following interpretation is offered: in the situation of horizontal cooperation, the global results should be improved if each individual algorithm provides a greater weight to algorithms that have the most comparable answers to the local solution (high $\tau^2 v,\, v'$ value for a particular NMF model v).

Otherwise, we find that the degree to which one algorithm should work with other collaborators who have answers that are different is proportional to the degree of normalization.

According to Eq. (28), each algorithm would cooperate most frequently with algorithms that have the most comparable answers. If many algorithms provide the identical most comparable results, they are assigned equal weight. The algorithms with the most comparable solutions would still be preferred for optimizing the global collaborative framework's cost function. However, solutions to algorithms with a lower degree of similarity would be considered locally.

Let us examine the mathematics used to solve the Karush-Kuhn-Tucker optimization problem in further depth. Given the constraint $Cv,\, v' \geq 0$, we are attempting to optimize the matrix $\{\tau v, v'\} V \times V$ as illustrated in the system below:

$$\begin{cases} \text{argmin}_{\tau} \sum_{v=1}^{V} \left(\mathcal{L}(F^{(v)}, G^{(v)}) + \sum_{v' \neq l} \tau_{v,v'}^2 \cdot C_{v,v'}(F^{(v)}, G^{(v)}) \right) \\ \forall v \quad \sum_{v' \neq v}^{V} (\tau_{v,v'})^2 = 1 \text{ and } \forall (v, v') \quad \tau_{v,v'} \geq 0. \end{cases} \tag{28}$$

Using Lagrange multipliers, we obtain the following KKT conditions from the prior system:

$$\forall (v, v'), l \neq v' \begin{cases} (a) & \tau_{v,v'} \geq 0 \\ (b) & \sum_{v' \neq l}^{V} (\tau_{v,v'})^2 = 1 \\ (c) & \lambda_{v,v'} \geq 0 \\ (d) & \tau_{v,v'} \cdot \lambda_{v,v'} = 0 \\ (e) & \nu_{v'} \cdot \left(2 \cdot (\tau_{v,v'})\right) - C_{v,v'} - \lambda_{v,v'} = 0 \end{cases} \tag{29}$$

Consider the following scenario: $\lambda_{v,v'} \neq 0$ in (29)(d). Then, we would have $\tau_{v,v'} = 0$ and with (29)(e): $C_{v,v'} = -\lambda_{v,v'} \leq 0$. Due to the fact that $C_{v,v'}$ has been specified as non-negative, this situation is not conceivable; hence, we shall analyze only the case $\tau_{v,v'} \neq 0$ and $\lambda_{v,v'} = 0$. Then, with (29)(e), we obtain:

$$C_{v,v'} = 2 \cdot \tau_{v,v'} \cdot \nu_{v'} \tag{30}$$

We obtain the following by tacking the square in Eq. (30):

$$C_{v,v'}^2 = 4 \cdot \tau_{v,v'}^2 \cdot \nu_{v'}^2. \tag{31}$$

By adding over v the Eq. (31) we get:

$$\sum_{v'' \neq v'} (C_{v'',v'})^2 = 4\nu_{v'}^2 \sum_{l'' \neq v'}^{V} (\tau_{l'',v'})^2 = 4\nu_{v'}^2, \tag{32}$$

in the last equality we have used Eq. (29)(b). By using $\nu_{v'}$ into Eq. (31) we get:

$$\tau_{v,v'}^2 = \frac{|C_{v,v'}|^2}{\left(\sum_{v''=1}^{V} |C_{v'',v'}|^2 \right)} \tag{33}$$

And thus we have proved the results of Eq. (27). After running V collaborative NMF models, the second algorithm $C + C - NMF$ (see Algorithm 4) establishes a collaboration between these V solutions, which is proceeded by a consensus between these V collaborative solutions.

The second algorithm is divided into three steps: Local is a stage that we have already discussed in greater detail. To obtain Vth collaborative solutions, we do the Collaborative step, in which we consider the V models with our suggested beta-optimization by utilizing Eq. (22) to obtain Vth collaborative solutions.

Consensus process is a phase in which we compute a collaborative agreement solution from the vth collaborative solutions in a collaborative solution set.

Algorithme 4: Multi-View NMF

Convex-initialization: Randomly set the cluster prototypes and P number of views.
For all realizations
Local phase:
forall the *methods* $\mathcal{A}^{(v)}$ **do**
| Optimiza the NMF cost function (20).
end
Collaboration phase:
Compute the optimized β **with Equation (27)**
forall the *methods* $\mathcal{A}^{(v)}$ **do**
| Estimate the partitions matrix of all views (25).
| Estimate the prototypes matrix of all views (26).
end

4 Experimental Results

To assess our proposed collaborative strategy, we employ a variety of datasets of varying size and complexity. To demonstrate the presented approach's premise, further details on the waveform dataset will be provided. The following datasets were used: waveform, SpamBase, NG5, Classic30, and Classic 300. We utilized the silhouette index and accuracy as criteria for validating our method, since the label is available for all datasets.

4.1 Purity Validation Method

The clustering result's purity (accuracy) is equivalent to the expected purity of all clusters.

A high degree of purity indicates a successful clustering process.

The purity of a cluster is defined as the proportion of data in the majority class.

Assuming that the data labels $L = l1, l2, \ldots, l|L|$ and the centroids $C = c1, c2, \ldots, c|C|$ are known, the following equation describes the purity of a clustering algorithm:

$$purity = \sum_{k=1}^{|C|} \frac{max_{i=1}^{|L|} |c_{ik}|}{|c_k|} \quad (34)$$

where $|c_k|$ represents the total amount of data associated with the cluster c_k, and $|c_{ik}|$ represents the total number of observations of class l_i associated with the cluster c_k.

4.2 Silhouette Validation Method

The Silhouette index (Rousseeuw, 1987 [23]) estimates the average silhouette width for each sample, cluster, and overall data. Using this method, each partition may be evaluated by a silhouette depending on tightness and separation. The average silhouette can be used to quantify the quality of the obtained clustering result. This index can also be an indicator for the number of clusters.

The silhouettes $S(i)$ are created using the formula:

$$S(i) = \frac{(b(i) - a(i))}{max\,\{a(i), b(i)\}} \quad (35)$$

where $a(i)$ represents the average dissimilarity of the ith item to all others in the same cluster, and $b(i)$ represents the average dissimilarity of the ith object to all other clusters (in the closest cluster).

The formula states for $-1 \leq s(i) \leq 1$. If the silhouette is close to 1, the corresponding data belongs to a good cluster (well clustered). If the silhouette is close to zero, the sample is equally far from both closest clusters. If the silhouette value is close to -1, the sample is misclassified and is between the clusters. An object's $S(i)$ is the average silhouette width for the whole dataset.

Basically, the optimal clustering is indicated by the biggest (number of clusters). The ideal number of clusters is the one with the highest overall average silhouette width.

4.3 Datasets

- *Waveform*—This dataset has 5000 occurrences classified into three categories. The initial base consisted of 40 variables, 19 of which are all noise characteristics with a mean of zero and a variation of one. Each class is created by combining two or three "basic" waves.
- *Spam Base*—This dataset contains 4601 observations that are characterized by 57 variables. Each attribute denoted an e-mail and its classification as spam or not-spam. The majority of the characteristics show how frequently a specific word

or character appeared in the e-mail. The run-length properties (55–57) specify the length of successive capital letter sequences.
- *Classic30 and Classic300*—The datasets were taken from Classic3, which originally had three classes designated Medline, Cisi, and Cranfield. Classic30 is a collection of 30 randomly chosen papers each explained in 1000 words, whereas Classic300 is a collection of 300 randomly chosen documents each detailed in 3625 words.
- *20-Newsgroup*—NG5 This dataset represents a subset of the 20-Newsgroup data and consists of 500 papers, each with a description of 2000 words, relating to politics field.

4.4 Illustration of the Method on the Waveform Dataset

In order to make the interpretation of the collaboration principle more straightforward, we will use the situation of a collaboration between two views.

We separated the 5000×40 basic waveform dataset into two datasets: the first is 5000×20 and contains all important variables; the second is 5000×20 and contains the noisy subset of variables.

Evidently, the first twenty variables in the waveform dataset relate to pertinent variables, whereas the latter twenty variables belong to noise variables.

We utilize these two sets of data to illustrate the entire process of horizontally collaboration.

The Fig. 2 shows the data projected in two dimensions using PCA (Principal Component Analysis), with the color shows the associated cluster for each data.

The original dataset is visualized in Fig. 3(a) using the true label.

The clustering shown in Fig. 3(b) was achieved using the NMF prior to collaboration on a subset including only pertinent features. The accuracy index achieved is 74.3%.

Following that, we used the suggested algorithm's second phase (the collaboration principle) to share clustering knowledge across all NMF clustering results without acceding to the the source data.

The Fig. 4(a), (b) demonstrate an instance of collaboration between the subsets of 1st and 2nd data (pertinent vs noisy). Following the collaboration of the first view with the noisy NMF2, the purity score dropped to 67.57% percent, owing to the fact that the NMF1 results (74.3%) incorporated data from a noisy clustering result (NMF2) with a small accuracy score (35.6%).

On the other hand, by reversing the collaboration phase, the purity index of the $NMF2 \rightarrow 1$ clustering rose to 58.63% as a consequence of collaboration with the pertinent NMF1 partition result.

Additionally, we calculated the Silhouette score to assess the data's clustering structure, obtaining a Silhouette score of 0.39 for the pertinent view and 0.042 for the second noisy view.

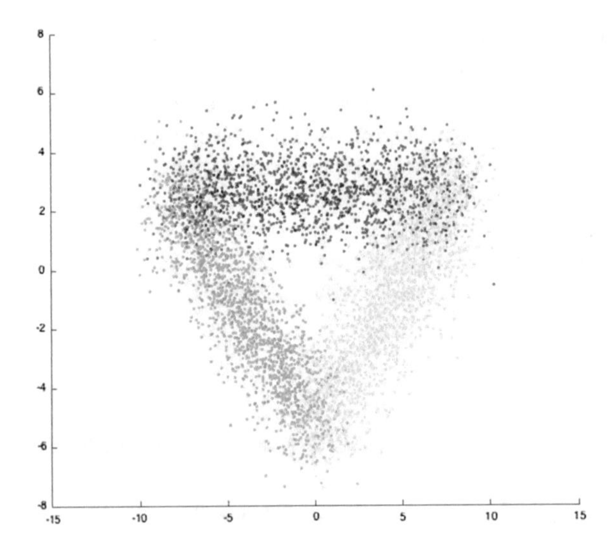

Fig. 2 Waveform data set

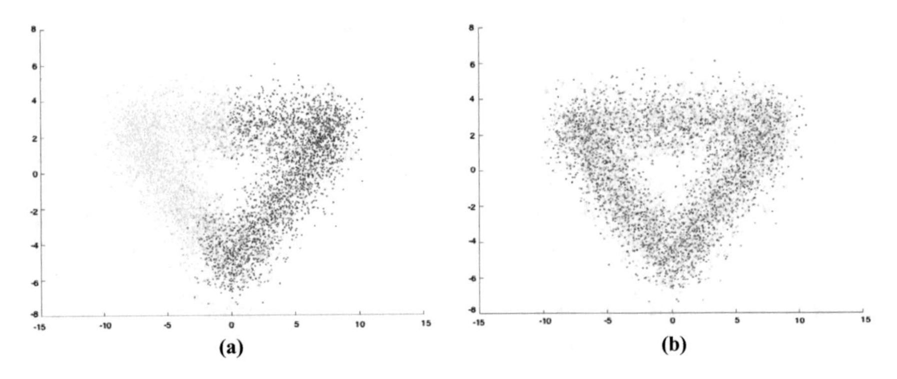

Fig. 3 Scatter of clustering results on relevant and noisy views prior collaboration

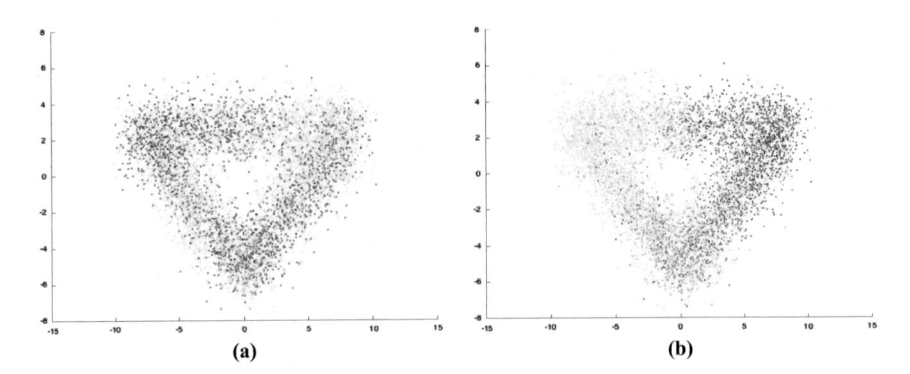

Fig. 4 Scatter of clustering results on relevant and noisy views after the collaboration

The Silhouette index increased to 0.058 following the collaboration $NMF_{1\to 2}$, and decreased to 0.26 following the collaboration $NMF_{2\to 1}$.

These results indicate that while collaboration improves the external quality of clustering (accuracy index), it has little effect on the internal quality index (Silhouette index), as collaboration considers the distant partition but does not alter the local structure of the dataset used by the internal indexes.

Collaboration between a noisy subset and a relevant subset results in an improvement in the quality of the noisy subset (the purity index).

4.5 Validation Using Additional Datasets

In this part, we used the NMF_{Collab} function between two views of distinct datasets and calculated the clustering accuracy before and after collaboration.

As shown in Table 1, following collaboration, the accuracy index increases in all situations where the distant collaborator has a stronger segmentation, but decreases in all cases where the distant dataset has the lowest clustering result.

Additionally, we can see that the collaboration has a negligible effect on the Silhouette index for all datasets, as the dataset structure remains same.

Table 1 Summary of the results of horizontal collaboration method on several datasets

Dataset	NMF	Horizontal collaboration	
		Purity	Silhouette score
SpamBase	NMF_1	81.20	0.46
	NMF_2	95.12	0.59
	$NMF_{1\to 2}$	81.39	0.48
	$NMF_{2\to 1}$	96.06	0.57
NG5	NMF_1	60.88	0.34
	NMF_2	62.64	0.39
	$NMF_{1\to 2}$	61.01	0.35
	$NMF_{2\to 1}$	63.57	0.38
Classic30	NMF_1	83.4	0.74
	NMF_2	82.52	0.68
	$NMF_{1\to 2}$	83.1	0.68
	$NMF_{2\to 1}$	82.86	0.72
Classic300	NMF_1	93.5	0.78
	NMF_2	94.1	0.81
	$NMF_{1\to 2}$	93.91	0.8
	$NMF_{2\to 1}$	93.38	0.79

5 Conclusions

We presented a novel technique in this study by collaborating with various Non-negative Matrix Factorization ($NMFCollab$). This collaboration allows different NMF to interact and disclose (identify) underlying structures and patterns in datasets.

We began by introducing the standard NMF technique and demonstrating that it is equal to k-means. Following that, we detailed our suggested horizontal collaboration technique, which is well-suited for collaboration between views that include the same objects but with different attributes.

The proposed method has been validated against a variety of datasets, and the experimental results indicate that it performs classical clustering NMF.

As part of our future work, we plan to develop a vertical collaborative method, which we will then merge with a horizontal one to create a hybrid approach.

6 Appendix

6.1 Gradients Computations

Notice that for all $v = 1, \dots V$ we have by construction:

$$[\nabla_G \mathcal{C}(F, G)]_\pm = \sum_{v=1}^{V} [\nabla_{G^{(v)}} \mathcal{C}(v, v')]_\pm \text{ and } [\nabla_F \mathcal{C}(F, G)]_\pm = \sum_{v=1}^{V} [\nabla_{F^{(v)}} \mathcal{C}(v, v')]_\pm. \tag{36}$$

By definition we have:

$$\begin{cases} \nabla_{F^{(v)}} \mathcal{C}(v, v') = \nabla_{F^{(v)}} \mathcal{L}(X^{(v)}, F^{(v)}, G^{(v)}) + \sum_{v' \neq l} \beta_{v,v'} \cdot \nabla_{F^{(v)}} \mathcal{C}(v, v')(F^{(v)}, G^{(v)}) \\ \nabla_{G^{(v)}} \mathcal{C}(v, v') = \nabla_{G^{(v)}} \mathcal{L}(X^{(v)}, F^{(v)}, G^{(v)}) + \sum_{v' \neq l} \beta_{v,v'} \cdot \nabla_{G^{(v)}} \mathcal{C}(v, v')(F^{(v)}, G^{(v)}) \end{cases} \tag{37}$$

We first recall the gradients of the local term.

$$\nabla_{F^{(v)}} (\|X^{(v)} - F^{(v)} G^{(v)}\|_F^2) = -2X^{(v)} (G^{(v)})^T + 2\, F^{(v)} G^{(v)} (G^{(v)})^T, \tag{38}$$

and that

$$\nabla_{G^{(v)}} (\|X^{(v)} - F^{(v)} G^{(v)}\|_F^2) = -2(F^{(v)})^T X^{(v)} + 2(F^{(v)})^T F^{(v)} G^{(v)}. \tag{39}$$

It remains to compute the gradient of $\mathcal{C}(v, v')(F^{(v)}, G^{(v)})$. Note that the gradient of the collaborative functional $\mathcal{C}_{l,l'}(F^{(v)}, G^{(v)})$ with respect to $F^{(v)}$ writes as follows:

$$\begin{aligned} \nabla_{F^{(v)}} \mathcal{C}_{l,l'}(F^{(v)}, G^{(v)}) = & -2X^{(v)} [(G^{(v)} - G^{(v')})^{\circ 2}]^T \\ & + 2\mathbf{1}_{M \times N} [(G^{(v)} - G^{(v')})^{\circ 2}]^T \circ F^{(v)}, \end{aligned} \tag{40}$$

and

$$\nabla_{G^{(v)}}\mathcal{J}(v) = -2(F^{(v)})^T X^{(v)} + 2(F^{(v)})^T F^{(v)} G^{(v)} \\ - 2\sum_{v' \neq v} \beta_{l,l'}[G^{(v)} - G^{(v')} \circ (D^{(v)})^{\circ 2}]. \quad (41)$$

By adding all these partial gradient computations and by separating the positive and negative arts of them we define the data-adaptative learning rates:

$$\eta_{F^{(v)}} = \frac{F^{(v)}}{F^{(v)}G^{(v)}\left((G^{(v)}\right)^T + \sum_{v' \neq v} \beta_{l,l'}\mathbf{1}_{M\times N}\left([(G^{(v)} - G^{(v')})^{\circ 2}\right]^T \circ F^{(v)}}, \quad (42)$$

and

$$\eta_{G^{(v)}} = \frac{G^{(v)}}{\left(F^{(v)}\right)^T F^{(v)}G^{(v)} + \sum_{v' \neq v} \beta_{l,l'}G^{(v')} \circ \left(D^{(v)}\right)^{\circ 2}}. \quad (43)$$

By using the recipe described by the Eqs. (26) and (25), we obtain the multiplicative rules:

$$F^{(v)} = F^{(v)} \circ \frac{X\left(G^{(v)}\right)^T + \sum_{v' \neq v} \beta_{l,l'} X\left[(G^{(v)} - G^{(v')})^{\circ 2}\right]^T}{F^{(v)}G^{(v)}\left(G^{(v)}\right)^T + \sum_{v' \neq v} \beta_{l,l'}\mathbf{1}_{M\times N}\left[(G^{(v)} - G^{(v')})^{\circ 2}\right]^T \circ F^{(v)}}, \quad (44)$$

and

$$G^{(v)} = G^{(v)} \circ \frac{\left(F^{(v)}\right)^T X + \sum_{v' \neq v} \beta_{l,l'}\left[G^{(v')} \circ (D^{(v)})^{\circ 2}\right]}{\left(F^{(v)}\right)^T F^{(v)}G^{(v)} + \sum_{v' \neq v} \beta_{l,l'}G^{(v)} \circ \left(D^{(v)}\right)^{\circ 2}}. \quad (45)$$

References

1. Andrzej, C., Rafal, Z., Anh Huy, P., Shun-ichi, A.: Nonnegative Matrix and Tensor Factorizations: Applications to Exploratory Multi-way Data Analysis and Blind Source Separation. Wiley (2009)
2. Coletta, L.F.S., Vendramin, L., Hruschka, E.R., Campello, R.J.G.B., Pedrycz, W.: Collaborative fuzzy clustering algorithms: some refinements and design guidelines. IEEE Trans. Fuzzy Syst. **20**(3), 444–462 (2012)
3. Cornuéjols, A., Wemmert, C., Gançarski, P., Bennani, Y.: Collaborative clustering: why, when, what and how. Inf. Fusion **39**, 81–95 (2018)
4. Depaire, B., Falcon, R. Vanhoof, K., Wets, G.: Pso driven collaborative clustering: a clustering algorithm for ubiquitous environments. Intell. Data Anal. **15**, 49–68 (2011)
5. Ding, C., He, X., Simon, H.D.: On the equivalence of nonnegative matrix factorization and spectral clustering. In: Proceedings of the 2005 SIAM International Conference on Data Mining, pp. 606–610. SIAM (2005)

6. Forestier, G., Gancarski, P., Wemmert, C.: Collaborative clustering with background knowledge. Data Knowl. Eng. **69**(2), 211–228 (2010)
7. Ghassany, M., Grozavu, N., Bennani, Y.: Collaborative clustering using prototype-based techniques. Int. J. Comput. Intell. Appl. **11**(03), 1250017 (2012)
8. Gillis, N., Glineur, F.: Accelerated multiplicative updates and hierarchical ALS algorithms for nonnegative matrix factorization. Neural Comput. **24**(4), 1085–1105 (2012)
9. Grozavu, N., Bennani, Y.: Topological collaborative clustering. Aust. J. Intell. Inf. Process. Syst. **12**(3) (2010)
10. Grozavu, N., Rogovschi, N., Lazhar, L.: Spectral clustering trought topological learning for large datasets. In: Neural Information Processing - 23rd International Conference, ICONIP, Proceedings, Part III, pp. 119–128 (2016)
11. Grozavu, N., Bennani, Y.: Topological collaborative clustering. Aust. J. Intell. Inf. Process. Syst. **12**(2) (2010)
12. Grozavu, N., Ghassany, M., Bennani, Y.: Learning confidence exchange in collaborative clustering. In: The 2011 International Joint Conference on Neural Networks (IJCNN), pp. 872–879. IEEE (2011)
13. Guillaume, C., Matthieu, E., Lionel, M., Jacques-Henri, S.: Cofkm: a centralized method for multiple-view clustering. In: ICDM 2009, The Ninth IEEE International Conference on Data Mining, Miami, Florida, USA, 6–9 December 2009, pp. 752–757 (2009)
14. Kim, J., Park, H.: Sparse nonnegative matrix factorization for clustering. Technical report, Georgia Institute of Technology (2008)
15. Kuhn, H.W., Tucker, A.W.: Nonlinear programming. In: Proceedings of 2nd Berkeley Symposium, pp. 481–492. Berkeley University of California Press (1951)
16. Lee, D.D., Seung, H.S.: Algorithms for non-negative matrix factorization. In: Advances in Neural Information Processing Systems, pp. 556–562 (2001)
17. LLoyd, S.P.: Least squares quantization in PCM. Special Issue on Quantization. IEEE Trans. Inf. Theory **28**(2), 129–137 (1982)
18. Nawaz, W., Khan, K.-U., Lee, Y.-K., Lee, S.: Intra graph clustering using collaborative similarity measure. In: Distributed and Parallel Databases, pp. 583–603 (2015)
19. Paatero, P., Tapper, U.: Positive matrix factorization: a non-negative factor model with optimal utilization of error estimates of data values. Environmetrics **5**(2), 111–126 (1994)
20. Pedrycz, W.: Collaborative fuzzy clustering. Pattern Recognit. Lett. **23**(14), 1675–1686 (2002)
21. Pedrycz, W.: Interpretation of clusters in the framework of shadowed sets. Pattern Recogn. Lett. **26**(15), 2439–2449 (2005)
22. Pedrycz, W., Hirota, K.: A consensus-driven fuzzy clustering. Pattern Recognit. Lett. **29**(9), 1333–1343 (2008)
23. Rousseeuw, R.J.: Silhouettes: a graphical aid to the interpretation and validation of cluster analysis. J. Comput. Appl. Math. **20**, 53–65 (1987)
24. Strehl, A., Ghosh, J., Cardie, C.: Cluster ensembles - a knowledge reuse framework for combining multiple partitions. J. Mach. Learn. Res. **3**, 583–617 (2002)
25. Xie, Y.-L., Hopke, P.K., Paatero, P.: Positive matrix factorization applied to a curve resolution problem. J. Chemom. **12**(6), 357–364 (1999)

A Graph-Based Multi-view Clustering Approach for Continuous Pattern Mining

Christoffer Åleskog, Vishnu Manasa Devagiri, and Veselka Boeva

Abstract Today's smart monitoring applications need machine learning models and data mining algorithms that are capable of analysing and mining the temporal component of data streams. These models and algorithms also ought to take into account the multi-source nature of the sensor data by being able to conduct multi-view analysis. In this study, we address these challenges by introducing a novel multi-view data stream clustering approach, entitled MST-MVS clustering, that can be applied in different smart monitoring applications for continuous pattern mining and data labelling. This proposed approach is based on the Minimum Spanning Tree (MST) clustering algorithm. This algorithm is applied for parallel building of local clustering models on different views in each chunk of data. The MST-MVS clustering transfers knowledge learnt in the current data chunk to the next chunk in the form of artificial nodes used by the MST clustering algorithm. These artificial nodes are identified by analyzing multi-view patterns extracted at each data chunk in the form of an integrated (global) clustering model. We further show how the extracted patterns can be used for post-labelling of the chunk's data by introducing a dedicated labelling technique, entitled Pattern-labelling. We study and evaluate the MST-MVS clustering algorithm under different experimental scenarios on synthetic and real-world data.

Keywords Data stream · Clustering analysis · Pattern mining · Minimum spanning tree

C. Åleskog · V. M. Devagiri · V. Boeva (✉)
Department of Computer Science, Blekinge Institute of Technology, Karlskrona, Sweden
e-mail: vbx@bth.se

C. Åleskog
e-mail: cck@bth.se

V. M. Devagiri
e-mail: vmd@bth.se

W. Pedrycz and S. Chen (eds.), *Recent Advancements in Multi-View Data Analytics*, Studies in Big Data 106, https://doi.org/10.1007/978-3-030-95239-6_8

1 Introduction

Data collected today in smart monitoring applications have a heterogeneous nature due to the fact that they are originated from multiple sources, more often providing information about different perspectives, views or perceptions of the monitored phenomenon, physical object, system. In addition, in many real-world applications (e.g., learning from sensor data streams) the availability of relevant labelled data is often low or even non-existing. In this context, one challenge is how the newly arriving information can be taken into account in the learning or monitoring phase and how the built model can be used for analysis and pattern mining of the temporal component of data streams. This problem however, cannot be considered and solved individually. It is also necessary to take into account the multi-source nature of the sensor data by developing machine learning (ML) techniques that can distribute model training and evaluation among multiple data sources or views. Some recent works in the data mining field address these challenges [11, 12, 30]. For example, MV Multi-Instance Clustering [11] is a novel multi-view stream clustering algorithm that handles the multi-source nature of today's data by clustering each data view separately, thereby inducing a parallel part to the process. The knowledge from different views is integrated using Formal Concept Analysis (FCA) [37]. MVStream [30], is another multi-view stream clustering approach. It combines the data points from different views by transferring them into a common kernel space and identifying the common support vectors for all views. More related studies are discussed in Sect. 2.3.

Inspired by the above discussed challenges and current need in the area, we propose a novel multi-view data stream clustering approach that can be applied in different smart monitoring applications for continuous pattern mining and data labelling. This approach is based on the Minimum Spanning Tree (MST) clustering algorithm, which is used at each data chunk for parallel building of local clustering models on different data views. Knowledge between chunks is transferred in the form of artificial nodes used by the MST clustering algorithm. The artificial nodes are identified by analyzing the global model that is built at each data chunk by integrating view patterns extracted from the local clustering models. The proposed approach, entitled MST-MVS clustering algorithm, can be considered as a communication-efficient and privacy-preserving solution, since it is not the chunk's entire data that is transferred for building the global model. The latter one consists of integrated multi-view patterns that can be used for post-labelling of the current chunk's data.

Different configurations of the MST-MVS clustering algorithm are studied and evaluated under different experimental scenarios on two types of data: synthetic and real-world. We study two different approaches for identifying artificial nodes that are used to seed the MST algorithm producing local clustering models at the next data chunk. In addition, we investigate how the knowledge transfer affects the performance of the proposed MST-MVS algorithm by comparing it with an algorithm version that does not use artificial nodes to seed the clustering at the next data chunk. We also propose a pattern-based labelling (Pattern-labelling) technique, which is evaluated and benchmarked to an approach relying on Convex Non-negative Matrix

Factorization (CNMF). Finally, the patterns extracted by the proposed algorithm from the sensor dataset used in our experiments are benchmarked to the results produced by other state-of-the-art algorithm, MV Multi-Instance Clustering, on the same dataset.

The evaluation of the MST-MVS clustering algorithm demonstrates a higher performance on the synthetic data than on the real-world data. This result is logical, since in comparison with the synthetic data, real-world data usually does not present a perfect clustering structure, but often has outliers and overlapping clusters. In addition, the labelled real-world dataset used in our experiments is not specifically designed for multi-view scenarios and this might have affected the performance of our multi-view clustering algorithm. The transfer of knowledge feature is shown to have a positive effect on the performance of MST-MVS algorithm. The proposed Pattern-labelling technique outperforms the CNMF-labelling algorithm in the conducted experiments both in terms of computational time and results obtained.

The rest of the paper is organised as follows. Section 2 presents the recent work done in the areas of multi-view clustering and stream clustering followed by the background in Sect. 3, which introduces the main concepts and methods used in the study. The proposed MST-MVS clustering algorithm is presented in Sect. 4. Then the data used along with the experimental settings are described in Sect. 5. This is followed by results and discussion in Sect. 6, and conclusion and future work in Sect. 7.

2 Related Work

2.1 Multi-view Clustering Algorithms

Multi-view clustering algorithms have been intensively studied in the recent years and many different algorithms and concepts have been proposed [5, 38, 45, 59]. Most of these algorithms cluster data from different views often into a consensus model while holding all data in memory.

Many research studies have proposed multi-view clustering algorithms based on Non-negative Matrix Factorization (NMF) [3, 38, 42, 58]. For example, Liu et al. combine the objective functions of different views (minimization problems) and add a consensus part to it [38]. They propose a novel way to use the l_1 normalization to solve the problem of multiple views with NMF by imposing a new constraint to the objective function. In [42], the authors propose a co-regularized multi-view NMF approach with correlation constraint for non-negative representation learning. The proposed approach imposes correlation constraint on the low-dimensional space to learn a common latent representation shared by different views.

Peng et al. [45] propose a novel multi-view clustering algorithm without parameter selections, entitled COMIC. The proposed algorithm projects data points into a space where the two properties, geometric consistency and cluster assignment consistency, are satisfied. The data are clustered without any parameters by enforcing a view consensus onto a connecting graph.

Bendechache and Kechadi [5] propose a distributed clustering algorithm using k-means. It clusters each view applying the k-means algorithm and merges overlapping clusters of different views. This is done multiple times with the same data until a specified level has been reached. This level is one of the algorithm parameters, additionally to the number of clusters for each view needed by the k-means algorithm.

Two graph-based multi-view clustering techniques have been recently published [32, 60]. The first study has proposed an affinity graph representation learning framework for modelling multi-view clustering tasks [32]. The approach consists of two separate stages. Namely, a robust graph Laplacian from each view is initially constructed and then those are fused using the proposed Consistent Affinity Graph Learning algorithm. In [60], multi-view subspace clustering networks using local and global graph information are introduced. The autoencoder networks are performed on different views simultaneously. In that way latent representations that conform to the linear subspace model can be achieved. Finally, a representation shared among the views is obtained by integrating the underlying graph information into the self-expressive layer of autoencoder networks.

2.2 *Stream Clustering Algorithms*

Ghesmoune et al. [20] discuss previous works on data stream clustering problems, and highlight the most relevant algorithms proposed in the literature to deal with these problems [1, 2, 8, 34].

Coa et al. [8] propose an iterative clustering algorithm with NMF, entitled ONMF. It uses the property in [13], that states that NMF is equivalent to k-means if the orthogonality constraint $HH^T = I$ is included in the minimization problem. The ONMF algorithm clusters temporally changing data streams, divided into chunks, where each chunk is clustered by applying the orthogonal NMF. Then the results are propagated to the next chunks' calculations to retain the knowledge found in the previous chunks. This process is continued until all data has been clustered.

In [2], the evolution of the data is addressed by dividing the clustering process into two components: online and offline. The online component periodically stores detailed summary statistics, while the offline component generates the final clusters. Similar solutions are proposed in [1, 34].

Wang et al. [57] has proposed an algorithm, entitled SVStream. It is a predecessor of MVStream [30], discussed in Sect. 2.3, and is similar to its successor by applying the Support Vector Clustering (SVC) to cluster the data. The difference in this algorithm when compared with its successor, excluding the multi-view part, is how SVStream merges the spheres created by the current and next data chunks. SVStream updates the kernel space, where the sphere is clustered after each new data chunk. It removes old support vectors that could affect the clustering and merging, and updates spheres accordingly, thereby transferring knowledge of the view's correlation to the next data chunk.

2.3 Multi-view Stream Clustering Algorithms

As it is discussed in the introduction, the research field that applies multi-view clustering analysis on streaming data is still relatively new. The study of Shao et al. published in [51] is seen as one of the first articles that has discussed the idea of sectioning data into different chunks for conducting multi-view clustering. In [36], NMF is used and showed that under certain constraints it can be used for clustering. A detail description can be found in [13]. The proposed version of the NMF algorithm is combined with itself to create a new minimization problem, where different views could be inputted, and a consensus matrix describing the clustering structure could be generated. The views are divided into chunks, and for each new chunk, the algorithm generates a new consensus matrix with the information from the consensus matrix of the previous chunk [13].

A new multi-view stream clustering algorithm, entitled MVStream, has been proposed in [30]. It is similar to the approach of Shao et al. [51], but uses the SVC algorithm [4], adjusted to unsupervised data with the Position Regularized Support Vector Domain Description (PSVDD) [56]. It works by combining and transforming all data points from the different views into a common kernel space (global model) and, from this space finding the common support vectors for all views. These support vectors are transformed back to each view's space, resulting in the contours of different arbitrary clusters. The algorithm also transfers knowledge between chunks by incorporating the previous chunks support vectors as data points in the current chunks views, thereby retaining the view's correlations in the previous chunks.

Devagiri et al. [11, 12] have proposed two of the recent algorithms in the field of multi-view stream clustering. The first algorithm [12], is entitled MV Split-Merge Clustering and is based on a previous work of the authors developing a stream clustering algorithm, called Split-Merge Clustering [7]. The MV Split-Merge Clustering algorithm works by updating the local (views') models in each chunk by applying the Split-Merge Clustering on the local models produced at the previous and current data chunks. Then the updated local models are integrated into a global model by applying Formal Concept Analysis (FCA) [37]. The algorithm proposed in [11], entitled MV Multi-Instance Clustering, is more effective by addressing the limitations of [12]. The MV Multi-Instance Clustering uses Bipartite Multi-Instance clustering algorithm to update the local models which proved to provide better results. In addition, it also uses closed patterns to mine the most frequent patterns or concepts from the formal context which is obtained using FCA. This reduces the size of the lattice generated in the global model making it easy to interpret and analyse.

3 Background

3.1 Minimum Spanning Tree Clustering

The *Cut-Clustering* algorithm is a graph-based clustering algorithm, based on minimum cut tree algorithms to cluster the input data [17]. It has been used in other clustering algorithms, e.g., [21, 49]. The input data is represented by a similarity (an undirected) graph, where each node is a data point and two nodes are connected if the similarity between the corresponding data points is positive, and the edge is weighted by the corresponding similarity score. The algorithm works by adding an artificial node to the existing graph and connecting it to all nodes in the graph with a value α. A minimum spanning tree is computed, and the artificial node is removed. The clusters consist of the nodes connected after the artificial node has been removed. The pseudo-code of Minimum Spanning Tree (MST) clustering algorithm is presented in Algorithm 1.

Algorithm 1 Minimum Spanning Tree Clustering

1: **procedure** MSTCLUSTERING($G(V, E), \alpha$)
2: Let $V' := V \cup t$.
3: **for** $v \in V$ **do**
4: Connect t to v with edge of weight α
5: Let $G'(V', E')$ be the expanded graph after connecting t to V
6: Calculate the minimum spanning tree T' of G'
7: Remove t from T'
8: **return** All connected components as the clusters of G

Lv et al. [40] propose another MST-based clustering algorithm, entitled CciMST. CciMST starts by finding a MST and calculating pairwise Euclidean and geodesic distances of all pairs of data points. The graph and distances are then used for finding the cluster centers based on the density and variances of the data points in the graph. The algorithm ends by determining which edges in the graph should be cut, producing and comparing two results, choosing the one where the clusters have a bigger gap between them.

Many different algorithms for finding a minimum spanning tree (MST) of the complete graph exist. The two well-known greedy algorithms for computing MST are Kruskal's and Prim's algorithms. The algorithm we use in our study is Kruskal's algorithm [35]. It works by sorting all the graph edges in increasing order of weights and choosing the shortest edge. The two nodes connected by the shortest edge are combined to the tree. The process repeats by finding a new shortest edge that does not create a cycle. These steps are continued until $m - 1$ (m is the number of graph nodes) edges are selected, i.e. a MST is found.

The value of α is an input parameter for the Cut-Clustering algorithm and it plays a crucial role in the quality of the produced clusters. Namely, the value of this parameter has an impact on how many clusters the algorithm will produce. It can be observed that as α goes to infinity, the Cut-Clustering algorithm will produce only one cluster, namely the entire graph. On the other extreme, as α goes to 0, the clustering algorithm will produce m trivial clusters (all singletons), where m is the number of graph nodes. For values of α between these two extremes the number of clusters will be between 1 and m, but the exact value depends on the graph structure and the distribution of the weights over the edges. What is important though, is that there is no direct correlation between the number of clusters and α value.

There are different ways to identify the α value. For example, a binary-like search algorithm is used in [17] to determine the optimal value for α. This is done by executing the algorithm for each loop and choosing the most stable α value. In this study, we use the mean α value that is calculated by the formula:

$$\alpha = \frac{1}{m}\sum_{i=1}^{m}\sum_{j\neq i}^{m}\frac{w_{ij}}{deg(v_i)}, \tag{1}$$

where m is the number of nodes in the graph, w_{ij} is the weight of the edge connecting nodes i and j, and $deg(v_i)$ is the degree of node i.

3.2 Non-negative Matrix Factorization

Traditional NMF [36] approximates two non-negative matrices $W \in \mathbb{R}_{+}^{n\times k}$ and $H \in \mathbb{R}_{+}^{k\times m}$ into a non-negative matrix $X \in \mathbb{R}_{+}^{n\times m}$. This results in a minimization problem with the objective:

$$X \approx WH. \tag{2}$$

Formerly used to save space in memory, the algorithm can also be used for clustering according to Ding et al. [13]. It is equivalent to k-means if the constraint $HH^T = I$ is additionally imposed on the objective function. Therefore, when NMF is used for clustering, k is the specified number of clusters to be found, columns of W holds the centroids of the clusters, and the rows with the maximum value in the columns of H describe which cluster a data point belongs to.

Convex Non-negative Matrix Factorization (CNMF) [14] is a modification of the traditional NMF that results in better centroids in W. CNMF approximates a mixed-sign data matrix X into a matrix W and non-negative matrix H. W is also divided into one mixed-sign data matrix S and a non-negative matrix L, where L is the labelling matrix and S is a data matrix. An extensive explanation of S and how it can be used for missing data can be found in [24]. However, since missing data is not considered in this study, $S \equiv X$. Then the factorization is in the following form:

$$F = X_{\pm}L_{+} \text{ and } X_{\pm} \approx FH_{+}. \tag{3}$$

CNMF imposes the constraint $\|L\|_1 = 1$ to lift the scale indeterminacy between L and H so that F_i is precisely a convex combination of the elements in X. The approximation is quantified by the use of a cost function that is constructed by distance measures. Usually the square of the Euclidean distance, also known as the Frobenius norm, is used [43]. The objective function is therefore, as follows:

$$\min_{W,H} \mathbb{L} = \|X - XLH^T\|_F^2 \text{ such that } L \geq 0, H \geq 0, \|L\|_1 = 1, \tag{4}$$

where $X \in \mathbb{R}^{n \times m}$ is the data to be clustered, $L \in \mathbb{R}^{m \times k}$ and $H \in \mathbb{R}^{k \times m}$ are non-negative matrices, n is the dimensionality of the feature space, m is the number of data points, and k is the dimension to reduce to. The symbols $\|.\|_F^2$ and $\|.\|^1$ denotes the Frobenius norm and Manhattan norm [10] of the expression it encases, respectively.

3.3 Cluster Validation Measures

Cluster validation measures have a very important role in cluster analysis by providing means for validation of clustering results. A range of different cluster validation measures are published in the data mining literature [23, 53]. They can be divided into two major categories: *external* and *internal* [33]. External validation measures have the benefit of providing an independent assessment of the clustering quality, since they evaluate the clustering result with respect to a pre-specified structure. Internal validation techniques on the other hand avoid the need for using such additional knowledge. Namely, they base their validation on the same information used to derive the clusters themselves. Internal measures can be split with respect to the specific clustering property they reflect and assess to find an optimal clustering scheme: *compactness*, *separation*, *connectedness*, and *stability* of the cluster partitions.

External validation measures can be of two types: *unary* and *binary* [26]. Some authors consider a third type of external validity approaches, namely *information theory* [29]. Unary external evaluation measures take a single clustering result as the input, and compare it with a known set of class labels to assess the degree of consensus between the two. Comprehensive measures like the F-measure provide a general way to evaluate this [50]. In addition to unary measures, the data-mining literature also provides a number of indices, which assess the consensus between a produced partitioning and the existing one based on the contingency table of the pairwise assignment of data items. Most of these indices are symmetric, and are therefore equally well-suited for the use as binary measures, i.e., for assessing the similarity of two different clustering results. Probably the best known such index is the Rand Index [46], which determines the similarity between two partitions as a function of positive and negative agreements in pairwise cluster assignments. A third class of indices is based on concepts from information theory [9]. Information theoretic indices assess the difference in shared information between two partitions. One of commonly used information theoretic indices is the adjusted mutual information [54].

The clustering algorithms usually do not perform uniformly well under all scenarios. Therefore, it is more reliable to use a few different cluster validation measures in order to be able to reflect various aspects of a partitioning. In this study, we use five different cluster validation measures to evaluate the clustering results generated in our experiments: one binary external measure (Adjusted Rand Index), three external measures related to information theory (Adjusted Mutual Information, Homogenity and Completeness) and one internal cluster validation measure (Silhouette Index). The definitions of the used evaluation measures are given hereafter.

Silhouette Index *Silhouette Index* (SI), introduced in [48], is a cluster validation index that is used to judge the quality of any clustering solution $C = C_1, C_2, \ldots, C_k$. Suppose a_i represents the average distance of object i from the other objects of its assigned cluster, and b_i represents the minimum of the average distances of object i from objects of the other clusters. Subsequently the Silhouette Index of object i can be calculated by:

$$s(i) = \frac{(b_i - a_i)}{\max\{a_i, b_i\}}. \tag{5}$$

The overall Silhouette Index for clustering solution C of m objects is defined as:

$$s(C) = \frac{1}{m} \sum_{i=1}^{m} \frac{(b_i - a_i)}{\max\{a_i, b_i\}}. \tag{6}$$

The values of Silhouette Index vary from -1 to 1, and higher values indicate better clustering results, while a negative value shows that there are wrongly placed data points within the clustering solution. As the Silhouette Index compares the distances from instances to its respective cluster against the distance to the nearest cluster, it assesses the separation and compactness between clusters.

The proposed MST-MVS clustering algorithm applies the SI analysis at each data chunk for assessing the quality of generated local clustering models. Only the clustering models that satisfy some preliminary defined threshold are used for building the integrated matrix of multi-view patterns (see Sect. 4).

Adjusted Rand Index *Adjusted Rand Index* (ARI) is used for evaluating a clustering solution and is adjusted for chance [31]. It works by calculating the Rand Index (RI) [46], i.e., computing the similarity measure between two clustering solutions by counting pairs of samples assigned to the same or different clusters. Essentially, it calculates the number of agreeing pairs divided by the total. Then adjusting it for chance, as shown in Eq. 7. A value of 1 indicates that the two clusterings are identical, while close to 0 shows random labeling independently of the number of clusters and samples.

$$ARI = \frac{\sum_{ij}\binom{n_{ij}}{2} - \left[\sum_i \binom{a_i}{2}\sum_j \binom{b_j}{2}\right]/\binom{n_{ij}}{2}}{\frac{1}{2}\left[\sum_i \binom{a_i}{2} + \sum_j \binom{b_j}{2}\right] - \left[\sum_i \binom{a_i}{2}\sum_j \binom{b_j}{2}\right]/\binom{n_{ij}}{2}}, \quad (7)$$

where $\sum_{ij}\binom{n_{ij}}{2}$ is the RI of samples i and j, a_i is the sum of the pair of samples w.r.t sample i, the same for b_j but for sample j. The minuend of the denominator is the expected RI, while the subtrahend is the maximum RI of samples i and j.

Adjusted Mutual Information *Adjusted Mutual Information* (AMI) score is an adjustment accounting for chance of the Mutual Information (MI) [54]. It is a measure of similarity between two clustering solutions, and the adjustment for it is similar to that of ARI. Replacing RI of samples i and j by MI, and the maximum value of RI by the maximum value of the entropy of the predicted and true labels. The calculation of the expected value is pretty complex, and therefore, it is not provided herein. A detailed description can be found in [54]. AMI scores a clustering solution from 0 to 1, where 1 stands for a perfect match and 0 shows a random partition.

Homogeneity and Completeness *Homogeneity* and *Completeness* are two other external cluster validation metrics, both components of the V-Measure [47]. The latter one is an entropy-based external cluster evaluation measure that introduces these two complementary criteria to capture desirable properties in clustering tasks. Homogeneity is satisfied by a clustering solution if all of its clusters consists of only data points which are members of a single class. Opposite, a clustering solution satisfies Completeness if all the data points that belong to a given class are assigned to the same cluster. Note that increasing the Homogeneity of a clustering solution often results in decreasing its Completeness. The calculations of these concepts are complex, and therefore, are not provided herein. A detailed explanation of how they work can be found in [47].

4 MST-MVS Clustering Algorithm

In this work, we study a streaming scenario, where a particular phenomenon (machine asset, patient, system etc.) is monitored under n different circumstances (views). Let us assume that the data arrives over time in chunks, i.e. a landmark window model according to the classification in [61]. In the landmark window model, the whole data between two landmarks form the current chunk are included in the processing. In our considerations each data chunk (window) t can contain different number of data points. Assume that chunk t contains N_t data points. In addition, in our multi-view context each chunk can be represented by a list of n different data matrices $D_t = \{D_{t1}, D_{t2}, \ldots, D_{tn}\}$, one per view. Each matrix D_{ti} $(i = 1, 2, \ldots, n)$ contains information about the data points in the current chunk t with respect to the corresponding view i. Further assume that $C_t = \{C_{t1}, C_{t2}, \ldots, C_{tn}\}$ is a set of clustering

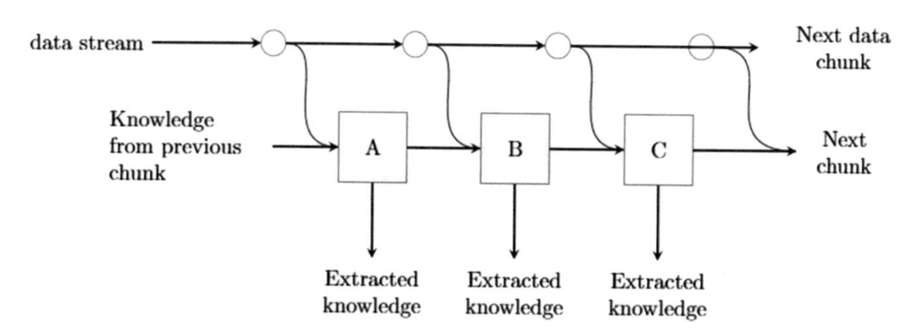

Fig. 1 A high level overview of the proposed MST-MVS clustering algorithm for three data chunks A, B and C. The figure depicts how the knowledge extracted from the previous data chunk is used in analyzing a new data chunk

solutions (local models), such that C_{ti} $(i = 1, 2, \ldots, n)$ represents the grouping of the data points in tth chunk with respect to ith view, i.e. a local model built on data set D_{ti}.

The proposed MST-MVS algorithm's main idea is schematically illustrated in Fig. 1. It has an initialization phase that is performed on the available historical data or on the first data chunk. Then on each subsequent data chunk the algorithm goes through a sequence of six different operational stages. At **Stage 1** the MST clustering algorithm is applied for parallel building of local clustering models on different data views. Then at the **Stage 2** the produced clustering solutions are evaluated and only ones satisfying a predefined evaluation criteria are considered in **Stage 3** for extracting individual view patterns. This makes our algorithm more robust to noisy and low-quality data, since the low-evaluated views' clustering solutions do not contribute to the global clustering model. The extracted patterns are used at the next stage to build an integrated matrix of multi-view profiles (**Stage 4**). In the proposed algorithm knowledge between chunks is transferred in the form of artificial nodes used by the MST clustering algorithm. The artificial nodes are identified at **Stage 6** by analyzing the global model that is built at **Stage 5** by using MST clustering algorithm on the integrated matrix. The six stages are described in more details hereafter and also illustrated by an example in Fig. 2.

The MST-MVS clustering algorithm goes through the following six operation stages (see Fig. 2) for each data chunk t $(t > 1)$:

Input: Newly arrived data D_t, set of artificial nodes A_{t-1} identified at chunk $t-1$, clustering quality threshold Θ_t.

1. **Views' clustering:** Produce a clustering solution C_{ti} on each data matrix D_{ti} $(i = 1, 2, \ldots, n)$ of tth chunk by applying MST clustering algorithm with a parameter $A_{(t-1)i}$, where $A_{(t-1)i}$ are artificial nodes' values corresponding to view i and chunk $t-1$.
2. **Cluster models' evaluation:** At this stage the quality of each produced clustering solution C_{ti} $(i = 1, 2, \ldots, n)$ is evaluated by applying SI analysis, i.e. $s(C_{ti})$

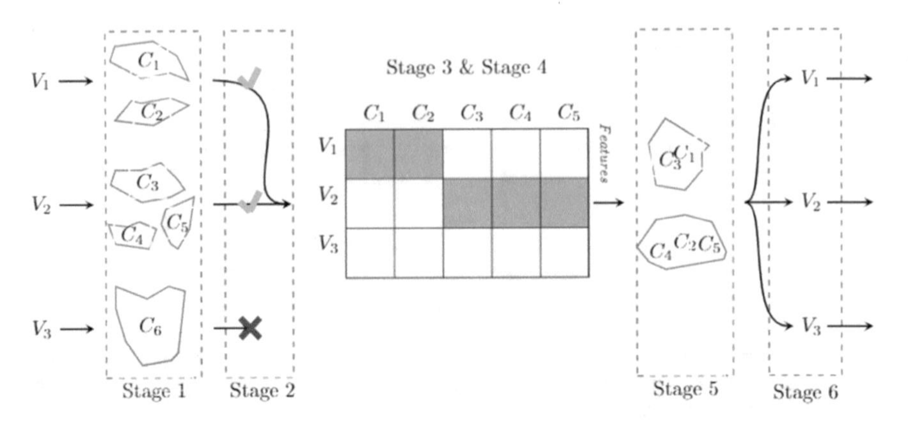

Fig. 2 A schematic illustration of different stages of the proposed MST-MVS clustering algorithm. The example presents a three-view scenario, where V_1, V_2, V_3 represent the views, and C_1, C_2, C_3, C_4, C_5, C_6 represent different clusters across views. Local clustering models in each view are built in Stage 1, followed by evaluating and selecting the local models in Stage 2, these are then used for pattern extraction and building integrated matrix in Stages 3 & 4. A global model is built in Stage 5 followed by identifying artificial nodes in Stage 6

is calculated for each view $i = 1, 2, \dots, n$. A set of selected clustering solutions $C_t^{'} = \{C_{ti} \mid s(C_{ti}) > \Theta_t\}$ $(C_t^{'} \subseteq C_t)$ satisfying the given threshold Θ_t is then built. Note that other internal cluster validation measures or an ensemble of few measures can be used in this stage.

3. **Pattern extraction from the individual views:** Consider only local clustering models which satisfy the given threshold Θ_t, i.e. the set $C_t^{'}$ built at the previous stage. The medoids of each $C_{ti} \in C_t^{'}$ are identified and extracted.
4. **Integrate the individual views' patterns:** The patterns extracted from the local models of $C_t^{'}$ are used at this stage to build an integrated matrix, denoted by M_t, that contains the multi-view profiles of the identified medoids. A detail explanation of the integration stage can be found in Sect. 4.1.
5. **Build a global model:** In order to build a global model the multi-view data points of M_t are clustered by applying MST clustering algorithm. The built global model, denoted by C_t^M, consists of a number of clusters of similar multi-view data points. Each cluster, as seen in Fig. 2, may be interpreted as a multi-view pattern that presents a relationship among individual views' patterns (further information about multi-view pattern extraction can be found in Sect. 4.2). This information, as it will be discussed further in this study, can be used for post-labelling of D_t (see Sect. 4.5).
6. **Identify artificial nodes:** The integrated matrix M_t and the built global model C_t^M are used to identify the set of artificial nodes A_t that will be used to seed the MST clustering algorithm at chunk $t + 1$. Two different techniques for finding the artificial nodes are developed and studied. Their descriptions and pseudo-codes can be found in Sect. 4.3.

Output: The set of multi-view patterns C_t^M, and the set of artificial nodes A_t.

The following sub-sections describe some of the complex steps of the algorithm, followed by an estimation of the computational complexity of the algorithm in Sect. 4.6. Section 4.1 explains the multi-view integration procedure conducted in stage 4 of the MST-MVS algorithm. Sections 4.2 and 4.3 supply with additional details about stages 5 and 6, respectively. Sections 4.4 and 4.5 are devoted to the two labelling (CNMF-based and Pattern-based) algorithms studied in this work and used in the evaluation of the algorithm performance.

4.1 Multi-view Data Integration

The integrated matrix M_t of chunk t ($t = 1, 2, \ldots$) is built using the medoids of the approved views' clustering solutions, i.e. ones in set $C_t^{'}$ built at stage 3 of the MST-MVS algorithm. A medoid is seen as a representative of a cluster and each clustering solution is summarized by its medoids. In that way, the proposed algorithm can be considered as a communication-efficient and privacy-preserving solution, since we do not need to transfer complete data from each view in order to build the global model C_t^M. The medoids of each clustering solution in $C_t^{'}$ are only transferred, i.e., privacy is preserved. The proposed algorithm can also be interpreted as a distributed clustering, since it clusters and evaluates multiple data views (sources) in parallel.

The integrated matrix M_t ($t = 1, 2, \ldots$) is built using the extracted data points from the views, where each column in the matrix is a medoid of a clustering solution in $C_t^{'}$. As one can see in Fig. 2, each medoid takes up one column, and each row presents one feature vector. For example, if C_1 in Fig. 3 is a medoid extracted from view 1 (the gray area in Fig. 3), the rest of the column (view 2 and view 3) contains the remaining features of this data point, namely ones extracted from the other two views. In general, if a medoid is extracted from one view, the data points corresponding to that medoid in the other views are also used to build the multi-view data point, i.e. to fill in the respective column in M_t. Evidently, all features of a medoid (in this case C_1) are placed in the same column. As we can see in Fig. 2 the third view (V_3)

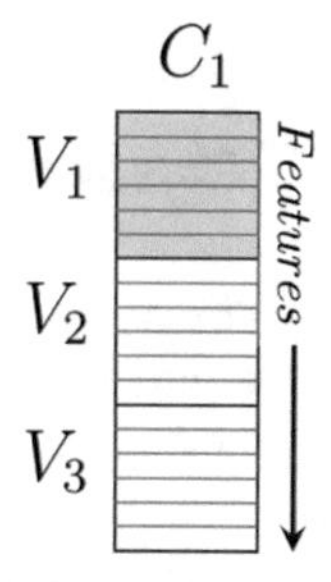

Fig. 3 Example of one column from the integrated matrix, where a multi-view data point (Column C_1 from Fig. 2) is built up by using the features' values from all the views that correspond to that data point

does not supply any medoids, since the view has not passed the evaluation criteria in stage 2.

4.2 Extraction of Multi-view Patterns

Remember that the matrix M_t of chunk t ($t = 1, 2, \ldots$) integrates the medoids of approved views' clustering solutions that are in $C_t^{'}$. This matrix is used in stage 5 of the MST-MVS clustering algorithm to build the global model C_t^M by clustering M_t with MST clustering algorithm. The built global model consists of clusters of similar multi-view data points. Note that each data point in M_t is a medoid in a view's clustering solution belonging to $C_t^{'}$, i.e. it is a pattern extracted from the respective view. Therefore, the clusters of the global model C_t^M present multi-view patterns. The latter ones may be interpreted as most typical relationships among the views supported by the data of the current chunk.

As it is mentioned above, MST clustering algorithm is used for clustering the integrated matrix M_t. However, one problem arises from this approach, namely, what value should α take? Note that the integrated matrix is usually very small, built just from the medoids of the approved views' clustering solutions. This implies that each cluster in the global model would often have only one medoid in it, representing a cluster from one view. As it was stated in Sect. 3.1, mean α values can be chosen. However, this choice depends more on the medoids themselves than the results from each view. Mean α is the average value α can take, producing often wrong numbers of clusters when there is an unbalance of the weights between nodes, focusing on the overall average.

In this work, we propose an alternative of the α value solution, entitled *Longest Edges* (LEdges). LEdges makes use of how the MST algorithm works. Instead of calculating α for an artificial node's edges, it builds a MST on the data points of the integrated matrix M_t and generates the global clustering model C_t^M by cutting the longest edges of the built tree. A detail explanation of LEdges is provided in the forthcoming section.

4.3 Transfer of Knowledge Through Artificial Nodes

At the initialization phase, first (or historical) data chunk of each view is clustered by applying the MST clustering algorithm with α as an input parameter. At each subsequent chunk t ($t > 1$) the integrated matrix M_t and the built global model C_t^M are used to identify data points that are transferred to data chunk $t + 1$. These are used as artificial nodes by the MST clustering algorithm that is applied to build the local clustering models of chunk $t + 1$. This knowledge transferred in the form of artificial nodes influences the $t + 1$ chunk's local clustering models. Particularly, the transferred artificial nodes seed and guide the clustering of the $t + 1$ chunk to be close/similar to the previous chunk's results, but still the clustering solutions can be modified since they are influenced by the data points in the $t + 1$ data chunk. In that way, the knowledge learnt through the processing of current data chunk (t) is used to support the clustering of the next chunk data ($t + 1$).

In the process of algorithm design, two different techniques for finding the artificial nodes are developed and studied: Boundary Nodes (BNodes) and LEdges. The main difference between BNodes and LEdges is the strategy that is applied for calculating the artificial nodes.

BNodes uses the centroids of global clustering model (C_t^M) built at stage 5 of the MST-MVS algorithm to calculate the set of artificial nodes (A_t). An artificial node is placed between two cluster centroids, i.e. it can be interpreted as a boundary node between two clusters. The pseudo-code of BNodes algorithm is shown in Algorithm 2. Note that this strategy of finding artificial nodes does not guarantee the optimal solution for all possible scenarios. For example, it can happen three clusters centroids to be in a straight line. Then one of the resulting artificial nodes will be placed on the middle centroid, splitting the central cluster in two. In practice, the possibility of this scenario to happen is very low.

Algorithm 2 Boundary Nodes

```
1: procedure BNODES(C_t^M)
2:     A_t := ∅.
3:     Pairs_of_centroids := All pairs of centroids of C_t^M
4:     for (u, v) ∈ Pairs_of_centroids do
5:         A_t ← Average(u, v)
6:     return A_t
```

As the name suggests, LEdges uses the longest edges of the MST generated from the integrated matrix (M_t) that is created at stage 4 of the MST-MVS algorithm. The artificial nodes are the middle points of the longest edges. However, this technique requires to know in advance how many longest edges to use. Each data view is already clustered and this information can be used to determine the number of artificial nodes, denoted by k_A. For example, it can be equal to the average number of clusters in the selected views, i.e. $k_A = \sum_{C_{ti} \in C_t'} | C_{ti} | / | C_t' |$, where $| . |$ represents the set cardinality. The LEdges algorithm pseudo-code is given in Algorithm 3.

As one can see in Algorithm 2, BNodes returns the artificial nodes by using the centroids of the global model C_t^M. The latter is built by clustering the integrated matrix M_t. In comparison LEdges, as a variation of MST clustering algorithm, finds not only the artificial nodes, but also clusters the integrated matrix M_t. Hence stages 5 and 6 are merged in the configuration of the MST-MVS algorithm using LEdges technique.

The artificial nodes returned by LEdges and BNodes are split into components corresponding to different views, e.g., the features belonging to view 1 are extracted from the artificial nodes and used in the clustering of view 1's data points in the next chunk. This allows MST clustering algorithm to be seeded with artificial nodes in each view.

Algorithm 3 Longest Edges

1: **procedure** LEDGES(M_t, k_A)
2: $A_t := \varnothing$
3: $removed\_edges := \varnothing$
4: $G(V, E) :=$ complete graph of M_t, i.e. $V \equiv M_t$
5: $MST :=$ Kruskal(G)
6: sorted(E) := sort edges of G in decreasing order
7: **for** ($e \in$ sorted(E)) $\wedge$ ($|$ $removed\_edges$ $| \leq k_A$) **do**
8: remove e from MST
9: $removed\_edges \leftarrow e$
10: **for** $e \in removed\_edges$ **do**
11: $A_t \leftarrow$ Average(u, v)
12: Build global clustering model C_t^M from MST
13: **return** (A_t, C_t^M)

4.4 CNMF-Based Labelling Algorithm

In this work, we propose and study two different techniques for post-labelling of the data points in chunk D_t ($t = 1, 2, \ldots$). The first technique uses CNMF, with fixed X and F matrices (see Eq. 3). CNMF is chosen over the traditional NMF as it allows negative values [14]. Let F consists of the centroids of the global model C_t^M built at stage 5 and $X \equiv D_t$ ($t = 1, 2, \ldots$). Then by applying Eq. 4, the rows with the maximum values of a column in H gives the label of each data point (columns in H).

Note that the CNMF method is slow and is not very accurate. CNMF initiates two matrices L and H with random values, resulting in different approximations for each execution. To rectify this, the CNMF algorithm is executed multiple times, and the instance which gives the lowest Frobenius norm $|X \approx FH_+^T|$ is used. However, this increases the time needed for labelling of the data points. Therefore, we propose in Sect. 4.5 a new algorithm for labelling the data chunk D_t, entitled Pattern-labelling. This labelling technique uses the cluster patterns identified by the global model C_t^M, and maps data points to the patterns they match.

4.5 Pattern-Based Labelling Algorithm

In this study, we propose a new algorithm for post-labelling of the data points in the current data chunk, entitled Pattern-labelling. The proposed algorithm uses the extracted patterns at each data chunk, which are the ones determined by the clusters of the global model. These patterns are mapped to the chunk's data points to identify matches. Each pattern's index (cluster label) is used as the label for all data points that match it.

In order to execute the Pattern-labelling algorithm on data chunk D_t it is necessary to format the global clustering model C_t^M. Remember that each cluster in C_t^M can be considered as a multi-view pattern and all data points in dataset D_t that match to it have to get its cluster label. Therefore each multi-view pattern (cluster) of C_t^M has to be presented as a sequence of its views' clustering labels. The Pattern-labelling algorithm pseudo-code is shown in Algorithm 4. Note that it is also necessary to translate each data point $D_t[i]$ ($i = 1, 2, \dots, N_t$) into a list of its views' clustering labels. In addition, assume that each multi-view pattern is denoted by $C_t^M[j]$, where $j = 1, 2, \dots, \mid C_t^M \mid$.

Algorithm 4 Pattern-labelling algorithm

```
1: procedure PATTERN_LABELLING(C_t^M, D_t)
2:     D_t^labelled := each data point in D_t is initialized to label −1
3:     for i ∈ {1, 2, ..., N_t} do
4:         D_t[i] is translated into a sequence of its views' clustering labels
5:         for j ∈ {1, 2, ..., | C_t^M |} do
6:             C_t^M [j] is presented by a sequence of its views' clustering labels
7:             if Match(D_t[i], C_t^M [j]) ≠ -1 then D_t^labelled [i] := Match(D_t[i], C_t^M [j])
8:     return D_t^labelled
```

If a data point does not match any of the patterns, it is seen as an outlier in the data. The definition of an outlier in this study is a data point that is hard to be grouped in any of the clusters. One view might assign it to one global cluster, and another view assigns the data point to another global cluster. For this reason, the data point will not match any pattern and is considered as an outlier, hence it is assigned the label -1.

4.6 *Computational Complexity*

The computational complexity of the proposed MST-MVS clustering algorithm is not easy to estimate due to its complexity involving different stages, variety of views, and dimensionality of the data. Therefore, we propose a separate estimation of the computational complexity for each stage of the core algorithm.

1. **Views' clustering:** The clustering of the views have a computational time complexity of $O(nN_t^2 + nN_t + nN_t \log N_t)$, where N_t is the number of data points in data chunk t and n is the number of views. The first part, $O(nN_t^2)$, is the identification of the medoids for all views. The second part, $O(nN_t)$, presents the creation of the complete graphs for all views. The third part stands for the complexity of Kruskal's algorithm for all views, i.e. $O(nN_t logN_t)$. The computational complexity of this stage of the algorithm can be approximated to $O(N_t^2)$, since $n << N_t$ and $N_t < N_t \log N_t < N_t^2$.
2. **Cluster models' evaluation:** The SI has a quadratic computational time complexity, which means that stage 2 of the algorithm has a computational time complexity of $O(nN_t^2)$. It can be approximated to $O(N_t^2)$, since $n << N_t$.

3. **Build of the integrated matrix:** This includes stages 3 and 4 of the algorithm. The computational complexity of these stages depend on how the creation of the integrated matrix is implemented. For the implementation used in this study, the computational time complexity is $O(n'n\sum_{i=1}^{n}(n_i \mid C_{ti} \mid))$, where n' is the number of approved views, $\mid C_{ti} \mid$ is the number of clusters in the local model of view i, and n_i is the number of features characterizing view i.
4. **Clustering of the integrated matrix:** The clustering of matrix M_t has different time complexity depending on which of the two methods (BNodes and LEdges) is used.

   – If BNodes is used then the computational time complexity is

   $$O(k_t^{(k_t-2)} + \mid M_t \mid + \mid M_t \mid log \mid M_t \mid)),$$

   where k_t is the number of centroids (clusters) of the global model C_t^M, and $\mid M_t \mid$ is the number of multi-view data points (columns) of matrix M_t. The first part in the above expression ($k_t^{(k_t-2)}$) presents the time complexity of BNodes, $\mid M_t \mid$ is proportional to the time needed for the creation of the complete graph, and the third part ($\mid M_t \mid log \mid M_t \mid$) stands behind the calculations of MST.

   – If LEdges is used then the computational time complexity is

   $$O(\mid M_t \mid + (\mid M_t \mid log \mid M_t \mid) + avg\_k_t),$$

   where avg_k_t is the average number of clusters found in each local model for chunk t. The above expression can be simplified to $O(\mid M_t \mid log \mid M_t \mid)$.

5. **Knowledge transfer:** The computational time complexity of stage 6 is $O(n)$. Due to the linearity of stage 6, it will not be included in the overall complexity evaluation.
6. **Pattern-labelling:** The computational complexity of the Pattern-labelling algorithm is approximated to

   $$O(\sum_{i=1}^{n} \mid C_{ti} \mid + \sum_{i=1}^{k_t} k_t^{(i)} + N_t \sum_{i=1}^{k_t} k_t^{(i)}),$$

   where $\mid C_{ti} \mid$ is the number of clusters in the local model of view i, k_t is the number of clusters in the global model, and $k_t^{(i)}$ is the number of data points in global cluster i. The first two parts in the above expression assess the mapping of the extracted patterns to the real medoids in each view, while the third part presents the matching of patterns for all data points.

Based on the above analysis the overall computational complexity of MST-MVS algorithm in a configuration using LEdges and Pattern-labelling is approximately (assuming all views are approved in stage 3):

$$O(N_t^2 + n^2 \sum_{i=1}^{n} n_i \mid C_{ti} \mid + \lambda(2 + N_t + \log \lambda)),$$

where n is the number of views, $\lambda = \sum_{i=1}^{n} \mid C_{ti} \mid$, and N_t is the number of data points in chunk t. The highest complexity is when all views are approved, resulting in $\sum_{i=1}^{n} \mid C_{ti} \mid = \sum_{i=1}^{k_t} k_t^{(i)} = \mid M_t \mid$.

5 Data and Experimental Settings

5.1 Data

We study and analyse the performance of the proposed MST-MVS clustering algorithm on synthetic and real-world data. We have used four different datasets: one synthetic and three real-world. These are listed in Table 1. A detailed description of the datasets used as well as their interpretation in a multi-view context are given below.

Dim32 dataset A synthetic dataset, Dim32 [18], is built to be used for evaluation of clustering algorithms. Each cluster in Dim32 dataset is well separated, even in higher dimensions. Initially, this dataset is used in [19] with variations of the number of attributes. The Dim32 dataset used in our work has 1024 instances with 32 features, and is divided into 16 different Gaussian clusters. This dataset is chosen to provide a controlled and well-known environment for studying different configurations of the algorithm. The experiments conducted on this dataset are presented in Sect. 5.3.

Dim32 has been interpreted as a two-view dataset in our experiments. The first 16 attributes are assigned to the first view and the rest to the second view.

Forest Cover-Type dataset The second dataset used in our experiments is a subset of the Cover-type dataset [6], available at the UCI repository [28]. The main motivation for selecting this dataset is the fact that it is well-known and used in many machine learning papers studying new clustering algorithms [7, 30, 39]. It has been created for classification tasks with 581012 instances and 54 attributes, divided into

Table 1 Datasets used in the experiments

Dataset	Type	#Samples	#Attributes	#Classes
Dim32	Synthetic	1024	32	16
Cover-type	Real-world	50000	14	7
One-year Sensor	Real-world	8664	8	Non-labelled
Two-year Sensor	Real-world	17544	8	Non-labelled

7 clusters. Cover-type is an unbalanced dataset with 283301 instances for the biggest cluster and 2747 instances for the smallest. From the total 54 attributes, 44 are binary values, and 40 of these indicate soil types, the remaining 4 indicate wilderness areas. These 44 binary attributes make the data sparse, since each instance has only 2 values of these 44, rest of them are 0.

In accordance to the experiments conducted in [7], the 40 binary soil types in the Cover-type dataset are not considered in the experiments due to their sparsity. In addition, a sample set of 50000 instances of the Cover-Type dataset is used in the experiments and is divided into four views. The first three attributes, Elevation, Aspect, and Slope, are assigned to the first view. The next three with the ninth attribute, i.e. horizontal and vertical distance to nearest surface water features, horizontal distance to nearest roadways, and wildfire ignition points are assigned to the second view. The third view includes attributes six, seven and eight in the dataset, corresponding to the hill shade at 9 am, noon, and 3 pm, respectively. The last view has been assigned to the remaining four binary attributes indicating Rawah, Neota, Comanche Peak, and Cache la Poudre wilderness areas.

Real-world sensor datasets The potential of the proposed algorithm is also demonstrated on real-world data from a company in the smart building domain. We have used two datasets: one covering a year (Jan 1st 2019 till Dec 27th 2019) and the other one containing measurements from two years (Jan 1st 2019 till Dec 31st 2020). The one-year dataset has been used in [15] for analysing and monitoring the control valve system behaviour. It has also been used in the evaluation of the MV Multi-Instance Clustering algorithm in [11].

In the smart building domain different types of metrics are collected from a wide range of sensors available for systems such as heating, ventilation, air conditioning, and refrigeration. The eight features listed in Table 2, used in [11] and seven of which also considered in [15], are used in our experiments.

The available data features are analysed and partitioned in three distinctive views: system operational behaviour parameters, performance indicators and contextual factors. The features SST, SRT, and PHL are selected to model the system typical operational behaviour. The system performance can be evaluated by these three indicators: VOM, VOS, and SE. Finally, the contextual factors are represented by the features: OTM and OTS. For each view, after removing the outliers using Hampel filter, averaged daily values of the corresponding features are calculated to build daily profiles.

5.2 Data Preparation

In this study, to identify and remove the outliers in each feature of the two Sensor datasets, Hampel filter [25], a method based on median absolute deviation (MAD) estimation is applied. In addition, each feature of the four used datasets are standardized using the z-score. Namely, each feature is subtracted by their mean

Table 2 Features included in the real-world sensor dataset

View	Id	Acronyms	Feature name	Units
Operation	1	SST	Secondary Supply Temperature	°C
	2	SRT	Secondary Return Temperature	°C
	3	PHL	Primary Heat Load	kW
Performance	4	VOM	Valve Openness Mean	%
	5	VOS	Valve Openness Standard Deviation	%
	6	SE	Sub-station Efficiency	%
Context	7	OTM	Outdoor Temperature Mean	°C
	8	OTS	Outdoor Temperature Standard Deviation	°C

value ($\bar{x}$) and divided by the standard deviation (σ), i.e.

$$z = \frac{x - \bar{x}}{\sigma}.$$

Ten different experimental datasets are generated from each of Cover-Type and Dim32 datasets and used in the conducted experiments. The obtained results reported and analysed in Sect. 6, are averaged values of the ones produced on these experimental datasets. These datasets are generated by randomly shuffling rows in the data, thereby generating new versions of the datasets. This is done to mitigate the biases in the results due to the data and in that way to provide with more objective evaluation of algorithm performance.

5.3 *Experiments and Validation*

The conducted experiments are categorised into three groups: (i) algorithm configuration, (ii) tuning of algorithm parameters, and (iii) evaluation of algorithm performance. In the first group, we have performed a series of experiments on Dim32 dataset to study algorithm properties and find its optimal configuration. This configuration, called standard configuration hereafter, has been used in our experiments evaluating algorithm performance on real-world datasets, i.e. in the third group of experiments. We have also conducted a number of experiments to fine tune the algorithm's parameters for each of the used dataset.

Algorithm configuration As described in Sect. 4.3, two different algorithms for calculating artificial nodes have been designed: BNodes and LEdges (see Algorithms 2 and 3, respectively). Two configurations of MST-MVS clustering algorithm, respectively using BNodes and LEdges to identify artificial nodes, are studied and compared on the Dim32 experimental datasets.

In this work, we have also proposed two different labelling techniques: CNMF-based labelling and Pattern-labelling (see Sects. 4.4 and 4.5, respectively). Those are also studied and validated on the experimental datasets of Dim32. Since Dim32 dataset is labelled, the results generated in both experiments are evaluated by the four external cluster validation measures discussed in Sect. 3.3.

Tuning of algorithm parameters We have conducted an experiment to find the optimal value of the threshold Θ_t, an algorithm parameter for each dataset used. We have explored each dataset by applying the MST-MVS algorithm on the dataset and plotting together the SI scores of the views' clustering solutions produced in each data chunk. The threshold is set as a trade-off value among the plotted scores.

The data chunk size is usually problem specific. However, Dim32 and Cover-type datasets are not considered in a concrete applied context. Therefore, we have performed an experiment to study and tune the chunk size of Cover-type dataset. The experiment executes the algorithm three times, each time using different chunk sizes. The data chunk size is chosen based on evaluating the results generated by the algorithm using the four external cluster validation measures described in Sect. 3.3.

In case of Dim32 dataset we have determined its data chunk size through reasoning, since it is a comparatively small dataset and an experiment is not needed. The Dim32 dataset is divided into two chunks containing 614 and 410 instances, respectively. As one can observe the first chunk is larger than the second. This is motivated by the algorithm working mechanism. Namely, the algorithm is initialized by mean α and transfers over knowledge from the current data chunk to the next. The quality of the initial clustering will have a significant impact on the upcoming data chunks. Therefore, the first chunk of Dim32 dataset is bigger, approximately 60% of the whole dataset, while the second have the remaining 40%.

Note that an experiment is not conducted for the Sensor datasets to determine the chunks' sizes. In case of one-year Sensor dataset we have simulated the experimental scenario used in the evaluation of MV Multi-Instance Clustering [11] to be able to compare the results. Thus the one-year Sensor dataset is divided into two chunks, 243 and 118, respectively. The first chunk contains the measures gathered during the months January to August, while the second chunk covers the period from September to December.

The two-year Sensor dataset is divided into two chunks, each one containing measurements from a year period. Our motivation behind this is to be able to compare the patterns extracted from two data chunks which are expected to capture similar seasonal behaviour modes. Note that there are missing values for a few days in the second year data. These have been removed from the dataset. Due to this the chunk sizes result in 363 and 334, respectively.

Evaluation of algorithm performance The standard configuration of the proposed MST-MVS algorithm has been studied and evaluated in a real-world context by using Cover-type and Sensor data. The standard configuration of the proposed algorithm is executed on the experimental datasets of Cover-type data and the obtained results are evaluated with the four external cluster validation measures discussed in Sect. 3.3.

We have investigated how the knowledge transfer affects the performance of the proposed MST-MVS algorithm by comparing it with an algorithm version that does not use artificial nodes to seed the clustering at the upcoming data chunk. This experiment has been conducted on the experimental datasets of Cover-type data. The generated results are evaluated with the four external cluster validation measures. In addition, the number of clusters of the global model have been compared to the ground-truth number of clusters to facilitate the interpretation of the results.

Our MST-MVS algorithm evaluates the quality of the clustering solutions produced on different views' data at each chunk and select the best ones to build the global clustering model. In order to study how quality of the data chunks impact the global result we have conducted an experiment on Cover-type data in which we skip the evaluation phase and build the global model using all views' clustering solutions and then compare the results with the one produced by the original version of the algorithm.

The one-year Sensor dataset is used to study and benchmark the patterns extracted by the proposed algorithm to ones identified by MV Multi-Instance Clustering in [11]. We have further studied and analysed the potential of MST-MVS algorithm in the pattern mining task by applying it on the two-year Sensor dataset.

In order to study how the algorithm's performance is affected by the quality of knowledge transferred we have applied the MST-MVS algorithm on the one-year Sensor dataset in a circular mode, i.e. the algorithm has been executed in 10 consecutive iterations by seeding the first data chunk with the artificial nodes extracted by the last chunk. The results are evaluated by calculating SI scores of the views' clustering models generated in each data chunk.

5.4 *Implementation and Availability*

The proposed MST-MVS clustering algorithm is implemented in Python [52]. The following libraries have been used:

- Scikit-Learn [44] for z-score, and external cluster validation measures: ARI, AMI, Completeness and Homogeneity.
- NetworkX [22] for Kruskal's algorithm, and creation of the complete graph.

Table 3 The cluster validation measures' scores generated on Dim32 dataset by Pattern-labelling (PL) and CNMF-labelling (CNMF). MST-MVS algorithm is executed with a threshold $\Theta = 0.0$ and LEdges method. The dataset is divided into two chunks containing 614 and 410 instances, respectively

Method	Minimum				Maximum				Mean			
	ARI	H	C	AMI	ARI	H	C	AMI	ARI	H	C	AMI
CNMF	0.82	0.91	0.98	0.94	0.84	0.92	0.98	0.94	0.83	0.92	0.98	0.94
PL	**1.00**	**1.00**	**1.00**	**1.00**	**1.00**	**1.00**	**1.00**	**1.00**	**1.00**	**1.00**	**1.00**	**1.00**

Note H and C stand for Homogenity and Completeness, respectively

- Pandas [41] for reading dataset from disk and manipulations of data.
- NumPy [27] for fast array manipulations, and operations.
- SciPy [55], utility functions for distance calculations, in this case, Euclidean.
- PyMF [16] for solving the CNMF algorithm, its implementation is based on [14].

Dim32 [18] and Cover-type [6] datasets are public. The Sensor datasets are provided by a company and are not publicly available. The algorithm code can be provided on request.

6 Results and Discussion

6.1 Algorithm Configuration

The Pattern-labelling technique introduced in Sect. 4.5 is benchmarked to CNMF-labelling[1] (see Sect. 4.4) on Dim32 dataset. As one can observe in Table 3, the Pattern-labeling has produced better results than the CNMF-labelling. This is logical and can be predicted due to the randomness embedded into the CNMF algorithm. The latter algorithm randomly initiates the H matrix and is executed multiple times to generate a good approximation. In comparison to it the Pattern-labelling is consistent, since randomness is not used in its calculations, as this is described in Sect. 4.5. Therefore, Pattern-labelling is chosen to be used in our experiments studying the MST-MVS algorithm performance.

In addition, we have compared the two algorithms for calculating artificial nodes (see Algorithms 2 and 3) on Dim32 dataset. As it can be observed in Table 4 both methods (BNodes and LEdges) have similar performance w.r.t. the used metrics. Note that BNodes, as stated before, has a small shortcoming, namely it would split a cluster if the clusters are ordered in a straight line. On the other hand, LEdges is more efficient by combining stages 5 and 6 of the algorithm. Due to this and the close

[1] Note that in our experiments we have used a CNMF-based, instead of NMF-based, version of the CNMF-labelling technique to allow for negative data.

Table 4 The cluster validation measures' scores generated on Dim32 dataset by LEdges and BNodes. MST-MVS algorithm is executed with a threshold $\Theta = 0.0$ and CNMF-labelling method. The dataset is divided into two chunks containing 614 and 410 instances, respectively

Method	Minimum				Maximum				Mean			
	ARI	H	C	AMI	ARI	H	C	AMI	ARI	H	C	AMI
BNodes	**0.83**	0.91	**0.99**	0.94	0.84	0.92	**0.99**	**0.95**	**0.84**	0.92	**0.99**	**0.95**
LEdges	0.82	0.91	0.98	0.94	0.84	0.92	0.98	0.94	0.83	0.92	0.98	0.94

Note H and C stand for Homogenity and Completeness, respectively

Table 5 The average SI scores of the views' clustering models produced in the two chunks of Dim32 dataset by applying the MST-MVS algorithm configuration using LEdges and CNMF-labelling

Chunk	Avg. SI	
	View 0	View 1
1	0.91	0.94
2	0.93	0.94

similarity between the results generated by the two methods LEdges is selected to be used in the experiments studying algorithm performance.

6.2 *Tuning of Algorithm Parameters*

The only parameter used in the algorithm, threshold Θ_t, is defined based on data exploration. This is done by running the MST-MVS algorithm on each used dataset and plotting the SI values of the views' clustering models calculated in each data chunk. For example, in case of Dim32 dataset, the threshold is set to 0.5. It could theoretically be higher due to the high SI scores produced on the views' clustering solutions, this can be observed in Table 5.

In case of Cover-type data we plot the SI values found for each view in Fig. 4. As one can observe they are relatively low, and most of the time below 0. Therefore, -0.2 is used as the threshold in the experiments conducted on Cover-type dataset. It could be argued that a positive threshold should be used, as negative SI denotes a poor clustering solution. However, most of the clusters of this dataset are overlapping and this affects the quality of the views' clustering solutions. For example, only the first chunk has produced positive SI values for all views. Hence, the threshold -0.2 is set to be a close average of the three lowest scoring views (views 0, 1, 2).

Similar to Cover-type data, the threshold on the Sensor dataset is set below 0 to include more than one view in the calculation of the global model. Specifically -0.4, a close average of the lowest and highest SI values with an added margin for error, as it can be observed in Table 6. Our motivation for this is the high difference between the values generated on view 2 and the other two views' values.

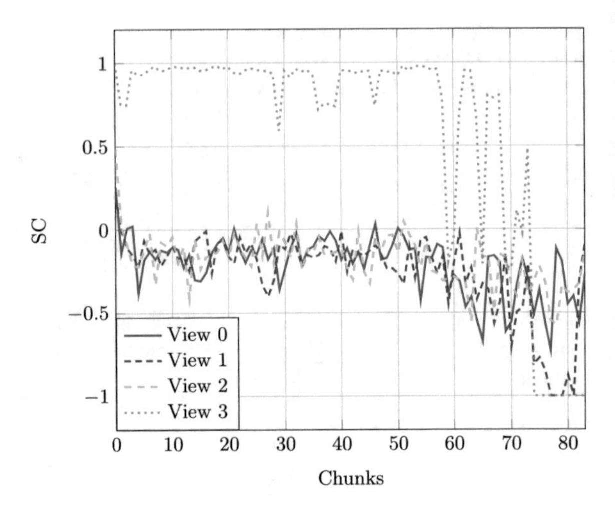

Fig. 4 The SI scores of the views' clustering models produced on Cover-Type dataset by applying the MST-MVS algorithm configuration using LEdges and Pattern-labelling

Table 6 The SI scores of the views' clustering models produced in the two chunks of one-year Sensor dataset by applying the MST-MVS algorithm configuration using LEdges and Pattern-labelling

	Avg. SI		
Chunk	View 0	View 1	View 3
1	−1.00	0.39	−1.00
2	−0.10	0.63	−0.11

Table 7 Average cluster validation measures' scores generated on the Cover-type experimental datasets using different chunk sizes

Chunk size	ARI	H	C	AMI
600	0.27	0.21	0.46	0.27
1000	0.25	0.20	0.49	0.26
2000	0.28	0.22	0.50	0.29

Note H and C stand for Homogeneity and Completeness, respectively

In order to determine the chunk size for the Cover-type dataset we have conducted experiments by studying different chunk sizes. Table 7 presents the produced cluster validation measures' scores for three different chunk sizes. As one can see they are very close. Therefore, we have selected 600 to be the chunk size, since the smallest size results in experimenting with more chunks. The latter provides more results for discussion and facilitates to get better insight into the working mechanism of the algorithm.

6.3 Evaluation of Algorithm Performance

Based on the experiments discussed in Sects. 6.1 and 6.2 a standard configuration of MST-MVS algorithm has been determined for each dataset. These standard configurations are used in the experiments discussed in this section. The standard configurations of the used datasets are shown in Table 8.

Note that in case of Cover-Type, the SI analysis for some chunks fails for all views, resulting in a skipped chunk. When plotting the results, the skipped chunk is not shown. The results from the last processed chunk are used instead. This can be observed in certain places in the graphs depicting the metrics' scores, e.g., see last chunks in Fig. 5, where the ARI scores plotted for MST-MVS algorithm (solid black line) between chunk 70 and 82 are constants, due to skipped chunks.

Cover-Type dataset The performance of the MST-MVS algorithm's standard configuration is evaluated on the Cover-Type dataset. The obtained results are reported in Table 9. It can be observed that the transfer of knowledge does have a positive impact on the resulting clustering solution's properties evaluated by the four external cluster validation measures. Note that the maximal values generated by the two versions of the algorithm are very similar. However, the version implementing the knowledge transfer generates higher minimal and average values then the other version.

It should also be noted that the evaluation of the views' clustering solutions does have a positive impact on the global models' clustering solutions, as supported by the results presented in Table 10. The difference between the obtained results, with and without evaluations of local clustering solutions is not significant, but in all conducted experiments except one (the minimum value of Completeness) the scores generated

Table 8 MST-MVS algorithm standard configurations for the used datasets

Dataset	Global model method	Labeling Method	Threshold	Chunk size(s)
Dim32	LEdges	Pattern-labelling	0.5	614(410)
Cover-Type	LEdges	Pattern-labelling	− 0.2	600
One-year Sensor	LEdges	Pattern-labelling	−0.4	243(118)
Two-year Sensor	LEdges	Pattern-labelling	−0.4	363(334)

Table 9 The cluster validation measures' scores generated on Cover-type dataset by the MST-MVS algorithm standard configuration with and without knowledge transfer between two consecutive data chunks

Transfer	Minimum				Maximum				Average			
	ARI	H	C	AMI	ARI	H	C	AMI	ARI	H	C	AMI
Yes	**0.10**	**0.12**	**0.11**	**0.12**	0.50	**0.48**	0.87	0.61	**0.28**	**0.23**	**0.50**	**0.29**
No	−0.02	0.00	0.03	0.00	0.50	0.46	**0.91**	0.61	0.17	0.14	0.45	0.19

Note H and C stand for Homogenity and Completeness, respectively

Table 10 The cluster validation measures' scores generated on Cover-type dataset by the MST-MVS algorithm standard configuration with and without evaluation of the views clustering solutions (Stage 2)

Evaluation	Minimum				Maximum				Average			
	ARI	H	C	AMI	ARI	H	C	AMI	ARI	H	C	AMI
Yes	**0.10**	**0.12**	0.11	**0.12**	**0.50**	**0.48**	**0.87**	**0.61**	**0.28**	**0.23**	**0.50**	**0.29**
No	−0.00	0.00	**0.19**	−0.00	0.38	0.37	0.81	0.48	0.24	0.19	0.48	0.25

Note H and C stand for Homogenity and Completeness, respectively

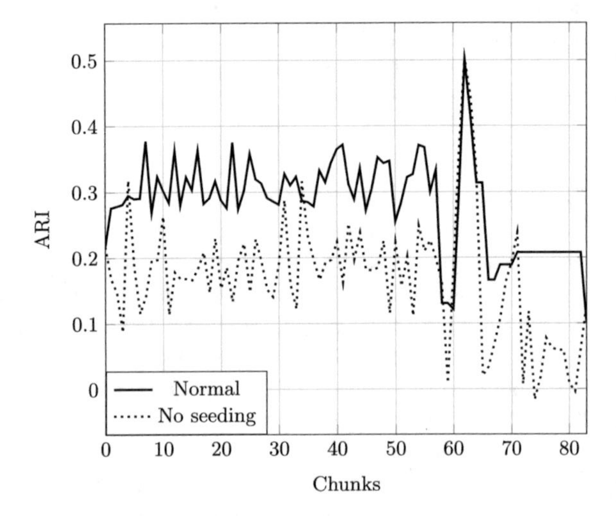

Fig. 5 ARI scores generated on Cover-type dataset by the MST-MVS algorithm standard configuration with and without knowledge transfer between two consecutive data chunks

in the former scenario, i.e. when the local clustering solutions are evaluated, are higher.

In order to get further insight into the results reported in Table 9, we have depicted the calculated ARI scores in Fig. 5. This plot indicates that the instability in the results is around and after the 58th chunk, i.e., something is different in the data after that chunk. Therefore, to better understand the data structure the number of clusters of the clustering solutions produced by the two versions are further benchmarked to that of the ground-truth clustering of the Cover-type dataset in Fig. 6.

As one can observe in Fig. 6 the number of clusters fluctuates after the 58th chunk, resulting in the instability in the algorithm's results after it. If the results after the 58th chunk are excluded, then as can be seen in Fig. 5, the knowledge transfer between chunks do have a positive impact on the results. This in turn indicates that the artificial nodes from the previous chunk really guide the clustering of the next chunk. This is additionally demonstrated in Fig. 6, where the two versions show different behaviour signatures w.r.t. the number of clusters. Despite the fact that the number of clusters produced by the version without implementing knowledge transfer is closer to the ground-truth number of clusters, its behaviour is more fluctuating in comparison to that of the other version.

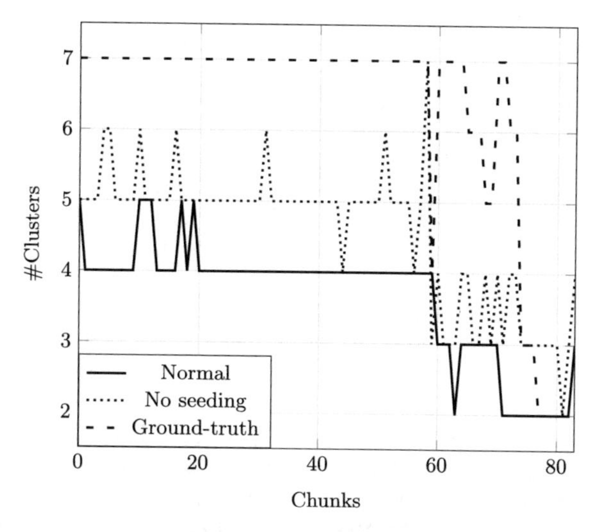

Fig. 6 Comparison of the number of clusters produced on Cover-type dataset by the MST-MVS algorithm standard configuration with and without knowledge transfer between two consecutive data chunks with the ground-true number of clusters

Table 11 Summary of the identified clusters in the first chunk (January–August) of the one-year Sensor dataset

Cluster	PHL	SST	SRT	VOM	VOS	SE	OTM	OTS	Months	Size	Concept
0	15.15	36.33	31.95	9.83	∓0.65	81	10.71	∓0.43	1–8	240	Unite concepts that share common trends for both heating and non-heating season
1	36.92	52.13	39.99	19.32	∓3.18	96	−2.89	∓0.36	1, 3	2	Concepts 12 & 13
2	11.42	34.33	31.43	11.77	∓7.28	84	11.29	∓0.44	5	1	Deviant behaviour
									Total	243	

Note The unit for PHL is kW and for SST, SRT, OTM, and OTS is °C. VOM, VOS, and SE are expressed in %. For the full form of each feature see Table 2

Sensor datasets Initially, the MST-MVS clustering algorithm has been applied on the one-year Sensor dataset by simulating the same experimental scenario as the one used in [11] to evaluate MV Multi-Instance Clustering. Namely, as described in Sect. 5.3, the created daily profiles (361 in total) are split into two parts in order to simulate two data chunks: the first one with 243 daily profiles (January–August) and the second chunk with 118 daily profiles (September–December).

Tables 11 and 12 present the clustering solutions generated by the MST-MVS algorithm on the first and second data chunks of the one-year Sensor dataset, respectively. These are benchmarked to 13 cluster concepts identified by MV Multi-Instance Clustering algorithm in [11]. The latter algorithm has extracted 8 concepts that links three views from the first chunk and 5 additional concepts from the second chunk. The last column *Concept*, in both Tables 11 and 12 present the concepts reported in [11] that based on our analysis match the clusters identified by applying the MST-MVS clustering algorithm.

Table 12 Summary of the identified clusters in the second chunk (September–December) of the one-year Sensor dataset

Cluster	PHL	SST	SRT	VOM	VOS	SE	OTM	OTS	Months	Size	Concept
0	14.27	36.86	33.44	10.62	∓0.59	87	9.79	∓0.30	9–12	65	Concept 6
−1	20.90	42.61	37.12	13.32	∓ 0.56	92	5.01	∓ 0.40	9–12	26	Concepts 10 & 11
2	20.51	43.48	38.26	13.25	∓ 0.45	94	5.00	∓ 0.35	10–12	23	Concepts 10 & 11
1	25.49	42.76	36.35	14.36	∓ 0.53	95	5.56	∓ 0.46	10	3	Concept 10
3	23.22	44.87	38.76	13.32	∓ 1.80	96	3.90	∓ 0.32	12	1	Concept 11
									Total	118	

Note The unit for PHL is kW and for SST, SRT, OTM, and OTS is °C. VOM, VOS, and SE are expressed in %. For the full form of each feature see Table 2

It can be observed that two concepts (concepts 12 and 13) extracted by MV Multi-Instance Clustering are very similar to cluster 1 in the first data chunk (see Table 11). Concepts 12 and 13 present the system behaviour typical for the heating season when the average outdoor temperature is close to zero. Cluster 1 differs from these concepts on having a negative mean value for the outdoor temperature (OTM), i.e. it groups together the profiles of the only two days in January and March when the temperature was below zero. Cluster 0 unites many of the concepts produced by the MV Multi-Instance Clustering in [11]. Evidently, it represents trends common for both heating and non-heating season. This result is mainly due to missing of previous knowledge. This is also confirmed by the improved results obtained when the algorithm is executed iteratively (see Fig. 7, and Tables 13 and 14). Cluster 2 is a singleton and can be considered as a pattern hinting deviating behaviour (from the one presented by cluster 0). It indicates a large deviation in the opening and closing of the valve in May. Therefore, it is worth to be further analysed and discussed with domain experts. In general, as it will be seen further in this section, clustering solutions produced by the proposed algorithm are more compact than the ones generated by MV Multi-Instance Clustering. This is due to the different working mechanisms of the two algorithms. Namely, the MV Multi-Instance Clustering algorithm integrates the clustering solutions that have been produced independently on two consecutive data chunks. Then it applies FCA analysis, and closed patterns to extract the integrated clustering result. This certainly implies the generation of clustering solutions that have more smaller-size clusters. In comparison with MV Multi-Instance Clustering, the proposed algorithm clusters each data chunk separately by applying a MST clustering algorithm seeded with artificial nodes from the previous chunk.

As one can see in Table 12, all clusters are linked to concepts identified by MV Multi-Instance Clustering in the second chunk of one-year Sensor dataset. We can notice the same trend as the one discussed above. Namely, more compact solution than the one produced by MV Multi-Instance Clustering. The five clusters (0, −1, 1, 2 and 3) identified by the MST-MVS algorithm are linked only to three concepts. It should be noted that cluster −1 indicates "outliers", i.e. data points that do not match any of the patterns extracted by the MST-MVS algorithm from the second data chunk. In comparison with clusters 1 and 2, cluster −1 seems to be a combination of them both.

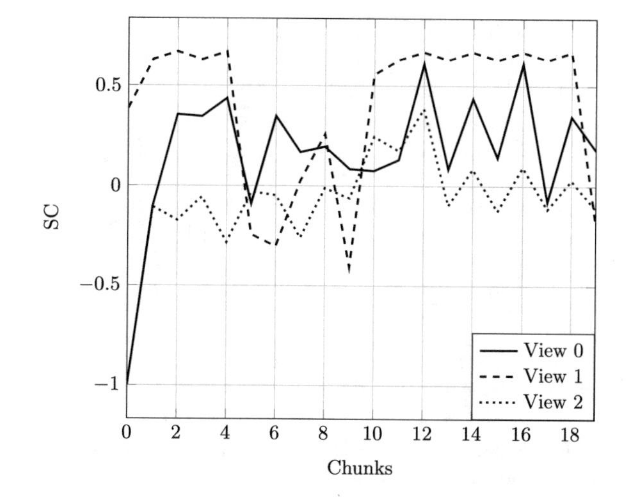

Fig. 7 The SI scores of local clustering models generated by the MST-MVS algorithm standard configuration applied 10 times on the one-year Sensor dataset in a circular mode. The threshold value is set to −0.4

This could therefore, be the reason why these data points are labelled as "outliers" by the algorithm. In addition, clusters 1 and 3 contain data points presenting behaviour deviating from the one modeled by cluster 2. This phenomenon is also noticed in the clustering solution summarized in Table 11 (see cluster 2) and clustering solutions generated on two-year Sensor data chunks (see Tables 15 and 16).

In order to study how the algorithm performance is affected by the quality of transfer knowledge we have applied the MST-MVS algorithm on the one-year Sensor dataset in a circular mode. This means that after the second chunk the algorithm is run on the first chunk again, but now it is seen as the next data chunk. This is repeated 10 times to observe the changes in SI score over time, see Fig. 7. It can be observed in the figure that the SI values stabilize after chunk 10 (i.e. after the 5th iteration), which results in an average SI value of 0.25 for all views' clustering solutions. In addition to this, we notice a significant increase in SI score in comparison to the starting value. All these indicate that the quality of the clustering solution of the initial chunk has an impact on the successive chunks' results, i.e. the quality of knowledge transferred is important.

We further study the clustering solutions produced on the first and second chunk after the 5th iteration of our algorithm on one-year Sensor dataset. They are presented in Tables 13 and 14, respectively. In comparison with the results reported in Tables 11 and 12 these clustering solutions obtained better capture the valve system behaviour. We can observe that the two clustering solutions have a very similar structure. Namely, they both have two bigger clusters (clusters 0 and 1 in Table 13 and clusters 2 and 0 in Table 14) that model the system behaviour typical for heating and non-heating seasons, respectively; a single cluster (cluster -1 in both clustering solutions) that presents behaviour fluctuating between heating and non-heating modes, i.e. the data points assigned to this cluster do not match any of the extracted patterns; and finally, a few smaller clusters that demonstrate behaviour deviating from the two typical modes. For example, clusters 3 and 4 in Table 13 are singletons that present

Table 13 Summary of the identified clusters in the first chunk (10th iterative chunk) after 5th iteration of the algorithm on the one-year Sensor dataset

Cluster	PHL	SST	SRT	VOM	VOS	SE	OTM	OTS	Months	Size
0	39.75	47.37	38.17	17.57	∓0.53	96	1.20	∓0.35	1–3, 5	73
3	19.81	50.32	39.17	18.59	∓4.18	98	−1.55	∓0.41	3	1
−1	6.59	37.38	32.89	12.43	∓0.73	89	9.34	∓0.44	3–7	66
1	3.40	27.89	26.96	2.65	∓0.72	67	18.31	∓0.49	3–8	101
4	11.42	34.34	31.44	11.77	∓7.28	85	11.30	∓0.44	3	1
2	10.40	32.73	30.08	11.18	∓1.09	81	13.00	∓0.47	3	1
									Total	243

Note The unit for PHL is kW and for SST, SRT, OTM, and OTS is °C. VOM, VOS, and SE are expressed in %. For the full form of each feature see Table 2

Table 14 Summary of the identified clusters in the second chunk (11th iterative chunk) after 5th iteration of the algorithm on the one-year Sensor dataset

Cluster	PHL	SST	SRT	VOM	VOS	SE	OTM	OTS	Months	Size
0	5.47	28.38	27.24	5.22	∓0.98	74	16.17	∓0.36	9	17
−1	11.28	34.37	31.41	11.54	∓0.65	85	11.76	∓0.36	9–10	10
1	12.45	34.94	32.08	11.81	∓0.47	88	11.10	∓0.32	9–10	10
2	21.11	43.21	37.88	13.29	∓0.47	95	4.94	∓0.33	9–12	80
3	23.23	44.88	38.76	13.33	∓1.80	97	3.90	∓0.32	12	1
									Total	118

Note The unit for PHL is kW and for SST, SRT, OTM, and OTS is °C. VOM, VOS, and SE are expressed in %. For the full form of each feature see Table 2

Table 15 Summary of the identified clusters in the first chunk (first year) of the two-year Sensor dataset

Cluster	PHL	SST	SRT	VOM	VOS	SE	OTM	OTS	Months	Size
0	17.41	38.73	33.87	11.66	∓0.71	87	8.36	∓0.39	1–12	328
1	11.42	34.34	31.44	11.77	∓7.28	85	11.30	∓0.44	5	1
2	2.43	26.27	26.07	0.00	∓0.00	58	21.50	∓0.47	6–9	34
									Total	363

Note The unit for PHL is kW and for SST, SRT, OTM, and OTS is °C. VOM, VOS, and SE are expressed in %. For the full form of each feature see Table 2

two days in March with a large deviation in the valve openness (VOS) in heating and non-heating season, respectively. Interestingly, the only day of cluster 2 is not assigned to cluster 4, mainly due to the fact that its VOS behaviour is normal. Evidently, the generated clustering models provide a lot of useful information that can facilitate the domain experts in analyzing of the system behaviour.

Tables 15 and 16 show the clustering solutions generated by the MST-MVS algorithm on the two chunks of the two-year Sensor dataset, respectively. We can see that there is a similarity between the two clustering solutions. In addition, the clustering

Table 16 Summary of the identified clusters in the second chunk (second year) of the two-year Sensor dataset

Cluster	PHL	SST	SRT	VOM	VOS	SE	OTM	OTS	Months	Size
1	11.67	38.63	33.80	12.46	∓ 0.57	−28	6.89	∓ 0.33	1–12	207
0	10.01	38.28	33.25	12.57	∓ 0.51	−31	7.23	∓ 0.42	2, 11	2
2	3.31	26.77	25.58	2.86	∓ 0.44	13	18.20	∓ 0.45	5–11	123
3	4.32	26.14	25.94	1.63	∓ 1.02	76	16.77	∓ 0.52	6	1
									Total	334

Note The unit for PHL is kW and for SST, SRT, OTM, and OTS is °C. VOM, VOS, and SE are expressed in %. For the full form of each feature see Table 2

produced on the first chunk (see Table 15) is similar to the one-year Sensor dataset's clustering results (see Tables 11 and 12). Namely, clusters 0 in the one-year Sensor dataset from both chunks have been combined to produce cluster 0 in the two-year dataset. Furthermore, cluster 1 from the two-year dataset is exactly the same as the deviant behaviour cluster from the one-year dataset (cluster 2 in Table 11). An interesting behaviour is presented by cluster 2 generated on the first chunk of the two-year Sensor dataset (see Table 15). We can see that all data points with VOM and VOS of 0.00 in a context of average outdoor temperature (OTM) above $20°C$ have been grouped together.

In Table 16, containing the clustering solution produced on the second chunk of two-year Sensor dataset, we can observe some interesting cluster patterns. First, clusters 0 and 1 in Table 16 have both negative values for the substation efficiency (SE). During a span in the second data chunk, that is from 23rd September 2020 till 14th December 2020 the system had faulty behaviour. The negative values for SE in these clusters are due to this. The primary supply temperature during these days is less than the primary return temperature which gave large negative SE values. Second, cluster 2 from the first chunk (see Table 15) seems to have been split into clusters 2 and 3 in the second chunk (see Table 16). Finally, clusters 0 and 1 contain the heating season patterns while clusters 2 and 3 present the two modes during the non-heating period. We can notice the clustering has the same structure as that of the clustering solutions presented in Tables 13 and 14. Namely, clusters 0 and 3 contain just two and one data points, respectively. These can be interpreted as ones presenting behaviours deviating from the typical for heating and non-heating season, respectively.

7 Conclusion and Future Work

In this work, we have proposed a novel multi-view graph-based clustering algorithm, entitled MST-MVS clustering, suitable for modelling multi-view streaming scenarios. We have developed different configurations of the MST-MVS clustering algorithm. They have been evaluated under different experimental scenarios on two types of data sets: synthetic and real-world. We have studied two different approaches for identifying artificial nodes that are used to seed the MST algorithm producing local clustering models at the upcoming data chunk. We have also investigated how the knowledge transfer affects the performance of the proposed MST-MVS algorithm. Finally, we have proposed two post-labelling techniques: Pattern-labelling and CNMF-labelling.

The MST-MVS clustering algorithm logically has shown a higher performance on the synthetic data than on the real-world data. This is due to the fact that synthetic data usually does not have the features which are characteristic for real-world data such as noise, missing values and overlapping clusters. The transfer of knowledge feature has shown to have a positive effect on the performance of MST-MVS algorithm. The Pattern-labelling technique has been demonstrated to outperform the CNMF-labelling algorithm in the conducted experiments both in terms of computational time and labelling accuracy. Our future plans include further evaluation of the proposed MST-MVS clustering algorithm on richer real-world datasets in different applied scenarios.

Acknowledgements We would like to thank *Farhad Basiri* for providing us the data from the smart building domain.

References

1. Ackermann, R.M., Märtens, M., Raupach, C., Swierkot, K., Lammersen, C., Sohler, C.: Streamkm++: a clustering algorithm for data streams. ACM J. Exp. Algorithmics **17**(1), 173–187 (2012)
2. Aggarwal, C., Han, J., Wang, J., Yu, P.: A framework for clustering evolving data streams. In: VLDB, vol. 7, pp. 81–92. VLDB Endowment, Berlin (2003)
3. Akata, Z., Thurau, C., Bauckhage, C.: Non-negative matrix factorization in multimodality data for segmentation and label prediction. In: 16th Computer Vision Winter Workshop (2011)
4. Ben-Hur, A., Horn, D., Siegelmann, H.T., Vapnik, V.: Support vector clustering. J. Mach. Learn. Res. **2**(Dec), 125–137 (2001)
5. Bendechache, M., Kechadi, M.T.: Distributed clustering algorithm for spatial data mining. In: 2015 2nd IEEE International Conference on Spatial Data Mining and Geographical Knowledge Services (ICSDM), pp. 60–65. IEEE (2015)
6. Blackard, J.A., Dean, D.J., Anderson, C.W.: UCI machine learning repository (1998). http://archive.ics.uci.edu/ml
7. Boeva, V., Angelova, M., Devagiri, V.M., Tsiporkova, E.: Bipartite split-merge evolutionary clustering. In: van den Herik, J., Rocha, A.P., Steels, L. (eds.) Agents and Artificial Intelligence, pp. 204–223. Springer, Cham (2019)
8. Cao, B., Shen, D., Sun, J.T., Wang, X., Yang, Q., Chen, Z.: Detect and track latent factors with online nonnegative matrix factorization. In: IJCAI, vol. 7, pp. 2689–2694 (2007)
9. Cover, T.M., Thomas, J.A.: Elements of information theory. In: Schilling, D. (ed.) Wiley Series in Telecommunications. Wiley, New York (1991)
10. Craw, S.: Manhattan Distance, pp. 790–791. Springer US, Boston, MA (2017)
11. Devagiri, V.M., Boeva, V., Abghari, S.: A Multi-View Clustering Approach for Analysis of Streaming Data, vol. AIAI 2021, IFIP AICT 627, pp. 169–183. Springer Nature Switzerland AG 2021 (2021)

12. Devagiri, V.M., Boeva, V., Tsiporkova, E.: Split-merge evolutionary clustering for multi-view streaming data. Procedia Comput. Sci. **176**, 460–469 (2020)
13. Ding, C., He, X., Simon, H.D.: On the equivalence of nonnegative matrix factorization and spectral clustering. In: Proceedings of the 2005 SIAM International Conference on Data Mining, pp. 606–610. SIAM (2005)
14. Ding, C.H., Li, T., Jordan, M.I.: Convex and semi-nonnegative matrix factorizations. IEEE Trans. Pattern Anal. Mach. Intell. **32**(1), 45–55 (2008)
15. Eghbalian, A., et al.: Multi-view data mining approach for behaviour analysis of smart control valve. In: Proceedings of the 19th IEEE ICMLA, pp. 1238–1245 (2020)
16. Erler, J., Ramos-Ceja, M.E., Basu, K., Bertoldi, F.: Introducing constrained matched filters for improved separation of point sources from galaxy clusters. ArXiv e-prints (2018)
17. Flake, G.W., Tarjan, R.E., Tsioutsiouliklis, K.: Graph clustering and minimum cut trees. Internet Math. **1**(4), 385–408 (2004)
18. Fränti, P., Sieranoja, S.: K-means properties on six clustering benchmark datasets (2018). http://cs.uef.fi/sipu/datasets/
19. Fränti, P., Virmajoki, O., Hautamäki, V.: Fast agglomerative clustering using a k-nearest neighbor graph. IEEE Trans. Pattern Anal. Mach. Intell. **28**(11), 1875–1881 (2006)
20. Ghesmoune, M., Lebbah, M., Azzag, H.: State-of-the-art on clustering data streams. Big Data Anal. **1**(1), 1–27 (2016)
21. Görke, R., Hartmann, T., Wagner, D.: Dynamic graph clustering using minimum-cut trees. In: Dehne, F., Gavrilova, M., Sack, J.R., Tóth, C.D. (eds.) Algorithms and Data Structures, pp. 339–350. Springer, Berlin, Heidelberg (2009)
22. Hagberg, A., Swart, P., S Chult, D.: Exploring network structure, dynamics, and function using networkx. Tech. rep., Los Alamos National Lab.(LANL), Los Alamos, NM (United States) (2008)
23. Halkidi, M., Batistakis, Y., Vazirgiannis, M.: On clustering validation techniques. J. Intell. Inf. Syst. **17**(2–3), 107–145 (2001)
24. Hamon, R., Emiya, V., Févotte, C.: Convex nonnegative matrix factorization with missing data. In: 2016 IEEE 26th International Workshop on Machine Learning for Signal Processing (MLSP), pp. 1–6. IEEE (2016)
25. Hampel, F.R.: A general qualitative definition of robustness. Ann. Math. Stat. **42**(6), 1887–1896 (1971). http://www.jstor.org/stable/2240114
26. Handl, J., Knowles, J., Kell, D.B.: Computational cluster validation in post-genomic data analysis. Bioinformatics **21**(15), 3201–3212 (2005)
27. Harris, C.R., Millman, K.J., van der Walt, S.J., Gommers, R., Virtanen, P., Cournapeau, D., Wieser, E., Taylor, J., Berg, S., Smith, N.J., Kern, R., Picus, M., Hoyer, S., van Kerkwijk, M.H., Brett, M., Haldane, A., Fernández del Río, J., Wiebe, M., Peterson, P., Gérard-Marchant, P., Sheppard, K., Reddy, T., Weckesser, W., Abbasi, H., Gohlke, C., Oliphant, T.E.: Array programming with NumPy. Nature **585**, 357–362 (2020)
28. Hettich, S., Bay, S.: The UCI KDD archive. University of California, Department of Information and Computer Science, Irvine, CA
29. van der Hoef, H., Warrens, M.J.: Understanding information theoretic measures for comparing clusterings. Behaviormetrika **46**, 353–370 (2019)
30. Huang, L., Wang, C.D., Chao, H.Y., Yu, P.S.: Mvstream: multiview data stream clustering. IEEE Trans. Neural Netw. Learn. Syst. **31**(9), 3482–3496 (2020)
31. Hubert, L., Arabie, P.: Comparing partitions. J. Classif. **2**(1), 193–218 (1985)
32. Jing, P., Su, Y., Li, Z., Nie, L.: Learning robust affinity graph representation for multi-view clustering. Inf. Sci. **544**, 155–167 (2021). https://doi.org/10.1016/j.ins.2020.06.068. https://www.sciencedirect.com/science/article/pii/S0020025520306575
33. Jain, A.K., Dubes, R.C.: Algorithms for Clustering Data. Prentice Hall, Englewood Cliffs, NJ (1988)
34. Kranen, P., Assent, I., Baldauf, C., et al.: The clustree: indexing micro-clusters for anytime stream mining. Knowl. Inf. Syst. **29**, 249–272 (2011)

35. Kruskal, J.B.: On the shortest spanning subtree of a graph and the traveling salesman problem. Proc. Am. Math. Soc. **7**(1), 48–50 (1956)
36. Lee, D.D., Seung, H.S.: Learning the parts of objects by non-negative matrix factorization. Nature **401**(6755), 788–791 (1999)
37. Lindig, C.: Fast concept analysis. Working with Conceptual Structures-Contributions to ICCS pp. 152–161 (2000)
38. Liu, J., Wang, C., Gao, J., Han, J.: Multi-view clustering via joint non-negative matrix factorization. In: Proceedings of the 2013 SIAM International Conference on Data Mining, pp. 252–260. SIAM (2013)
39. Lughofer, E.: A dynamic split-and-merge approach for evolving cluster models. Evolving Syst. **3**(3), 135–151 (2012)
40. Lv, X., Ma, Y., He, X., Huang, H., Yang, J.: Ccimst: a clustering algorithm based on minimum spanning tree and cluster centers. Mathematical Problems in Engineering (2018)
41. Wes McKinney: Data structures for statistical computing in Python. In: van der Walt, S., Millman, J. (eds.) Proceedings of the 9th Python in Science Conference, pp. 56–61 (2010)
42. Ou, W., Long, F., Tan, Y., Yu, S., Wang, P.: Co-regularized multiview nonnegative matrix factorization with correlation constraint for representation learning. Multimedia Tools Appl. **77**(10), 12955–12978 (2018)
43. Paatero, P., Tapper, U.: Positive matrix factorization: a non-negative factor model with optimal utilization of error estimates of data values. Environmetrics **5**(2), 111–126 (1994)
44. Pedregosa, F., Varoquaux, G., Gramfort, A., Michel, V., Thirion, B., Grisel, O., Blondel, M., Prettenhofer, P., Weiss, R., Dubourg, V., Vanderplas, J., Passos, A., Cournapeau, D., Brucher, M., Perrot, M., Duchesnay, E.: Scikit-learn: machine learning in Python. J. Mach. Learn. Res. **12**, 2825–2830 (2011)
45. Peng, X., Huang, Z., Lv, J., Zhu, H., Zhou, J.T.: Comic: multi-view clustering without parameter selection. In: International Conference on Machine Learning, pp. 5092–5101. PMLR (2019)
46. Rand, W.M.: Objective criteria for the evaluation of clustering methods. J. Am. Stat. Assoc. **66**(336), 846–850 (1971)
47. Rosenberg, A., Hirschberg, J.: V-measure: a conditional entropy-based external cluster evaluation measure. In: Proceedings of the 2007 Joint Conference on Empirical Methods in Natural Language Processing and Computational Natural Language Learning (EMNLP-CoNLL), pp. 410–420 (2007)
48. Rousseeuw, P.: Silhouettes: a graphical aid to the interpretation and validation of cluster analysis. J. Comput. Appl. Math. **20**, 53–65 (1987)
49. Saha, B., Mitra, P.: Dynamic algorithm for graph clustering using minimum cut tree. In: Sixth IEEE International Conference on Data Mining - Workshops (ICDMW'06), pp. 667–671 (2006)
50. Schwarz, G.: Estimating the dimension of a model. Ann. Stat. **6**(2), 461–464 (1978)
51. Shao, W., He, L., Lu, C., Yu, P.S.: Online multi-view clustering with incomplete views. In: 2016 IEEE International Conference on Big Data (Big Data), pp. 1012–1017 (2016)
52. Van Rossum, G., Drake, F.L.: Python 3 Reference Manual (2009)
53. Vendramin, L., Campello, R., Hruschka, E.: Relative clustering validity criteria: a comparative overview. Stat. Anal. Data Min. **3**, 209–235 (2010)
54. Vinh, N.X., Epps, J., Bailey, J.: Information theoretic measures for clusterings comparison: is a correction for chance necessary? In: Proceedings of the 26th Annual International Conference on Machine Learning, ICML'09, pp. 1073–1080 (2009)
55. Virtanen, P., Gommers, R., Oliphant, T.E., Haberland, M., Reddy, T., Cournapeau, D., Burovski, E., Peterson, P., Weckesser, W., Bright, J., van der Walt, S.J., Brett, M., Wilson, J., Millman, K.J., Mayorov, N., Nelson, A.R.J., Jones, E., Kern, R., Larson, E., Carey, C.J., Polat, İ., Feng, Y., Moore, E.W., VanderPlas, J., Laxalde, D., Perktold, J., Cimrman, R., Henriksen, I., Quintero, E.A., Harris, C.R., Archibald, A.M., Ribeiro, A.H., Pedregosa, F., van Mulbregt, P., SciPy 1.0 Contributors: SciPy 1.0: Fundamental Algorithms for Scientific Computing in Python. Nature Methods **17**, 261–272 (2020)
56. Wang, C.D., Lai, J.: Position regularized support vector domain description. Pattern Recognit. **46**(3), 875–884 (2013)

57. Wang, C.D., Lai, J.H., Huang, D., Zheng, W.S.: Svstream: a support vector-based algorithm for clustering data streams. IEEE Trans. Knowl. Data Eng. **25**(6), 1410–1424 (2011)
58. Wang, J., Tian, F., Yu, H., Liu, C.H., Zhan, K., Wang, X.: Diverse non-negative matrix factorization for multi-view data representation. IEEE Trans. Cybern. **48**(9), 2620–2632 (2017)
59. Yang, Y., Wang, H.: Multi-view clustering: a survey. Big Data Min. Anal. **1**(2), 83–107 (2018)
60. Zheng, Q., Zhu, J., Ma, Y., Li, Z., Tian, Z.: Multi-view subspace clustering networks with local and global graph information. Neurocomputing **449**, 15–23 (2021). https://doi.org/10.1016/j.neucom.2021.03.115. https://www.sciencedirect.com/science/article/pii/S0925231221005075
61. Zubaroglu, A., Atalay, V.: Data stream clustering: a review. Artif. Intell. Rev. **54**, 1201–1236 (2021)

Learning Shared and Discriminative Information from Multiview Data

Jia Chen, Hongjie Cao, Alireza Sadeghi, and Gang Wang

Abstract With the advent of the Internet-of-Things (IoTs) and Big Data movements, the rate of gathering as well as accumulating information nowadays is growing dramatically. Huge amounts of data that abound with heterogeneous features representing distinct perspectives of the same underlying patterns arise in various scientific fields. For instance, LIDAR signals, radar signals, and camera videos can be seen as three different views of a particular self-driving car. In signal processing, machine learning, statistics, and data science, multiview learning (a.k.a., data fusion/integration) is a popular field with well-documented analytical tools and wide-range application domains. In this chapter, we will introduce two algorithms which seek to find the shared and discriminative information from multiview data. First, graph multiview canonical correlation analysis (GMCCA) models will be introduced to unravel such shared information, which aims at searching for the low-dimensional representations from multiview data, while incorporating some graph prior information of shared latent components of the multiview data. Capitalizing on kernel methods, a generalization of GMCCA to capture the latent components of nonlinear data data is further established. A theoretical analysis of the GMCCA is provided to estimate the generalization error bound. Second, we will introduce discriminative principal component analysis (dPCA) to learn the unique subspace of

The work was supported in part by the National Key R&D Program of China under Grant 2021YFB1714800, the National Natural Science Foundation of China under Grants 62173034, 61925303, 62088101, and the Chongqing Natural Science Foundation under Grant 2021ZX4100027. The first three authors contributed equally.

J. Chen
Department of Electrical and Computer Engineering, University of California, Riverside, CA 92521, USA

H. Cao · G. Wang (✉)
State Key Lab of Intelligent Control and Decision of Complex Systems and School of Automation, Beijing Institute of Technology, Beijing 100081, China
e-mail: gangwang@bit.edu.cn

Beijing Institute of Technology Chongqing Innovation Center, Chongqing 401120, China

A. Sadeghi
Department of Electrical and Computer Engineering, University of Minnesota, Minneapolis, MN 55455, USA

W. Pedrycz and S. Chen (eds.), *Recent Advancements in Multi-View Data Analytics*, Studies in Big Data 106, https://doi.org/10.1007/978-3-030-95239-6_9

one dataset (a.k.a., target data) relative to the other dataset (a.k.a., background data). Under some mild assumptions, dPCA can be shown to be optimal in terms of finding the discriminative subspace of the target data with respect to the background data. Moreover, some representative applications are provided to validate the effectiveness of the aforementioned models in discovering shared or discriminative knowledge of multiview data.

1 Introduction

Over the past decade, it has become easier to access or generate heterogeneous data capturing multiple aspects, views, or modalities of the same information entities. For example, in social media, Facebook and Twitter posts can be understood as two views of the same set of users; and various news channels offer complementary and shared information of the same events. In transportation systems for instance, a traffic network indicating the traffic volume among cities and a distance network can be viewed as two views of the cities. Extracting the knowledge of heterogeneous data is known as multiview learning, which is gaining increasing attention in signal processing, machine learning, and data science fields with well-established novel algorithms and a broad spectrum of applications [15, 39, 40, 53].

Aiming at combining the strength of different types of data, multiview learning estimates the function of each individual view and optimizes all the functions jointly improving the generalization performance. This is realized by enforcing the agreement or disagreement among different views in the new space defined by the sought-after view-specific function, which correspond to learning shared and discriminative knowledge of multiview data, respectively. Representative models to find such shared information include canonical correlation analysis (CCA) variants [2, 4, 9, 24, 29, 31], multiview support vector machine (SVM), e.g., SVM-2K [16], multi-view twin SVM [58], and co-Laplacian SVM [51], multiple kernel learning [21, 52, 62], tensor methods [22, 43, 50], and matrix factorization-based methods [37, 41, 64, 65]. Learning discriminative information of multiview data is closely related to the paradigm of contrastive learning. Linear discriminant analysis (LDA) is one of the classical and supervised contrastive learning methods, which looks for subspcace directions along which the data within a cluster are close and among clusters are as far as possible [18]. Such discriminative concepts are adopted in kernel LDA [42], discriminative vanishing component analysis [30] and re-constructive and discriminative subspaces [17]. Recently, in the field of self-supervised representation learning, contrastive learning is showing appealing performance in downstream machine learning tasks [12, 25, 27, 36, 54]. For example, both unsupervised and semi-supervised multiview contrastive learning approaches are proposed in [54] to learn the influence of different views by reducing the mutual information between views.

Focusing on latent component discovery, canonical correlation analysis (CCA) is a 'workhorse' multiview representation learning method which projects two views to the same space and maximizes the correlation or minimizes the distance between the projected views [29]. Generalizations of the standard CCA include multiview CCA allowing more than two views [35], kernel or deep CCA dealing with nonlinear data [2, 14, 55, 56, 59], and nonnegative or sparse CCA promoting structural priors [8, 13, 31, 44, 57]. Building upon the maximum variance (MAXVAR) multiview CCA [28] framework, graph-regularized multiview (GM) CCA is developed in [9]. Later in this chapter, we will discuss GMCCA in detail. Given a extra graph view as prior information besides multiple feature datasets, GMCCA learns a shared and lower-dimensional representations of the multiple views while enforcing the graph smoothness. Introducing such graph regularizers has proven to be successful in improving the performance of various machine learning tasks, e.g., clustering, classification, data compression and reconstruction [11, 33, 47, 48]. A very related work reported in [5, 61] differs from the GMCCA by introducing different graph regularizers for different views in semi-supervised setup. Generalization error analysis of the GMCCA for unseen data samples will also be introduced from theoretical perspective, which measures the upperbound of the average dissimilarities among the projected multiview data. Relying on kernel method, a more general model of GMCCA, namely GKMCCA, will also be introduced to capture the nonlinearility of multiview data. The effectiveness of GMCCA and GKMCCA will be validated in several real-world data tests.

Most of the existing multiview learning frameworks are focused on the discovery of common patterns. Nonetheless, in some real-world applications, it is appealing to extract the discriminative knowledge of one view (a.k.a., target data) relative to the other view (a.k.a., background data). For example, consider the target data as gene-expression measurements of cancer patients from across different geographical areas and genders, and the background data as the gene-expression levels of healthy volunteers and the object is to cluster cancer patients based on the molecular sub-types. It is most likely that performing single view analysis such as principal component analysis (PCA) on any single dataset yields principal components corresponding to the shared information between target and background data [20]. Addressing such challenge, contrastive PCA is proposed in [1] to extract target data-specific information often missed by PCA. Inspired by contrastive PCA, we developed discriminative (d) PCA in [10], which searches for directions along which the variance of target data is maximized while the variance of background data is minimized. This boils down to solving a ratio trace maximization problem, similar to LDA. Despite the similarity in the formulation, dPCA is fundamentally from LDA due to the fact that there are unlabeled data. Furthermore, we will show the least-squares (LS) optimality of dPCA when there is one projection vector under some mild assumptions.

The remainder of this chapter is structured as follows. We begin in Sect. 2 by introducing the basis of CCA and MCCA, GMCCA and GKMCCA formulations, along with the generalization error bounds of GMCCA, and discussing representative applications of G(K)MCCA. Section 3 revisits the standard PCA and contrastive PCA, laying out the formulation and establishing under certain assumptions the optimality of dPCA, and ends with applications of dPCA. Section 4 concludes this chapter. The technical material that constitutes this chapter was mainly adapted from our previous works on G(K)MCCA [9] and dPCA [10].

2 Joint Knowledge Discovery from Multiview Data

2.1 Canonical Correlation Analysis

We consider N common sources $\{\check{\mathbf{s}}_n \in \mathbb{R}^{\rho}\}_{n=1}^{N}$ with M views expressed by M centered datasets $\{\mathbf{X}_m \in \mathbb{R}^{D_m \times N}\}_{m=1}^{M}$ where possibly we have $\rho \ll \min_m \{D_m\}_{m=1}^{M}$. For notional simplicity, we stack the N common source vectors and the m-th view vectors into the columns of data matrices $\check{\mathbf{S}} \in \mathbb{R}^{\rho \times N}$ and $\mathbf{X}_m$, respectively. The standard CCA is applicable when $M = 2$, and it seeks for two subspace matrices $\mathbf{U}_1 \in \mathbb{R}^{D_1 \times d}$ and $\mathbf{U}_2 \in \mathbb{R}^{D_2 \times d}$ with $d \leq \rho$, such that the two views become as close as possible in the projected space $\mathbb{R}^d$, which is mathematically realized by solving [24]

$$\min_{\mathbf{U}_1, \mathbf{U}_2} \quad \left\| \mathbf{U}_1^\top \mathbf{X}_1 - \mathbf{U}_2^\top \mathbf{X}_2 \right\|_F^2 \tag{1a}$$

$$\text{s. to} \quad \mathbf{U}_m^\top \left(\mathbf{X}_m \mathbf{X}_m^\top \right) \mathbf{U}_m = \mathbf{I}, \quad m = 1,\ 2 \tag{1b}$$

where the projection vectors, i.e., the columns of $\mathbf{U}_m$, are also known as the loading vectors. Tackling (1) is tantamount to solving a generalized eigenvalue decomposition problem, as initially noted in the seminal contribution [29]. Enforcing the projected data $\{\mathbf{U}_m^\top \mathbf{X}_m\}_{m=1}^{2}$ to be uncorrelated in the new dimensions, see (1b), prevents the trivial solution $\{\mathbf{U}_m = \mathbf{0}\}_{m=1}^{2}$. Here, the projections $\{\mathbf{U}_m^\top \mathbf{X}_m\}_{m=1}^{2}$ are oftentimes termed canonical variables and they can be interpreted as the estimates of the common sources.

Generalizing the standard CCA formulation to deal with more than 2 views ($M > 2$) is commonly referred to as multiview (M) CCA with the following two most popular formulations well-accepted in the literature: the sum-of-correlations (SUMCOR) MCCA [7] and the maximum-variance (MAXVAR) MCCA [35]. Both formulations linearly transform the M views from their original space to a common space $\mathbb{R}^d$, and minimize the Euclidean distances among the views in the common space. Specifically, the SUMCOR MCCA addresses the following constrained minimization problem

$$\min_{\{\mathbf{U}_m\}_{m=1}^{M}} \quad \sum_{m=1}^{M-1} \sum_{m' > m}^{M} \left\| \mathbf{U}_m^\top \mathbf{X}_m - \mathbf{U}_{m'}^\top \mathbf{X}_{m'} \right\|_F^2 \tag{2a}$$

$$\text{s. to} \quad \mathbf{U}_m^\top \left(\mathbf{X}_m \mathbf{X}_m^\top \right) \mathbf{U}_m = \mathbf{I}, \quad m = 1, \ldots, M \tag{2b}$$

whereas, the MAXVAR MCCA introduces an auxiliary common source vector $\mathbf{S} \in \mathbb{R}^{d \times N}$ and minimizes the summed distances from all projected views to the common source

$$\min_{\{\mathbf{U}_m\}_{m=1}^{M}, \mathbf{S}} \quad \sum_{m=1}^{M} \left\| \mathbf{U}_m^\top \mathbf{X}_m - \mathbf{S} \right\|_F^2 \tag{3a}$$

$$\text{s. to} \quad \mathbf{S}\mathbf{S}^\top = \mathbf{I}. \tag{3b}$$

It has been proven that the SUMCOR CCA problem is NP-hard in general [45]. When the covariance matrices from all the views are non-singular, the columns of $\hat{\mathbf{S}}^\top$ coincide with the first d eigenvectors of the 'normalized' data matrix $\sum_{m=1}^{M} \mathbf{X}_m^\top (\mathbf{X}_m \mathbf{X}_m^\top)^{-1} \mathbf{X}_m$ that correspond to its top d eigenvalues, and $\{\hat{\mathbf{U}}_m = (\mathbf{X}_m \mathbf{X}_m^\top)^{-1} \mathbf{X}_m \hat{\mathbf{S}}^\top\}_{m=1}^{M}$ [35].

2.2 *Graph-Regularized Multiview CCA*

The common source vectors $\{\check{\mathbf{s}}_i\}_{i=1}^{N}$, in an emerging set of contemporary applications including citation, social, power, transportation, and gene regulatory networks, reside on N connected nodes that can be (approximately) understood as forming a graph or network. For instance, in ResearchIndex networks, besides keywords, titles, abstracts, and introductions of collected articles, one usually has also access to the citation network capturing the contextual relationship among the papers. Such a graph/network of inter-dependent sources can be learned from the underlying physics, or it can be given a priori by some 'expert' in the field, or, it can be inferred/identified from extra (e.g., historical) views of the data. Such structural prior information can be leveraged along with the given (limited number of) multiview datasets to further enhance the MCCA performance. Specifically, here we will advocate a graph regularization term to extract this extra knowledge, which is judiciously accounted for in the low-dimensional space through the common source estimates.

In the following, we will use the tuple $\mathcal{G} := \{\mathcal{N}, \mathcal{W}\}$ to stand for the graph formed by the N common sources, in which $\mathcal{N} := \{1, \ldots, N\}$ denotes the vertex set, and $\mathcal{W} := \{w_{ij}\}_{(i,j) \in \mathcal{N} \times \mathcal{N}}$ collects all edge weights $\{w_{ij}\}$ over all vertex pairs (i, j). The so-termed weighted adjacency matrix $\mathbf{W} \in \mathbb{R}^{N \times N}$ is constructed with w_{ij} being its (i, j)-th entry. Here we only consider undirected graphs, for which $\mathbf{W} = \mathbf{W}^\top$ holds. Upon defining $d_i := \sum_{j=1}^{N} w_{ij}$ and $\mathbf{D} := \text{diag}(\{d_i\}_{i=1}^{N}) \in \mathbb{R}^{N \times N}$, the Laplacian matrix of graph $\mathcal{G}$ can be found as below

$$\mathbf{L}_{\mathcal{G}} := \mathbf{D} - \mathbf{W}. \tag{4}$$

Next, we will delineate a relationship between canonical correlations as well as graph regularization. To start with, we posit that the source vectors $\{\check{\mathbf{s}}_i\}_{i=1}^{N}$ are 'smooth' over the graph $\mathcal{G}$. In other words, the two source vectors $(\check{\mathbf{s}}_i, \check{\mathbf{s}}_j)$ rooted on the two connected nodes $i, j \in \mathcal{N}$ are also 'close' to each other in terms of Euclidean distance. As alluded earlier, vectors $\mathbf{s}_i$ and $\mathbf{s}_j$ are the d-dimensional approximants of $\check{\mathbf{s}}_i$ and $\check{\mathbf{s}}_j$, respectively. Building on this observation, a meaningful regularizer can thus be selected as the weighted sum of distances among all pairs of common source estimates $\mathbf{s}_i$ and $\mathbf{s}_j$ over $\mathcal{G}$, namely

$$\operatorname{Tr}\left(\mathbf{S}\mathbf{L}_{\mathcal{G}}\mathbf{S}^{\top}\right) = \frac{1}{2}\sum_{i=1}^{N}\sum_{j=1}^{N} w_{ij}\left\|\mathbf{s}_i - \mathbf{s}_j\right\|_2^2. \tag{5}$$

This regularization promotes that source vectors $\mathbf{s}_i$ and $\mathbf{s}_j$ residing on adjacent nodes $i, j \in \mathcal{N}$ with large weights w_{ij} should be similar (in the sense that their Euclidean distance should be small) to each other as well. To leverage this property, the quadratic term (5) is advocated as a regularizer in the standard MAXVAR MCCA formulation, leading to the following graph-regularized (G) MCCA formulation

$$\min_{\substack{\{\mathbf{U}_m\}\\ \mathbf{S}}} \sum_{m=1}^{M}\left\|\mathbf{U}_m^{\top}\mathbf{X}_m - \mathbf{S}\right\|_F^2 + \gamma\operatorname{Tr}\left(\mathbf{S}\mathbf{L}_{\mathcal{G}}\mathbf{S}^{\top}\right) \tag{6a}$$

$$\text{s. to} \quad \mathbf{S}\mathbf{S}^{\top} = \mathbf{I} \tag{6b}$$

with the coefficient $\gamma \geq 0$ trading off between minimizing the distance between the canonical variables and their corresponding common source estimates as well as promoting the smoothness of common source estimates over the graph $\mathcal{G}$. Specifically, in case when $\gamma = 0$, the proposed GMCCA boils down to the classical MCCA in (3); and, in case when γ increases, the GMCCA aggressively capitalizes on the graph knowledge to extract the canonical variables.

To find the optimal solution of the GMCCA problem, we first postulate that all per-view sample covariance matrices, i.e., $\{\mathbf{X}_m\mathbf{X}_m^{\top}\}$ have full rank. Next, by taking the partial derivative of the objective function in (6a) with respect to each variable $\mathbf{U}_m$, we arrive at the solver $\hat{\mathbf{U}}_m = (\mathbf{X}_m\mathbf{X}_m^{\top})^{-1}\mathbf{X}_m\mathbf{S}^{\top}$. The following eigenvalue problem (cf. (7)) will be brought about via substituting $\mathbf{U}_m$ by $\hat{\mathbf{U}}_m$ as well as neglecting some constant terms

$$\max_{\mathbf{S}} \ \operatorname{Tr}\Big[\mathbf{S}\Big(\sum_{m=1}^{M}\mathbf{X}_m^{\top}(\mathbf{X}_m\mathbf{X}_m^{\top})^{-1}\mathbf{X}_m - \gamma\mathbf{L}_{\mathcal{G}}\Big)\mathbf{S}^{\top}\Big] \tag{7a}$$

$$\text{s. to} \quad \mathbf{S}\mathbf{S}^{\top} = \mathbf{I}. \tag{7b}$$

Interestingly, akin to the standard MCCA, the solver $\hat{\mathbf{S}}$ of (7) coincides with the d leading eigenvectors of the processed data matrix

Algorithm 1: Graph-regularized MCCA.

1 **Input:** $\{\mathbf{X}_m\}_{m=1}^M$, d, γ, and $\mathbf{W}$.
2 **Build** $\mathbf{L}_{\mathcal{G}}$ using (4).
3 **Construct** $\mathbf{C} = \sum_{m=1}^M \mathbf{X}_m^\top \left(\mathbf{X}_m \mathbf{X}_m^\top\right)^{-1} \mathbf{X}_m - \gamma \mathbf{L}_{\mathcal{G}}$.
4 **Perform** eigendecomposition
5 on $\mathbf{C}$ to obtain the d eigenvectors associated with the d largest eigenvalues, which are collected as columns of $\hat{\mathbf{S}}^\top$.
6 **Compute** $\left\{\hat{\mathbf{U}}_m = \left(\mathbf{X}_m \mathbf{X}_m^\top\right)^{-1} \mathbf{X}_m \hat{\mathbf{S}}^\top\right\}_{m=1}^M$.
7 **Output:** $\{\hat{\mathbf{U}}_m\}_{m=1}^M$ and $\hat{\mathbf{S}}$.

$$\mathbf{C} := \sum_{m=1}^{M} \mathbf{X}_m^\top (\mathbf{X}_m \mathbf{X}_m^\top)^{-1} \mathbf{X}_m - \gamma \mathbf{L}_{\mathcal{G}}. \tag{8}$$

At the optimum, it is straightforward to check that the following holds

$$\sum_{m=1}^{M} \left\| \hat{\mathbf{U}}_m^\top \mathbf{X}_m - \hat{\mathbf{S}} \right\|_F^2 + \gamma \mathrm{Tr}\left(\hat{\mathbf{S}} \mathbf{L}_{\mathcal{G}} \hat{\mathbf{S}}^\top\right) = Md - \sum_{i=1}^{d} \lambda_i$$

where λ_i denotes the i-th largest eigenvalue of $\mathbf{C}$ in (8).

A detailed description of the GMCCA algorithm is summarized in Algorithm 1. We make some remarks at this point.

Remark 1 So far, we have developed a two-view graph-regularized CCA scheme by capitalizing upon the SUMCOR MCCA formulation. Nonetheless, the original cost was surrogated in [11] by its lower bound in order to derive an analytical solution, which cannot be readily generalized to deal with multi-view datasets with $M \geq 3$. In contrast with that scheme, the proposed GMCCA scheme in (6) is able to afford an analytical solution for any number $M \geq 2$ of views.

Remark 2 In general, when $N \gg D_m$, it is likely that $\mathbf{X}_m \mathbf{X}_m^\top$ has full rank. Even if it is singular, one could replace $\mathbf{X}_m \mathbf{X}_m^\top$ with $\mathbf{X}_m \mathbf{X}_m^\top + c_m \mathbf{I}$ by adding a small share of the identity matrix to ensure invertibility, where $c_m > 0$ is a small constant.

Remark 3 Different from our single graph regularizer in (6), the proposals in [5] and [61] rely on M different regularizers $\{\mathbf{U}_m^\top \mathbf{X}_m \mathbf{L}_{\mathcal{G}_m} \mathbf{X}_m^\top \mathbf{U}_m\}_m$ to exploit the extra graph knowledge, for view-specific graphs $\{\mathbf{L}_{\mathcal{G}_m}\}_m$ on data $\{\mathbf{X}_m\}_m$. However, the formulation in [61] does not admit an analytical solution, and convergence of iterative solvers for the resulting nonconvex problem can be guaranteed only to a stationary point. The approach in [5] focuses on semi-supervised learning tasks, in which cross-covariances of pair-wise datasets are not fully available. In contrast, the single graph Laplacian regularizer in (6) is effected on the common sources, to extract the pair-wise similarities of the N common sources. This is of practical significance when one has prior knowledge about the common sources in addition to the M given datasets. What is more, our proposed GMCCA approach comes with simple analytical solutions.

Remark 4 There are a number of approaches for selecting the hyper parameter γ, among which two are: (i) the cross-validation method, where some labeled training data are given (for supervised learning tasks), and γ is fixed to the one that yields the optimal empirical performance on the validation data; or, (ii) the spectral clustering method, that automatically chooses the best γ from a given set of candidates; see, for instance, [10].

2.3 Generalization Bound of GMCCA

The finite-sample performance of the proposed GMCCA is analyzed. In particular, the following analysis builds on the regression formulation introduced in [49, Ch. 6.5], which belongs to the alternating conditional expectation method proposed in [6]. The analysis establishes a generalization error bound of GMCCA, by leveraging the Rademacher's complexity.

The ultimate objective of MCCA is to seek for the common low-dimensional representation of the M-view data. To effectively measure the closeness between the estimated M low-dimensional sources, we consider the next error function

$$g(\check{\mathbf{s}}) := \sum_{m=1}^{M-1} \sum_{m'>m}^{M} \left\| \mathbf{U}_m^\top \psi_m(\check{\mathbf{s}}) - \mathbf{U}_{m'}^\top \psi_{m'}(\check{\mathbf{s}}) \right\|_F^2 \tag{9}$$

where the underlying source vector $\check{\mathbf{s}} \in \mathbb{R}^\rho$ is assumed to follow some fixed but unknown distribution denoted by $\mathcal{D}$, and there exists a linear mapping $\psi_m(\cdot)$ that projects a source vector from space $\mathbb{R}^\rho$ to the m-the view in $\mathbb{R}^{D_m}$, for $m = 1, \ldots, M$.

A generalization bound is established by evaluating the empirical average of $g(\check{\mathbf{s}})$ over a number, say N, of training data samples as follows

$$\begin{aligned}
\bar{g}_N(\check{\mathbf{s}}) &:= \frac{1}{N} \sum_{n=1}^{N} \sum_{m=1}^{M-1} \sum_{m'>m}^{M} \left\| \mathbf{U}_m^\top \psi_m(\check{\mathbf{s}}_n) - \mathbf{U}_{m'}^\top \psi_{m'}(\check{\mathbf{s}}_n) \right\|_F^2 \\
&= \frac{1}{N} \sum_{n=1}^{N} \sum_{m=1}^{M-1} \sum_{m'>m}^{M} \Big[\psi_m^\top(\check{\mathbf{s}}_n) \mathbf{U}_m \mathbf{U}_m^\top \psi_m(\check{\mathbf{s}}_n) - 2\psi_m^\top(\check{\mathbf{s}}_n) \\
&\quad \times \mathbf{U}_m \mathbf{U}_{m'}^\top \psi_{m'}(\check{\mathbf{s}}_n) + \psi_{m'}^\top(\check{\mathbf{s}}_n) \mathbf{U}_{m'} \mathbf{U}_{m'}^\top \psi_{m'}(\check{\mathbf{s}}_n) \Big].
\end{aligned}$$

If the terms are quadratic, it can be easily seen that

$$\psi_m^\top(\check{\mathbf{s}}) \mathbf{U}_m \mathbf{U}_m^\top \psi_m(\check{\mathbf{s}}) = \left\langle \mathrm{vec}(\mathbf{U}_m \mathbf{U}_m^\top),\ \mathrm{vec}(\psi_m(\check{\mathbf{s}}) \psi_m^\top(\check{\mathbf{s}})) \right\rangle \tag{10}$$

$$\psi_m^\top(\check{\mathbf{s}}) \mathbf{U}_m \mathbf{U}_{m'}^\top \psi_{m'}(\check{\mathbf{s}}) = \left\langle \mathrm{vec}(\mathbf{U}_m \mathbf{U}_{m'}^\top),\ \mathrm{vec}(\psi_m(\check{\mathbf{s}}) \psi_{m'}^\top(\check{\mathbf{s}})) \right\rangle. \tag{11}$$

Define the next two $\sum_{m=1}^{M-1} \sum_{m'>m}^{M} (D_m^2 + D_{m'}^2 + D_m D_{m'}) \times 1$ vectors

$$\psi(\check{\mathbf{s}}) := \left[\psi_{11}^\top(\check{\mathbf{s}}) \;\cdots\; \psi_{1M}^\top(\check{\mathbf{s}})\; \psi_{23}^\top(\check{\mathbf{s}}) \;\cdots\; \psi_{M,M-1}^\top(\check{\mathbf{s}})\right]^\top$$
$$\mathbf{u} := \left[\mathbf{u}_{11}^\top \;\cdots\; \mathbf{u}_{1M}^\top\; \mathbf{u}_{23}^\top \;\cdots\; \mathbf{u}_{M,M-1}^\top\right]^\top$$

where the two $(D_m^2 + D_{m'}^2 + D_m D_{m'}) \times 1$ vectors $\psi_{mm'}(\check{\mathbf{s}})$ and $\mathbf{u}_{mm'}$ are defined as below

$$\psi_{mm'} := \left[\text{vec}^\top(\psi_m \psi_m^\top)\; \text{vec}^\top(\psi_{m'}\psi_{m'}^\top)\; \sqrt{2}\text{vec}^\top(\psi_m \psi_{m'}^\top)\right]^\top$$
$$\mathbf{u}_{mm'} := \left[\text{vec}^\top(\mathbf{U}_m \mathbf{U}_m^\top)\; \text{vec}^\top(\mathbf{U}_{m'}\mathbf{U}_{m'}^\top)\; -\sqrt{2}\text{vec}^\top(\mathbf{U}_m \mathbf{U}_{m'}^\top)\right]^\top$$

for $m = 1, \ldots, M-1$ and $m' = 2, \ldots, M$.

Substituting Eqs. (10) and (11) into (9), one can see that $g(\check{\mathbf{s}})$ can be simplified to

$$g(\check{\mathbf{s}}) = \left\langle \mathbf{u},\; \psi(\check{\mathbf{s}}) \right\rangle. \tag{12}$$

with the norm of $\mathbf{u}$ given by

$$\|\mathbf{u}\|_2^2 = \sum_{m=1}^{M-1} \sum_{m'>m}^{M} \left\| \mathbf{U}_m^\top \mathbf{U}_m + \mathbf{U}_{m'}^\top \mathbf{U}_{m'} \right\|_F^2 .$$

Starting from (12), we establish next an upper bound on the expected value of $g(\check{\mathbf{s}})$ using (12). This is important, mainly because the expectation is carried over not only the N training samples, but also more importantly over the unseen samples.

Theorem 1 *Assume that (i) the N common source vectors $\{\check{\mathbf{s}}_n\}_{n=1}^N$ are drawn independent and identically distributed from some distribution $\mathcal{D}$; (ii) the M transformations $\{\psi_m(\cdot)\}_{m=1}^M$ of vectors $\{\check{\mathbf{s}}_n\}_{n=1}^N$ are bounded; and, iii) subspaces $\{\mathbf{U}_m \in \mathbb{R}^{D_m \times d}\}_{m=1}^M$ satisfy $\sum_{m=1}^{M-1}\sum_{m'>m}^{M} \|\mathbf{U}_m^\top \mathbf{U}_m + \mathbf{U}_{m'}^\top \mathbf{U}_{m'}\|_F^2 \le B^2$ $(B > 0)$ and $\{\mathbf{U}_m\}_{m=1}^M$ are the optimizers of (6). If we obtain low-dimensional representations of $\{\psi_m(\check{\mathbf{s}})\}_{m=1}^M$ specified by subspaces $\{\mathbf{U}_m \in \mathbb{R}^{D_m \times d}\}_{m=1}^M$, it holds with probability at least $1 - p$ that*

$$\mathbb{E}[g(\check{\mathbf{s}})] \le \bar{g}_N(\check{\mathbf{s}}) + 3RB\sqrt{\frac{\ln(2/p)}{2N}}$$
$$+ \frac{4B}{N}\sqrt{\sum_{n=1}^{N}\sum_{m=1}^{M-1}\sum_{m'>m}^{M}\left[\kappa_m(\check{\mathbf{s}}_n,\, \check{\mathbf{s}}_n) + \kappa_{m'}(\check{\mathbf{s}}_n,\, \check{\mathbf{s}}_n)\right]^2} \tag{13}$$

where $\kappa_m(\check{\mathbf{s}}_n, \check{\mathbf{s}}_n) := \left\langle \psi_m(\check{\mathbf{s}}_n), \psi_m(\check{\mathbf{s}}_n)\right\rangle$ for $n = 1, \ldots, N$, and $m = 1, \ldots, M$, while the constant R is given by

$$R := \max_{\check{\mathbf{s}} \sim \mathcal{D}} \sqrt{\sum_{m=1}^{M-1}\sum_{m'>m}^{M}\left[\kappa_m(\check{\mathbf{s}}, \check{\mathbf{s}}) + \kappa_{m'}(\check{\mathbf{s}}, \check{\mathbf{s}})\right]^2}.$$

Proof Equation (12) suggests that $g(\check{\mathbf{s}})$ belongs to the function class

$$\mathcal{F}_B := \left\{\check{\mathbf{s}} \rightarrow \langle \mathbf{u},\ \psi(\check{\mathbf{s}})\rangle : \|\mathbf{u}\| \leq B\right\}.$$

Consider the function class

$$\mathcal{H} = \left\{h : \check{\mathbf{s}} \rightarrow 1/(RB) f(\check{\mathbf{s}}) \big| f(\cdot) \in \mathcal{F}_B\right\} \subseteq \mathcal{A} \circ \mathcal{F}_B$$

where the function $\mathcal{A}$ is defined as

$$\mathcal{A}(x) = \begin{cases} 0, & \text{if } x \leq 0 \\ \frac{x}{RB}, & \text{if } 0 \leq x \leq RB \\ 1, & \text{otherwise} \end{cases}.$$

It can be shown that the function $\mathcal{A}(\cdot)$ is Lipschitz with Lipschitz constant $1/(RB)$, and that the range of functions in $\mathcal{H}$ is $[0,\ 1]$. Calling for [49, Th. 4.9], it can be further deduced that with probability at least $1-p$, the following holds true

$$\begin{aligned} \mathbb{E}[h(\check{\mathbf{s}})] &\leq \frac{1}{N}\sum_{n=1}^{N} h(\mathbf{s}_n) + R_N(\mathcal{H}) + \sqrt{\frac{\ln 2/p}{2N}} \\ &\leq \frac{1}{N}\sum_{n=1}^{N} h(\check{\mathbf{s}}_n) + \hat{R}_N(\mathcal{H}) + 3\sqrt{\frac{\ln 2/p}{2N}} \end{aligned} \tag{14}$$

where $\mathbb{E}[h(\check{\mathbf{s}})]$ denotes the expected value of $h(\cdot)$ on a new common source $\check{\mathbf{s}}$; and the Rademacher complexity $R_N(\mathcal{H})$ of $\mathcal{H}$ along with its empirical $\hat{R}_N(\mathcal{H})$ is defined as

$$\begin{aligned} R_N(\mathcal{H}) &:= \mathbb{E}_{\check{\mathbf{s}}}[\hat{R}_N(\mathcal{H})] \\ \hat{R}_N(\mathcal{H}) &:= \mathbb{E}_{\delta}\Big[\sup_{h\in\mathcal{H}}\Big|\frac{2}{N}\sum_{n=1}^{N} \delta_n h(\check{\mathbf{s}}_n)\Big| \,\big|\check{\mathbf{s}}_1,\ \check{\mathbf{s}}_2,\ \ldots,\ \check{\mathbf{s}}_N\Big] \end{aligned}$$

where $\boldsymbol{\delta} := \{\delta_n\}_{n=1}^N$ collects independent random variables drawn from the Rademacher distribution, i.e., $\{\Pr(\delta_n = 1) = \Pr(\delta_n = -1) = 0.5\}_{n=1}^N$. Further, $\mathbb{E}_{\delta}[\cdot]$ and $\mathbb{E}_{\check{\mathbf{s}}}[\cdot]$ denote the expectation with respect to $\boldsymbol{\delta}$ and $\check{\mathbf{s}}$, respectively.

Since $\mathcal{A}(\cdot)$ is a Lipschitz function with constant $1/(RB)$ obeying $\mathcal{A}(0) = 0$, the result in [3, Th. 12] asserts that

$$\hat{R}_N(\mathcal{H}) \leq 2/(RB)\hat{R}_N(\mathcal{F}_B). \tag{15}$$

Applying [49, Th. 4.12] leads to

$$\hat{R}_N(\mathcal{F}_B) \leq 2B/N\sqrt{\mathrm{Tr}(\mathbf{K})} \tag{16}$$

here the (i, j)-th entry of $\mathbf{K} \in \mathbb{R}^{N\times N}$ is $\langle \psi(\check{\mathbf{s}}_i), \psi(\check{\mathbf{s}}_j)\rangle$, for $i, j = 1, \ldots, N$. One can also confirm that

$$\mathrm{Tr}(\mathbf{K}) = \sum_{n=1}^{N}\sum_{m=1}^{M-1}\sum_{m'>m}^{M}\left[\kappa_m(\check{\mathbf{s}}_n, \check{\mathbf{s}}_n) + \kappa_{m'}(\check{\mathbf{s}}_n, \check{\mathbf{s}}_n)\right]^2. \tag{17}$$

Substituting (16) and (17) to (15) yields

$$\hat{R}_N(\mathcal{H}) \leq \frac{4}{RN}\sqrt{\sum_{n=1}^{N}\sum_{m=1}^{M-1}\sum_{m'>m}^{M}\left[\kappa_m(\check{\mathbf{s}}_n, \check{\mathbf{s}}_n) + \kappa_{m'}(\check{\mathbf{s}}_n, \check{\mathbf{s}}_n)\right]^2}.$$

Multiplying (14) by RB along with the last equation results in (13). □

Theorem 1 demonstrates that the empirical expectation of $g(\cdot)$ stays close to its ensemble $\mathbb{E}(g(\check{\mathbf{s}}))$, if $\{\|\mathbf{U}_m\|_F\}_m$ is controlled. As a consequence, it is prudent to trade off maximization of correlations among the M datasets with the norms of the resultant loading vectors.

2.4 Graph-Regularized Kernel MCCA

Unfortunately, the described GMCCA approach is only amendable to linear data transformations. Nonetheless, in practice, complex nonlinear data correlations are prevailing. To take care of nonlinear data, a graph-regularized kernel (GK) MCCA formulation is developed here to capture the nonlinear relationships among M datasets $\{\mathbf{X}_m\}_m$ through the kernel-based method. Specifically, the idea of GKM-CCA is to map the data vectors $\{\mathbf{X}_m\}_m$ to higher or even infinite dimensional feature spaces using M nonlinear functions, such that the task becomes linear in the new feature space, and thus one can apply the proposed GMCCA to learn the shared low-dimensional canonical variables.

To that aim, let ϕ_m represent the non-linear mapping from $\mathbb{R}^{D_m}$ to $\mathbb{R}^{L_m}$ for all m, where the dimension L_m can even be infinity. By mapping all data vectors $\{\mathbf{x}_{m,i}\}_{i=1}^N$ into $\{\phi_m(\mathbf{x}_{m,i})\}_{i=1}^N$, the linear similarities $\{\langle \mathbf{x}_{m,i}, \mathbf{x}_{m,j}\rangle\}_{i,j=1}^N$ is replaced with the mapped nonlinear similarities $\{\langle \phi_m(\mathbf{x}_{m,i}), \phi_m(\mathbf{x}_{m,j})\rangle\}_{i,j=1}^N$. Upon carefully selecting kernel function κ^m such that $\kappa^m(\mathbf{x}_{m,i}, \mathbf{x}_{m,j}) := \langle \phi_m(\mathbf{x}_{m,i}), \phi_m(\mathbf{x}_{m,j})\rangle$, the (i, j)-th entry of the kernel matrix $\bar{\mathbf{K}}_m \in \mathbb{R}^{N\times N}$ is given by $\kappa^m(\mathbf{x}_{m,i}, \mathbf{x}_{m,j})$, for all i, j, and m. In the sequel, centering $\{\phi_m(\mathbf{x}_{m,i})\}_{i=1}^N$ is realized by centering the kernel matrix for data $\mathbf{X}_m$ as

Algorithm 2: Graph-regularized kernel MCCA.

1 **Input:** $\{\mathbf{X}_m\}_{m=1}^M$, $\{\epsilon_m\}_{m=1}^M$, γ, $\mathbf{W}$, and $\{\kappa^m\}_{m=1}^M$.
2 **Construct** $\{\mathbf{K}_m\}_{m=1}^M$ using (18).
3 **Build** $\mathbf{L}_{\mathcal{G}}$ using (4).
4 **Form** $\mathbf{C}_g = \sum_{m=1}^M (\mathbf{K}_m + \epsilon_m \mathbf{I})^{-1} \mathbf{K}_m - \gamma \mathbf{L}_{\mathcal{G}}$.
5 **Perform** eigendecomposition
6 on $\mathbf{C}_g$ to obtain the d eigenvectors associated with the d largest eigenvalues, which are collected as columns of $\hat{\mathbf{S}}^\top$.
7 **Compute** $\{\hat{\mathbf{A}}_m = (\mathbf{K}_m + \epsilon_m \mathbf{I})^{-1} \hat{\mathbf{S}}^\top\}_{m=1}^M$.
8 **Output:** $\{\hat{\mathbf{A}}_m\}_{m=1}^M$ and $\hat{\mathbf{S}}$.

$$\mathbf{K}_m(i,\ j) := \bar{\mathbf{K}}_m(i,\ j) - \frac{1}{N}\sum_{k=1}^N \bar{\mathbf{K}}_m(k,\ j) - \frac{1}{N}\sum_{k=1}^N \bar{\mathbf{K}}_m(i, k) + \frac{1}{N^2}\sum_{i,j=1}^N \bar{\mathbf{K}}_m(i,\ j) \tag{18}$$

for $m = 1, \ldots, M$.

Replacing $\{\mathbf{U}_m\}_m$ with $\{\mathbf{X}_m\mathbf{A}_m\}_m$ and $\{\mathbf{X}_m^\top \mathbf{X}_m\}_m$ with centered kernel matrices $\{\mathbf{K}_m\}_m$ in the GMCCA formulation (6) gives our GKMCCA

$$\min_{\{\mathbf{A}_m\},\mathbf{S}} \quad \sum_{m=1}^M \left\|\mathbf{A}_m^\top \mathbf{K}_m - \mathbf{S}\right\|_F^2 + \gamma \mathrm{Tr}\left(\mathbf{S}\mathbf{L}_{\mathcal{G}}\mathbf{S}^\top\right) + \sum_{m=1}^M \epsilon_m \mathrm{Tr}\left(\mathbf{A}_m^\top \mathbf{K}_m \mathbf{A}_m\right) \tag{19a}$$

$$\text{s. to} \quad \mathbf{S}\mathbf{S}^\top = \mathbf{I}. \tag{19b}$$

Upon selecting nonsingular matrices $\{\mathbf{K}_m\}_{m=1}^M$, the columns of the optimizer $\hat{\mathbf{S}}^\top$ become the first d principal eigenvectors of $\mathbf{C}_g := \sum_{m=1}^M (\mathbf{K}_m + \epsilon_m \mathbf{I})^{-1}\mathbf{K}_m - \gamma \mathbf{L}_{\mathcal{G}} \in \mathbb{R}^{N\times N}$, and the solver $\hat{\mathbf{A}}_m$ can be obtained as $\hat{\mathbf{A}}_m = (\mathbf{K}_m + \epsilon_m \mathbf{I})^{-1}\hat{\mathbf{S}}^\top$. For implementation purposes, GKMCCA is shown in details in Algorithm 2.

Remark 5 When the (non)linear maps $\phi_m(\cdot)$ needed to form the kernel matrices $\{\mathbf{K}_m\}_{m=1}^M$ in (19) are not given in advance, a multi-kernel approach can be advocated (see e.g., [63]). Concretely, one presumes each $\mathbf{K}_m$ is a linear combination of some pre-selected kernel matrices P, namely $\mathbf{K}_m = \sum_{p=1}^P \beta_m^p \mathbf{K}_m^p$, where $\{\mathbf{K}_m^p\}_{p=1}^P$ represent view-specific kernel matrices for data $\mathbf{X}_m$. The unknown coefficients $\{\beta_m^p \geq 0\}_{m,p}$ are then jointly optimized along with $\{\mathbf{A}_m\}_m$ and $\mathbf{S}$ in (19).

Remark 6 When more than one type of connectivity information on the common sources are available, our single graph-regularized MCCA schemes can be generalized to accommodate multiple or multi-layer graphs. Specifically, the single graph-based regularization term $\gamma \mathrm{Tr}(\mathbf{S}\mathbf{L}_{\mathcal{G}}\mathbf{S}^\top)$ in (6) and (19) can be replaced with

$\sum_{i=1}^{I} \gamma_i \mathrm{Tr}(\mathbf{S}\mathbf{L}_{\mathcal{G}i}\mathbf{S}^\top)$ with possibly unknown yet learnable coefficients $\{\gamma_i\}_i$, where $\mathbf{L}_{\mathcal{G}i}$ denotes the graph Laplacian matrix of the i-th graph, for $i = 1, \ldots, I$.

2.5 Applications

2.5.1 UCI Data Clustering

The clustering performance of GMCCA can be verified by performing the techniques on the Multiple Features Data Set from the UCI machine learning repository,[1] which contains handwritten numerals from a collection of Dutch utility maps. The dataset which consists of 200 patterns for each of the class 0 to class 9, is represented in terms of 6 feature sets. And each feature set which is shown in Table 1 contains 2000 patterns. The views $\{\mathbf{X}_m \in \mathbb{R}^{D_m \times 1,400}\}_{m=1}^{6}$ consist of seven data sample groups comprising digits 1, 2, 3, 4, 7, 8, and 9, whose dimension respectively are different from each other and can be seen from Table 1 specifically. The figure is constructed by the view $\mathbf{X}_3$ and the (i, j)-th item of the graph adjacency matrix $\mathbf{W} \in \mathbb{R}^{N \times N}$ with $N = 1,400$ is

$$w_{ij} := \begin{cases} K^t(i, j), \ i \in \mathcal{N}_{k_1}(j) \text{ or } j \in \mathcal{N}_{k_1}(i) \\ 0, \qquad\qquad \text{otherwise} \end{cases} \tag{20}$$

where $\mathbf{K}^t$ represents a Gaussian kernel matrix of $\mathbf{X}_3$ whose bandwidth is tantamount to the average Euclidean distance, and $\mathcal{N}_{k_1}(j)$ signifies a set containing the $\mathbf{K}^t$'s column indices corresponding to k_1 nearest neighbors of j-th column. The data vectors were concatenated in order to apply GPCA and PCA techniques, which makes the data vectors dimension $\sum_{m=1}^{6} D_m$. Meanwhile we applied the K-means method to the dataset by utilizing either $\hat{\mathbf{S}}$, or the principal components with $\gamma = 0.1$ and $d = 3$.

Two different metrics, namely the clustering accuracy and the scatter ratio, are employed to assess the clustering performance. Clustering accuracy refers to the proportion of correctly clustered samples among the total samples. In this chapter, we define the scatter ratio as $C_t / \sum_{i=1}^{7} C_i$. Here, C_t refers to the total scatter value given by $C_t := \|\hat{\mathbf{S}}\|_F^2$, and C_i denotes the within-cluster scatter value given by $C_i := \sum_{j \in \mathcal{C}_i} \|\hat{\mathbf{s}}_j - \frac{1}{|\mathcal{C}_i|} \sum_{\ell \in \mathcal{C}_i} \hat{\mathbf{s}}_\ell\|_2^2$. In particular, $\mathcal{C}_i$ represents data vectors from the i-th cluster and the cardinality of $\mathcal{C}_i$ can be defined as $|\mathcal{C}_i|$ (Table 2).

The clustering capability of the techniques mentioned above under multiple k_1 values is exhibited in Table 3. As is shown in the table, both the clustering accuracy and scatter ratio of GMCCA demonstrate the highest value among all methods. The first two dimensions of the common source estimates obtained by (G)MCCA under $k_1 = 50$ are plotted in the Fig. 1, as well as the first two principal components of (G)PCA. Seven different colors indicate seven different clusters.It can be inferred from the scatter diagrams that GMCCA has the optimal separation effect on 7 clusters,

[1] See http://archive.ics.uci.edu/ml/datasets/Multiple+Features for the Multiple Features Data Set.

Table 1 Six features sets of handwritten digits

mfeat-fou	76-dim. Fourier coefficients of character shapes features
mfeat-fac	216-dim. profile correlations features
mfeat-kar	64-dim. Karhunen-Love coefficients features
mfeat-pix	240-dim. pixel averages in 2 × 3 windows
mfeat-zer	47-dim. Zernike moments features
mfeat-mor	6-dim. morphological features

Table 2 Clustering performance comparison

k_1	Clustering accuracy		Scatter ratio	
	GMCCA	GPCA	GMCCA	GPCA
10	0.8141	0.5407	9.37148	4.9569
20	0.8207	0.5405	11.6099	4.9693
30	0.8359	0.5438	12.2327	4.9868
40	0.8523	0.5453	12.0851	5.0157
50	0.8725	0.5444	12.1200	5.0640
MCCA	0.8007		5.5145	
PCA	0.5421		4.9495	

Table 3 Performance comparison between dPCA and PCA

d	Clustering error		Scatter ratio	
	dPCA	PCA	dPCA	PCA
1	0.1660	0.4900	2.0368	1.0247
2	0.1650	0.4905	1.8233	1.0209
3	0.1660	0.4895	1.6719	1.1327
4	0.1685	0.4885	1.4557	1.1190
5	0.1660	0.4890	1.4182	1.1085
10	0.1680	0.4885	1.2696	1.0865
50	0.1700	0.4880	1.0730	1.0568
100	0.1655	0.4905	1.0411	1.0508

that is, the data points within the clusters are relatively more centralized, but the data points across the clusters seem to be farther apart.

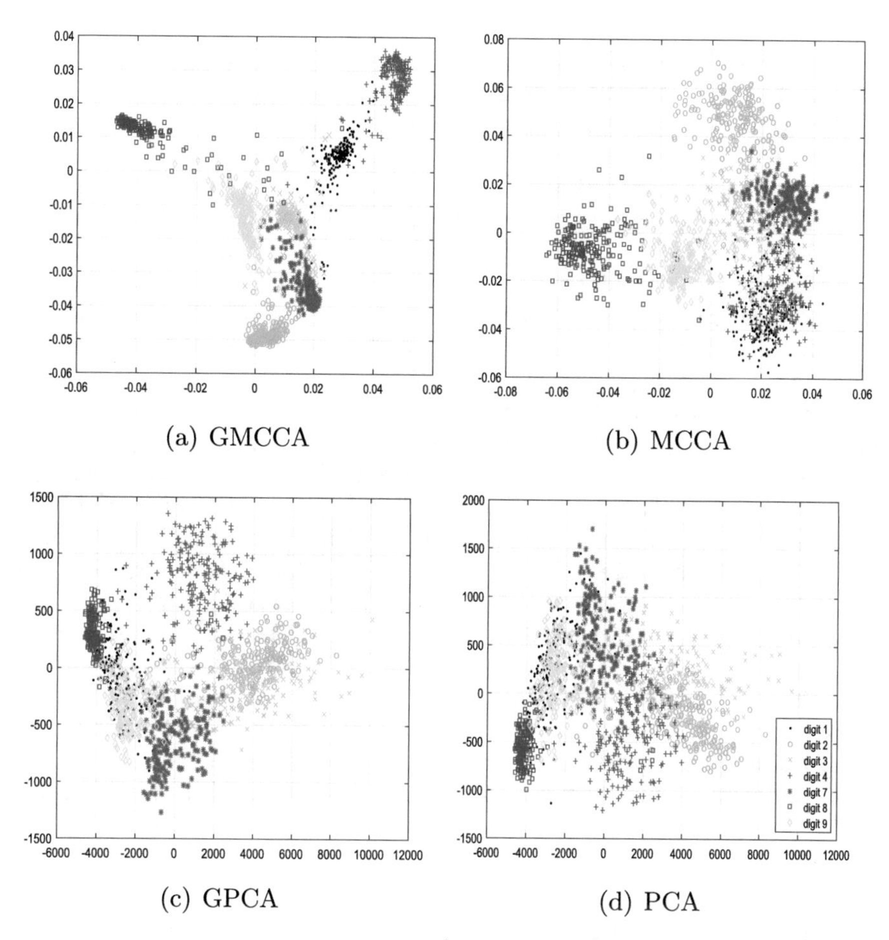

(a) GMCCA (b) MCCA

(c) GPCA (d) PCA

Fig. 1 Scatter diagrams of the first two rows of $\hat{\mathbf{S}}$ or principal components

2.5.2 Image Data Classification

The advantage of GKMCCA in performing classification tasks is evaluated within the MNIST database,[2] which contains 70,000 images of handwritten 28×28 digit images spread over 10 classes. In each MC test, three sets of N_{tr} images were randomly selected from each class for training, validation and testing respectively. In order to construct $\{\mathbf{X}_m \in \mathbb{R}^{196 \times 10N_{\text{tr}}}\}_{m=1}^3$, $\{\mathbf{X}_m^{\text{tu}} \in \mathbb{R}^{196 \times 10N_{\text{tr}}}\}_{m=1}^3$, and $\{\mathbf{X}_m^{\text{te}} \in \mathbb{R}^{196 \times 10N_{\text{tr}}}\}_{m=1}^3$, we adopted the method for producing three different views data, namely training data, tuning data and testing data which was described in Section VIII-E of [9], with the only difference being that all the pixel size of the data samples were adjusted to 14×14 pixels.

[2] Downloaded from http://yann.lecun.com/exdb/mnist/.

We employed Gaussian kernels for the training images $\{\mathbf{X}_m\}_{m=1}^3$ which can be expressed as $\{\mathbf{o}_i \in \mathbb{R}^{196\times 1}\}_{i=1}^{10N_{\text{tr}}}$ after resizing and vectorization operations, where the bandwidth parameters were specified as their average Euclidean distances correspondingly.

Drawing on the method of constructing graph adjacency matrix demonstrated in (20), the graph adjacency matrix was created by utilizing the kernel matrix of $\{\mathbf{o}_i\}$ which is represented by $\mathbf{K}_o \in \mathbb{R}^{10N_{\text{tr}}\times 10N_{\text{tr}}}$ except that $K^t(i,\ j)$ was substituted by the $(i,\ j)$-th entry of $\mathbf{K}_o$ as well as N was substituted by $10N_{\text{tr}}$. In the article that proposed the graph Laplacian regularized kernel method [5], three graph adjacency matrices were used, which were calculated using (20) after substituting $K^t(i,\ j)$ with the $(i,\ j)$-th entry of $\{\mathbf{K}_m\}_{m=1}^3$. By employing (20), the graph adjacency matrices were constructed for implementing graph dual (GD) MCCA (which is a special case of GKMCCA by using linear kernels) and graph dual (GD) PCA. We set $k_2 = N_{\text{tr}} - 1$ for all tests in this subsection. A set of 30 logarithmically spaced values between 10^{-3} and 10^3 are selected to determine the best hyperparameters of GKMCCA, KMCCA, GKPCA and many other methods which achieves optimal classification performance. By employing GKMCCA, KMCCA, GKPCA and other similar methods, we obtain ten obtain ten projection vectors, which are subsequently utilized to generate low-dimensional representations of $\{\mathbf{X}_m^{\text{te}}\}_{m=1}^3$. Based on the average of 30 MC runs, the classification accuracies of GKMCCA, KMCCA, GKPCA, KPCA, LKMCCA, GDMCCA, DMCCA, GDPCA, and DPCA methods are all presented.

Figures 1, 2, and 3 show the classification accuracy of the 10-dimensional representations of $\mathbf{X}_1^{\text{te}}$, $\mathbf{X}_2^{\text{te}}$, and $\mathbf{X}_3^{\text{te}}$. As clearly demonstrated in the figures, the competitive advantage of GKMCCA remains notable, no matter which test data view is considered.

3 Discriminative Knowledge Discovery from Multiview Data

3.1 (Contrastive) Principal Component Analysis Revisit

Suppose there exists two different datasets, namely the target dataset $\{\mathbf{x}_i \in \mathbb{R}^D\}_{i=1}^m$ that we are interested to analyze, and a background dataset $\{\mathbf{y}_j \in \mathbb{R}^D\}_{j=1}^n$ that contains latent background-related vectors that also exist in the target dataset. Assuming that the datasets are performed centralize processing in advance in order to maintain generality; namely, the sample mean $m^{-1}\sum_{i=1}^m \mathbf{x}_i$ $(n^{-1}\sum_{j=1}^n \mathbf{y}_j)$ should be subtracted from $\mathbf{x}_i$ $(\mathbf{y}_j)$. The following sections outline some PCA and cPCA basics to provide motivation for our novel approaches.

Only one dataset can be analyzed at a time with standard PCA. Through maximizing the variances of $\{\chi_i\}_{i=1}^m$ [34], standard PCA finds low-dimensional representations $\{\chi_i \in \mathbb{R}^d\}_{i=1}^m$ of $\{\mathbf{x}_i\}_{i=1}^m$ with $d < D$ being linear projections of $\{\mathbf{x}_i\}_{i=1}^m$. In particular, under $d = 1$, the normal (linear) PCA yields $\chi_i := \hat{\mathbf{u}}^\top \mathbf{x}_i$, with the vector

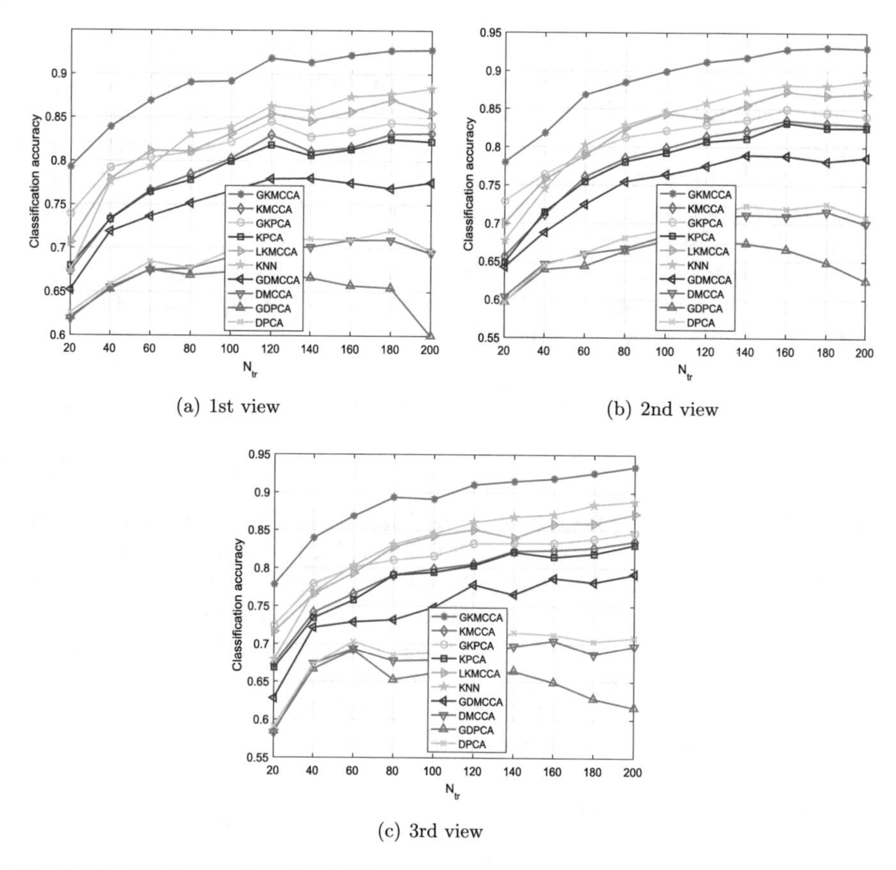

(a) 1st view

(b) 2nd view

(c) 3rd view

Fig. 2 Classification results using MNIST data

Fig. 3 Superimposed images

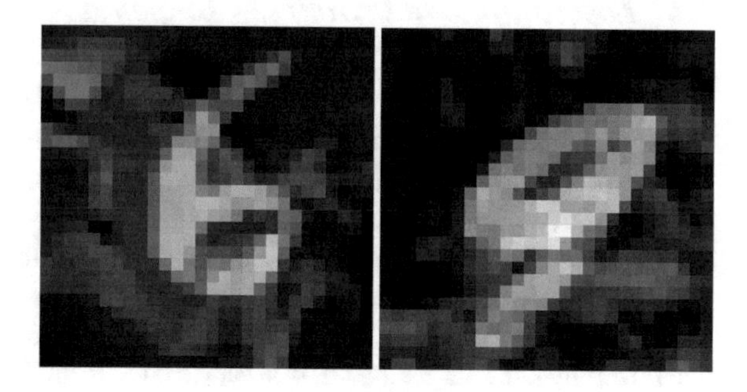

$\hat{\mathbf{u}} \in \mathbb{R}^D$ found by

$$\hat{\mathbf{u}} := \arg\max_{\mathbf{u}\in\mathbb{R}^D} \quad \mathbf{u}^\top \mathbf{C}_{xx}\mathbf{u} \quad \text{s. to} \quad \mathbf{u}^\top\mathbf{u} = 1 \tag{21}$$

where $\mathbf{C}_{xx} := (1/m)\sum_{i=1}^{m} \mathbf{x}_i\mathbf{x}_i^\top \in \mathbb{R}^{D\times D}$ represents the $\{\mathbf{x}_i\}_{i=1}^m$'s sample covariance matrix. Solving for (21) gives $\hat{\mathbf{u}}$, which is the top one normalized eigenvector of

$\mathbf{C}_{xx}$. The first principal component (PC) corresponding to the data points from the target dataset is composed of the obtained projections $\{\chi_i = \hat{\mathbf{u}}^\top \mathbf{x}_i\}_{i=1}^m$. If $d > 1$, PCA searches for $\{\mathbf{u}_i \in \mathbb{R}^D\}_{i=1}^d$, which is derived from the d eigenvectors of $\mathbf{C}_{xx}$ corresponding to the top d eigenvalues arranged in descending order. According to Sect. 1, PCA utilized on $\{\mathbf{x}_i\}_{i=1}^m$ only, or on the combined datasets $\{\{\mathbf{x}_i\}_{i=1}^m, \{\mathbf{y}_j\}_{j=1}^n\}$ can barely capture the specific features or patterns uniquely existed in target data compared to the background.

As an alternative, the recently proposed contrastive PCA technique finds a vector $\mathbf{u} \in \mathbb{R}^D$ along which large variations are maintained for the target data while small variations are kept for the background data, by solving [1]

$$\max_{\mathbf{u} \in \mathbb{R}^D} \quad \mathbf{u}^\top \mathbf{C}_{xx} \mathbf{u} - \alpha \mathbf{u}^\top \mathbf{C}_{yy} \mathbf{u} \tag{22a}$$

$$\text{s. to} \quad \mathbf{u}^\top \mathbf{u} = 1. \tag{22b}$$

Here $\mathbf{C}_{yy} := (1/n) \sum_{j=1}^n \mathbf{y}_j \mathbf{y}_j^\top \in \mathbb{R}^{D \times D}$ denotes the $\{\mathbf{y}_j\}_{j=1}^n$'s sample covariance matrix, and the contrastive parameter $\alpha \geq 0$ denotes the trade-off between maximizing the target data variance in (22a) and minimizing the background data variance. For a given α value, one can tackle the problem (22) by calculating the eigenvector corresponding to the largest eigenvalue of $\mathbf{C}_{xx} - \alpha \mathbf{C}_{yy}$, and finding the data projections along this eigenvector to form the first contrastive (c) PC. However, how to choose α remains an issue to be solved. Based on the spectral clustering method, α can be automatically selected from a series of potential candidates [1], although its search method tends to be brute-force and brings excessively high computational overhead when applied on large datasets.

3.2 *Discriminative Principal Component Analysis*

In contrast to PCA, linear discriminant analysis (LDA) is a supervised algorithm which take into account the information of class labels while learning from data to achieve dimensionality reduction. It seeks for linear projections that identify a new feature space, while maximizing the separability between classes and minimizing the variation within the same classes [18]. Namely, this technique maximizes the ratio of the between-classes variance to the within-classes variance.

Suppose that in an unsupervised situation similar to the above problem, consider two different data sets, a target data set and a background data set. Our task is to extract vectors that are meaningful in representing $\{\mathbf{x}_i\}_{i=1}^m$, but not $\{\mathbf{y}_j\}_{j=1}^n$. Therefore, it is natural to consider an approach that aims at maximizing the ratio of the projected target data variance to the background data variance. Our *discriminative (d) PCA* technique gives

$$\hat{\mathbf{u}} := \arg\max_{\mathbf{u} \in \mathbb{R}^D} \frac{\mathbf{u}^\top \mathbf{C}_{xx} \mathbf{u}}{\mathbf{u}^\top \mathbf{C}_{yy} \mathbf{u}} \tag{23}$$

We call the solution in (23) the discriminant subspace vector, and the set of projections $\{\hat{\mathbf{u}}^\top \mathbf{x}_i\}_{i=1}$ the first discriminative (d) PC. The solution of (23) will be discussed below.

The solution mentioned in (23) can be derived by calculating the $\mathbf{C}_{yy}^{-1}\mathbf{C}_{xx}$'s right eigenvector, according to the celebrated Lagrangian duality theory. The solution of (23) coincides with the right eigenvector of $\mathbf{C}_{yy}^{-1}\mathbf{C}_{xx}$ corresponding to its largest eigenvalue. In order to demonstrate this, we can rewrite (23) as the following formulation equivalently

$$\hat{\mathbf{u}} := \arg \max_{\mathbf{u}\in\mathbb{R}^D} \quad \mathbf{u}^\top \mathbf{C}_{xx}\mathbf{u} \tag{24a}$$

$$\text{s. to} \quad \mathbf{u}^\top \mathbf{C}_{yy}\mathbf{u} = 1. \tag{24b}$$

Assuming that λ represents the dual variable associated with the constraint (24b), the Lagrangian of (24) can be expressed as

$$\mathcal{L}(\mathbf{u};\ \lambda) = \mathbf{u}^\top \mathbf{C}_{xx}\mathbf{u} + \lambda\left(1 - \mathbf{u}^\top \mathbf{C}_{yy}\mathbf{u}\right). \tag{25}$$

When $(\hat{\mathbf{u}};\ \hat{\lambda})$ reaches the optimum, the KKT conditions imply that

$$\mathbf{C}_{xx}\hat{\mathbf{u}} = \hat{\lambda}\mathbf{C}_{yy}\hat{\mathbf{u}}. \tag{26}$$

Unlike standard eigen-equations, the equation above is a generalized eigenvalue problem, of which the solution $\hat{\mathbf{u}}$ is $(\mathbf{C}_{xx},\ \mathbf{C}_{yy})$'s generalized eigenvector associated with the generalized eigenvalue $\hat{\lambda}$. Left-multiplying both sides of the Eq. (26) by $\hat{\mathbf{u}}^\top$, we can get $\hat{\mathbf{u}}^\top \mathbf{C}_{xx}\hat{\mathbf{u}} = \hat{\lambda}\hat{\mathbf{u}}^\top \mathbf{C}_{yy}\hat{\mathbf{u}}$, which proves that the objective value of (24a) reached optimum while $\hat{\lambda} := \lambda_1$ is the largest. In addition, there are well-documented solvers based on e.g., the Cholesky's factorization [46] that can efficiently solve (26).

Moreover, if we assume that $\mathbf{C}_{yy}$ is nonsingular, Eq. (26) can be written as follows

$$\mathbf{C}_{yy}^{-1}\mathbf{C}_{xx}\hat{\mathbf{u}} = \hat{\lambda}\hat{\mathbf{u}} \tag{27}$$

which suggests that $\hat{\mathbf{u}}$ in (24) is the right eigenvector of $\mathbf{C}_{yy}^{-1}\mathbf{C}_{xx}$ associated with the largest eigenvalue $\hat{\lambda} = \lambda_1$.

When the dimension of subspace is more than one ($d \geq 2$), namely $\{\mathbf{u}_i \in \mathbb{R}^D\}_{i=1}^d$ that form $\mathbf{U} := [\mathbf{u}_1 \ \cdots \ \mathbf{u}_d] \in \mathbb{R}^{D\times d}$, in (23) with $\mathbf{C}_{yy}$ being nonsingular, can be generalized as follows (cf. (23))

$$\hat{\mathbf{U}} := \arg \max_{\mathbf{U}\in\mathbb{R}^{D\times d}} \operatorname{Tr}\left[\left(\mathbf{U}^\top \mathbf{C}_{yy}\mathbf{U}\right)^{-1}\mathbf{U}^\top \mathbf{C}_{xx}\mathbf{U}\right]. \tag{28}$$

Obviously, (28) is a *ratio trace* maximization problem; see e.g., [32], whose solution will be given in Theorem 2 (See [19, p. 448] for the proof).

Theorem 2 *Given centered data* $\{\mathbf{x}_i \in \mathbb{R}^D\}_{i=1}^m$ *and* $\{\mathbf{y}_j \in \mathbb{R}^D\}_{j=1}^n$ *with sample covariance matrices* $\mathbf{C}_{xx} := (1/m)\sum_{i=1}^m \mathbf{x}_i\mathbf{x}_i^\top$ *and* $\mathbf{C}_{yy} := (1/n)\sum_{j=1}^n \mathbf{y}_j\mathbf{y}_j^\top \succ \mathbf{0}$, *the i-th column of the dPCA optimal solution* $\hat{\mathbf{U}} \in \mathbb{R}^{D \times d}$ *in* (28) *is given by the right eigenvector of* $\mathbf{C}_{yy}^{-1}\mathbf{C}_{xx}$ *associated with the i-th largest eigenvalue, where* $i = 1, \ldots, d$.

Our dPCA technique for discriminative analytics of two datasets is outlined in Algorithm 3.

Four remarks are made below.

Remark 7 If the background data doesn't exist, it is obvious to see that $\mathbf{C}_{yy} = \mathbf{I}$, and in this case, dPCA becomes the standard PCA.

Remark 8 The following are several examples of possible target and background dataset combinations: (i) measurement data from healthy group $\{\mathbf{y}_j\}$ and diseased group $\{\mathbf{x}_i\}$, where the former and the latter have similar variations at the population level, but there are distinct variations between the two datasets due to the existence of different subtypes of diseases; (ii) 'before-treatment' data $\{\mathbf{y}_j\}$ and 'after-treatment' $\{\mathbf{x}_i\}$ datasets, in which the latter contains variations caused by treatment of diseases and the former contains additional measurement noise; and (iii) signal-free $\{\mathbf{y}_j\}$ and signal recordings $\{\mathbf{x}_i\}$, where the former consists of only noise.

Remark 9 Consider the eigenvalue decomposition $\mathbf{C}_{yy} = \mathbf{U}_y \Sigma_{yy} \mathbf{U}_y^\top$. With $\mathbf{C}_{yy}^{1/2} := \Sigma_{yy}^{1/2}\mathbf{U}_y^\top$, and the definition $\mathbf{v} := \mathbf{C}_{yy}^{\top/2}\mathbf{u} \in \mathbb{R}^D$, (24) can be expressed as

$$\hat{\mathbf{v}} := \arg\max_{\mathbf{v}\in\mathbb{R}^D} \quad \mathbf{v}^\top \mathbf{C}_{yy}^{-1/2}\mathbf{C}_{xx}\mathbf{C}_{yy}^{-\top/2}\mathbf{v} \tag{29a}$$

$$\text{s. to} \quad \mathbf{v}^\top \mathbf{v} = 1 \tag{29b}$$

where $\hat{\mathbf{v}}$ corresponds to the leading eigenvector of $\mathbf{C}_{yy}^{-1/2}\mathbf{C}_{xx}\mathbf{C}_{yy}^{-\top/2}$. Subsequently, $\hat{\mathbf{u}}$ in (24) is recovered as $\hat{\mathbf{u}} = \mathbf{C}_{yy}^{-\top/2}\hat{\mathbf{v}}$. This indeed suggests that discriminative analytics of $\{\mathbf{x}_i\}_{i=1}^m$ and $\{\mathbf{y}_j\}_{j=1}^n$ using dPCA can be viewed as PCA of the 'denoised' or 'background-removed' data $\{\mathbf{C}_{yy}^{-1/2}\mathbf{x}_i\}$, followed by an 'inverse' transformation to map the obtained subspace vector of the $\{\mathbf{C}_{yy}^{-1/2}\mathbf{x}_i\}$ data to $\{\mathbf{x}_i\}$ that of the target data. In this sense, $\{\mathbf{C}_{yy}^{-1/2}\mathbf{x}_i\}$ can be seen as the data obtained after removing the dominant 'background' subspace vectors from the target data.

Remark 10 The principal eigenvectors in (27) can be computed by running the inexpensive and easy-to-implement power method or Lanczos iterations [46].

Take another look at (24). On the basis of Lagrange duality theory, when choosing $\alpha = \hat{\lambda}$ in (22), where $\hat{\lambda}$ is the maximum eigenvalue of $\mathbf{C}_{yy}^{-1}\mathbf{C}_{xx}$, cPCA maximizing $\mathbf{u}^\top(\mathbf{C}_{xx} - \hat{\lambda}\mathbf{C}_{yy})\mathbf{u}$ is equivalent to $\max_{\mathbf{u}\in\mathbb{R}^D} \mathcal{L}(\mathbf{u}; \hat{\lambda}) = \mathbf{u}^\top(\mathbf{C}_{xx} - \hat{\lambda}\mathbf{C}_{yy})\mathbf{u} + \hat{\lambda}$, which coincides with (25) at the optimal $\lambda = \hat{\lambda}$. This suggests that when α in cPCA is equal to the optimal dual variable $\hat{\lambda}$ of our dPCA in (24) the optimizers of cPCA and dPCA share the same direction. This equivalence between dPCA and cPCA with appropriate α can also be derived from the following.

Algorithm 3: Discriminative PCA.

1 **Input:** Nonzero-mean target and background data $\{\mathring{\mathbf{x}}_i\}_{i=1}^m$ and $\{\mathring{\mathbf{y}}_j\}_{j=1}^n$; number of dPCs d.
2 **Exclude** the means from $\{\mathring{\mathbf{x}}_i\}$ and $\{\mathring{\mathbf{y}}_j\}$ to obtain centered data $\{\mathbf{x}_i\}$, and $\{\mathbf{y}_j\}$. Construct $\mathbf{C}_{xx}$ and $\mathbf{C}_{yy}$.
3 **Perform** eigendecomposition on $\mathbf{C}_{yy}^{-1}\mathbf{C}_{xx}$ to obtain the d right eigenvectors $\{\hat{\mathbf{u}}_i\}_{i=1}^d$ associated with the d largest eigenvalues.
4 **Output** $\hat{\mathbf{U}} = [\hat{\mathbf{u}}_1 \ \cdots \ \hat{\mathbf{u}}_d]$.

Theorem 3 ([23, Theorem 2]) *For real symmetric matrices* $\mathbf{C}_{xx} \succeq \mathbf{0}$ *and* $\mathbf{C}_{yy} \succ \mathbf{0}$, *the following holds*

$$\check{\lambda} = \frac{\check{\mathbf{u}}^\top \mathbf{C}_{xx}\check{\mathbf{u}}}{\check{\mathbf{u}}^\top \mathbf{C}_{yy}\check{\mathbf{u}}} = \max_{\|\mathbf{u}\|_2=1} \frac{\mathbf{u}^\top \mathbf{C}_{xx}\mathbf{u}}{\mathbf{u}^\top \mathbf{C}_{yy}\mathbf{u}}$$

if and only if

$$\check{\mathbf{u}}^\top(\mathbf{C}_{xx} - \check{\lambda}\mathbf{C}_{yy})\check{\mathbf{u}} = \max_{\|\mathbf{u}\|_2=1} \mathbf{u}^\top(\mathbf{C}_{xx} - \check{\lambda}\mathbf{C}_{yy})\mathbf{u}.$$

In order to explore the connection between dPCA and cPCA, we furthermore assume that $\mathbf{C}_{xx}$ and $\mathbf{C}_{yy}$ can be diagonalized simultaneously; namely, we can find a unitary matrix $\mathbf{U} \in \mathbb{R}^{D\times D}$ such that the following hold at the same time

$$\mathbf{C}_{xx} := \mathbf{U}\Sigma_{xx}\mathbf{U}^\top, \quad \text{and} \quad \mathbf{C}_{yy} := \mathbf{U}\Sigma_{yy}\mathbf{U}^\top$$

where diagonal matrices Σ_{xx} and $\Sigma_{yy} \succ \mathbf{0}$ hold accordingly eigenvalues $\{\lambda_x^i\}_{i=1}^D$ of $\mathbf{C}_{xx}$ and $\{\lambda_y^i\}_{i=1}^D$ of $\mathbf{C}_{yy}$ on their main diagonals. Although the two datasets may share some subspace vectors, $\{\lambda_x^i\}_{i=1}^D$ and $\{\lambda_y^i\}_{i=1}^D$ are in general not the same. It is straightforward to check that $\mathbf{C}_{yy}^{-1}\mathbf{C}_{xx} = \mathbf{U}\Sigma_{yy}^{-1}\Sigma_{xx}\mathbf{U}^\top = \mathbf{U}\text{diag}\big(\{\frac{\lambda_x^i}{\lambda_y^i}\}_{i=1}^D\big)\mathbf{U}^\top$. Seeking the top d latent subspace vectors is the same as taking the d columns of $\mathbf{U}$ associated with the d largest values among $\{\frac{\lambda_x^i}{\lambda_y^i}\}_{i=1}^D$. Alternatively, cPCA given a fixed α, finds the first d latent subspace vectors of $\mathbf{C}_{xx} - \alpha\mathbf{C}_{yy} = \mathbf{U}(\Sigma_{xx} - \alpha\boldsymbol{\Sigma}_{yy})\mathbf{U}^\top = \mathbf{U}\text{diag}\big(\{\lambda_x^i - \alpha\lambda_y^i\}_{i=1}^D\big)\mathbf{U}^\top$, which is equivalent to taking the d columns of $\mathbf{U}$ associated with the d largest values in $\{\lambda_x^i - \alpha\lambda_y^i\}_{i=1}^D$. This further confirms that when α is sufficiently large (small), cPCA returns the d columns of $\mathbf{U}$ associated with the d largest λ_y^i's (λ_x^i's). In the case that α is not chosen properly, there will be cases that cPCA is unable to extract the most contrastive feature in target data compared to background data. In contrast, dPCA does not have this problem simply because it makes you not bother to tune parameters.

3.3 Optimality of dPCA

In this section, we demonstrate that only when data obey a certain affine model will dPCA reach the optimal. Along the similar way, a factor analysis model is employed by PCA to demonstrate the non-centered background data $\{\mathring{\mathbf{y}}_j \in \mathbb{R}^D\}_{j=1}^n$ as

$$\mathring{\mathbf{y}}_j = \mathbf{m}_y + \mathbf{U}_b \boldsymbol{\psi}_j + \mathbf{e}_{y,j}, \quad j = 1, \ldots, n \tag{30}$$

where $\mathbf{m}_y \in \mathbb{R}^D$ represents the unknown location (mean) vector; $\mathbf{U}_b \in \mathbb{R}^{D\times k}$ has orthonormal columns with $k < D$; $\{\boldsymbol{\psi}_j \in \mathbb{R}^k\}_{j=1}^n$ are some unknown coefficients with covariance matrix $\boldsymbol{\Sigma}_b := \text{diag}(\lambda_{y,1}, \lambda_{y,2}, \ldots, \lambda_{y,k}) \in \mathbb{R}^{k\times k}$; and the modeling errors $\{\mathbf{e}_{y,j} \in \mathbb{R}^D\}_{j=1}^n$ are assumed zero-mean with covariance matrix $\mathbb{E}[\mathbf{e}_{y,j}\mathbf{e}_{y,j}^\top]=\mathbf{I}$. Adopting the least-squares (LS) criterion, the unknowns $\mathbf{m}_y$, $\mathbf{U}_b$, and $\{\boldsymbol{\psi}_j\}$ can be estimated by [60]

$$\min_{\substack{\mathbf{m}_y, \{\boldsymbol{\psi}_j\} \\ \mathbf{U}_b}} \sum_{j=1}^{n} \|\mathring{\mathbf{y}}_j - \mathbf{m}_y - \mathbf{U}_b \boldsymbol{\psi}_j\|_2^2 \quad \text{s. to} \quad \mathbf{U}_b^\top \mathbf{U}_b = \mathbf{I}$$

we find at the optimum $\hat{\mathbf{m}}_y := (1/n)\sum_{j=1}^n \mathring{\mathbf{y}}_j$, $\{\hat{\boldsymbol{\psi}}_j := \hat{\mathbf{U}}_b^\top(\mathring{\mathbf{y}}_j - \hat{\mathbf{m}}_y)\}$, with $\hat{\mathbf{U}}_b$ columns given by the first k leading eigenvectors of $\mathbf{C}_{yy} = (1/n)\sum_{j=1}^n \mathbf{y}_j\mathbf{y}_j^\top$, in which $\mathbf{y}_j := \mathring{\mathbf{y}}_j - \hat{\mathbf{m}}_y$. It is clear that $\mathbb{E}[\mathbf{y}_j\mathbf{y}_j^\top] = \mathbf{U}_b\boldsymbol{\Sigma}_b\mathbf{U}_b^\top + \mathbf{I}$. Let matrix $\mathbf{U}_n \in \mathbb{R}^{D\times(D-k)}$ with orthonormal columns satisfying $\mathbf{U}_n^\top\mathbf{U}_b = \mathbf{0}$, and $\mathbf{U}_y := [\mathbf{U}_b\ \mathbf{U}_n] \in \mathbb{R}^{D\times D}$ with $\boldsymbol{\Sigma}_y := \text{diag}(\{\lambda_{y,i}\}_{i=1}^D)$, where $\{\lambda_{y,k+\ell} := 1\}_{\ell=1}^{D-k}$. Therefore, $\mathbf{U}_b\boldsymbol{\Sigma}_b\mathbf{U}_b^\top + \mathbf{I} = \mathbf{U}_y\boldsymbol{\Sigma}_y\mathbf{U}_y^\top$. As $n \to \infty$, the strong law of large numbers asserts that $\mathbf{C}_{yy} \to \mathbb{E}[\mathbf{y}_j\mathbf{y}_j^\top]$; that is, $\mathbf{C}_{yy} = \mathbf{U}_y\boldsymbol{\Sigma}_y\mathbf{U}_y^\top$ as $n \to \infty$.

Our assumption is that the target data $\{\mathring{\mathbf{x}}_i \in \mathbb{R}^D\}_{i=1}^m$ and data $\{\mathring{\mathbf{y}}_j\}$ share the background related matrix $\mathbf{U}_b$. Nevertheless, the target data have d additional vectors which are specific to the target data and are not present in the background data. In real-world scenarios, this assumption is well justified. Take another look at the example discussed in Sect. 1, the gene expression data of patients and normal people have some common patterns due to differences in geographic environment and gender differences; yet there exist some specific latent subspace vectors corresponding to their disease in the gene expression data of patients. For brevity on $d = 1$, $\{\mathring{\mathbf{x}}_i\}$ can be modeled as

$$\mathring{\mathbf{x}}_i = \mathbf{m}_x + [\mathbf{U}_b\ \ \mathbf{u}_s]\begin{bmatrix}\chi_{b,i} \\ \chi_{s,i}\end{bmatrix} + \mathbf{e}_{x,i}, \quad i = 1, \ldots, m \tag{31}$$

where $\mathbf{m}_x \in \mathbb{R}^D$ represents the location of $\{\mathring{\mathbf{x}}_i\}_{i=1}^m$; $\{\mathbf{e}_{x,i}\}_{i=1}^m$ account for zero-mean modeling errors; $\mathbf{U}_x := [\mathbf{U}_b\ \mathbf{u}_s] \in \mathbb{R}^{D\times(k+1)}$ collects orthonormal columns, where $\mathbf{U}_b$ is the shared latent subspace vectors corresponding to background data, and $\mathbf{u}_s \in \mathbb{R}^D$ is a latent subspace vector which is unique to the target data instead of the background data. We devote to extracting this discriminative subspace $\mathbf{u}_s$ under given $\{\mathring{\mathbf{x}}_i\}_{i=1}^m$ and $\{\mathring{\mathbf{y}}_j\}_{j=1}^n$ condition.

Similarly, given $\{\mathring{\mathbf{x}}_i\}$, the unknowns $\mathbf{m}_x$, $\mathbf{U}_x$, and $\{\chi_i := [\chi_{b,i}^\top, \chi_{s,i}]^\top\}$ can be estimated by solving the following problem

$$\max_{\substack{\mathbf{m}_x, \{\chi_i\} \\ \mathbf{U}_x}} \sum_{i=1}^{m} \left\|\mathring{\mathbf{x}}_i - \mathbf{m}_x - \mathbf{U}_x\chi_i\right\|_2^2 \quad \text{s. to} \quad \mathbf{U}_x^\top\mathbf{U}_x = \mathbf{I}$$

which gives rise to $\hat{\mathbf{m}}_x := (1/m)\sum_{i=1}^m \mathring{\mathbf{x}}_i$, $\hat{\chi}_i := \hat{\mathbf{U}}_x^\top\mathbf{x}_i$ with $\mathbf{x}_i := \mathring{\mathbf{x}}_i - \hat{\mathbf{m}}_x$, where $\hat{\mathbf{U}}_x$ has columns the $(k+1)$ principal eigenvectors of $\mathbf{C}_{xx} = (1/m)\sum_{i=1}^m \mathbf{x}_i\mathbf{x}_i^\top$. When $m \to \infty$, it follows that $\mathbf{C}_{xx} = \mathbf{U}_x\boldsymbol{\Sigma}_x\mathbf{U}_x^\top$, with $\boldsymbol{\Sigma}_x := \mathbb{E}[\chi_i\chi_i^\top] = \text{diag}(\lambda_{x,1}, \lambda_{x,2}, \ldots, \lambda_{x,k+1}) \in \mathbb{R}^{(k+1)\times(k+1)}$.

Let $\boldsymbol{\Sigma}_{x,k} \in \mathbb{R}^{k\times k}$ denote the submatrix of $\boldsymbol{\Sigma}_x$, which is composed by its top k rows and columns. While $\mathbf{C}_{yy}$ is nonsingular as well as $m, n \to \infty$, one can express $\mathbf{C}_{yy}^{-1}\mathbf{C}_{xx}$ as follows

$$\begin{aligned}
&\mathbf{U}_y\boldsymbol{\Sigma}_y^{-1}\mathbf{U}_y^\top\mathbf{U}_x\boldsymbol{\Sigma}_x\mathbf{U}_x^\top \\
&= [\mathbf{U}_b\ \mathbf{U}_n]\begin{bmatrix}\boldsymbol{\Sigma}_b^{-1} & \mathbf{0} \\ \mathbf{0} & \mathbf{I}\end{bmatrix}\begin{bmatrix}\mathbf{I} & \mathbf{0} \\ \mathbf{0} & \mathbf{U}_n^\top\mathbf{u}_s\end{bmatrix}\begin{bmatrix}\boldsymbol{\Sigma}_{x,k} & \mathbf{0} \\ \mathbf{0} & \lambda_{x,k+1}\end{bmatrix}\begin{bmatrix}\mathbf{U}_b^\top \\ \mathbf{u}_s^\top\end{bmatrix} \\
&= [\mathbf{U}_b\ \mathbf{U}_n]\begin{bmatrix}\boldsymbol{\Sigma}_b^{-1}\boldsymbol{\Sigma}_{x,k} & \mathbf{0} \\ \mathbf{0} & \lambda_{x,k+1}\mathbf{U}_n^\top\mathbf{u}_s\end{bmatrix}\begin{bmatrix}\mathbf{U}_b^\top \\ \mathbf{u}_s^\top\end{bmatrix} \\
&= \mathbf{U}_b\boldsymbol{\Sigma}_b^{-1}\boldsymbol{\Sigma}_{x,k}\mathbf{U}_b^\top + \lambda_{x,k+1}\mathbf{U}_n\mathbf{U}_n^\top\mathbf{u}_s\mathbf{u}_s^\top.
\end{aligned}$$

It becomes obvious that the rank of the top two summands is respectively k and 1, and we can infer that the rank of $\mathbf{C}_{yy}^{-1}\mathbf{C}_{xx}$ is at most $k+1$. Assuming that $\mathbf{u}_{b,i}$ represents the i-th column of $\mathbf{U}_b$ that is orthogonal to $\{\mathbf{u}_{b,j}\}_{j=1,j\neq i}^k$ and $\mathbf{u}_s$, then we can right-multiply $\mathbf{C}_{yy}^{-1}\mathbf{C}_{xx}$ by $\mathbf{u}_{b,i}$ and arrive at

$$\mathbf{C}_{yy}^{-1}\mathbf{C}_{xx}\mathbf{u}_{b,i} = (\lambda_{x,i}/\lambda_{y,i})\,\mathbf{u}_{b,i}$$

for $i = 1, \ldots, k$, which hints that $\{\mathbf{u}_{b,i}\}_{i=1}^k$ are k eigenvectors of $\mathbf{C}_{yy}^{-1}\mathbf{C}_{xx}$ associated with the eigenvalues $\{\lambda_{x,i}/\lambda_{y,i}\}_{i=1}^k$. Again, right-multiplying $\mathbf{C}_{yy}^{-1}\mathbf{C}_{xx}$ by $\mathbf{u}_s$ leads to

$$\mathbf{C}_{yy}^{-1}\mathbf{C}_{xx}\mathbf{u}_s = \lambda_{x,k+1}\mathbf{U}_n\mathbf{U}_n^\top\mathbf{u}_s\mathbf{u}_s^\top\mathbf{u}_s = \lambda_{x,k+1}\mathbf{U}_n\mathbf{U}_n^\top\mathbf{u}_s. \tag{32}$$

Here are three facts that we will utilize to proceed: Fact (i) $\mathbf{u}_s$ is orthogonal to all columns of $\mathbf{U}_b$; Fact (ii) columns of $\mathbf{U}_n$ are orthogonal to those of $\mathbf{U}_b$; and, Fact (iii) $[\mathbf{U}_b\ \mathbf{U}_n]$ has full rank. Based on Facts (i)–(iii), it is easy to conclude that $\mathbf{u}_s$ can be uniquely expressed as a linear combination of columns of $\mathbf{U}_n$; that is, $\mathbf{u}_s := \sum_{i=1}^{D-k} p_i\mathbf{u}_{n,i}$, where $\{p_i\}_{i=1}^{D-k}$ are some unknown coefficients, and $\mathbf{u}_{n,i}$ denotes the i-th column of $\mathbf{U}_n$. One can further express $\mathbf{U}_n\mathbf{U}_n^\top\mathbf{u}_s$ in (32) as follows

$$\begin{aligned}
\mathbf{U}_n\mathbf{U}_n^\top \mathbf{u}_s &= [\mathbf{u}_{n,1} \ \cdots \ \mathbf{u}_{n,D-k}] \begin{bmatrix} \mathbf{u}_{n,1}^\top \mathbf{u}_s \\ \vdots \\ \mathbf{u}_{n,D-k}^\top \mathbf{u}_s \end{bmatrix} \\
&= \mathbf{u}_{n,1}\mathbf{u}_{n,1}^\top \mathbf{u}_s + \cdots + \mathbf{u}_{n,D-k}\mathbf{u}_{n,D-k}^\top \mathbf{u}_s \\
&= p_1 \mathbf{u}_{n,1} + \cdots + p_{D-k}\mathbf{u}_{n,D-k} \\
&= \mathbf{u}_s
\end{aligned}$$

leading to $\mathbf{C}_{yy}^{-1}\mathbf{C}_{xx}\mathbf{u}_s = \lambda_{x,k+1}\mathbf{u}_s$; that is, $\mathbf{u}_s$ is the $(k+1)$-st eigenvector of $\mathbf{C}_{yy}^{-1}\mathbf{C}_{xx}$ corresponding to the eigenvalue $\lambda_{x,k+1}$.

Before moving on, we make the following two standing assumptions.

Assumption 1 The background and target data are generated according to the models (30) and (31), respectively, with the background data sample covariance matrix being nonsingular.

Assumption 2 It holds for all $i = 1, \ldots, k$ that $\lambda_{x,k+1}/\lambda_{y,k+1} > \lambda_{x,i}/\lambda_{y,i}$.

Assumption 2 basically suggests that $\mathbf{u}_s$ is discriminative enough in the target data compared to the background data. After combining Assumption 2 and the fact that $\mathbf{u}_s$ is an eigenvector of $\mathbf{C}_{yy}^{-1}\mathbf{C}_{xx}$, it can be inferred that the eigenvector of $\mathbf{C}_{yy}^{-1}\mathbf{C}_{xx}$ associated with the largest eigenvalue is $\mathbf{u}_s$. Under these two assumptions, we formally state the optimality of dPCA next.

Theorem 4 *Under Assumptions 1 and 2 with $d = 1$, as $m, n \to \infty$, the solution of* (23) *recovers the subspace vector specific to target data relative to background data, namely* $\mathbf{u}_s$.

3.4 Applications

To examine the validity of dPCA, we employed a new dataset that contains Semi-synthetic target $\{\mathring{\mathbf{x}}_i \in \mathbb{R}^{784}\}_{i=1}^{2,000}$ and background images $\{\mathring{\mathbf{y}}_j \in \mathbb{R}^{784}\}_{j=1}^{3,000}$ by superimposing. The images were selected randomly from the MNIST[3] and CIFAR-10 [38] datasets. Specifically, we created the target data $\{\mathbf{x}_i \in \mathbb{R}^{784}\}_{i=1}^{2,000}$ by randomly superimposing 2,000 handwritten digits 6 and 9 (1,000 for each) on the top of 2,000 frog images from the CIFAR-10 database [38], and the sample mean from each data point are removed. The schematic diagrams are shown in Fig. 3. We resized all the images to 28×28 pixels. The zero-mean background data $\{\mathbf{y}_j \in \mathbb{R}^{784}\}_{j=1}^{3,000}$ were constructed using 3,000 cropped frog images, which were randomly chosen from the remaining frog images in the CIFAR-10 database.

The dPCA Algorithm 3 was executed on $\{\mathring{\mathbf{x}}_i\}$ and $\{\mathring{\mathbf{y}}_j\}$ with $d = 2$, while PCA was performed on $\{\mathring{\mathbf{x}}_i\}$ only. The first two PCs and dPCs are presented in the left

[3] Downloaded from http://yann.lecun.com/exdb/mnist/.

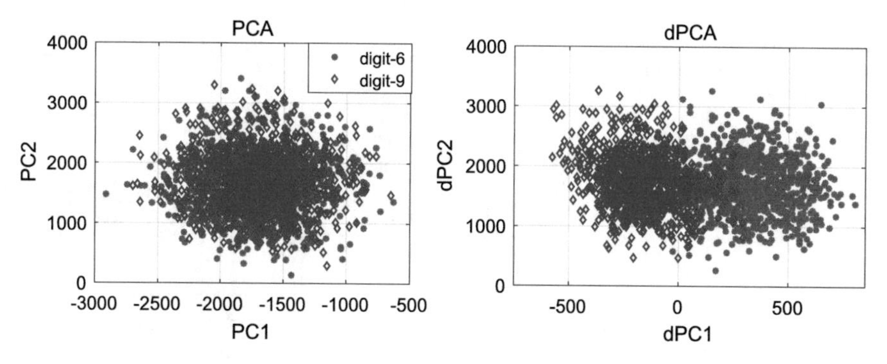

Fig. 4 dPCA versus PCA on semi-synthetic images

and right panels of Fig. 4, respectively. Obviously, dPCA reveals the discriminative information of the target data describing digits 6 and 9 relative to the background data, so as to realize the clustering of numbers 6 and 9 subtypes. Conversely, PCA cannot capture the patterns that associated with digits 6 and 9, but only the patterns corresponding to the generic background.

In order to further evaluate the performance of dPCA and PCA, K-means is employed by utilizing the resulting low-dimensional representations of the target data. There exists two metrics to evaluate the clustering performance: one is called clustering error defined as the ratio of the number of incorrectly clustered data vectors over m. And the other one is called scatter ratio which verifies the cluster separation and scatter ratio is defined as $S_t / \sum_{i=1}^{2} S_i$, where S_t given by $S_t := \sum_{j=1}^{2,000} \|\hat{\mathbf{U}}^\top \mathbf{x}_j\|_2^2$ represent the total scatter value, and $\{S_i\}_{i=1}^{2}$ given by and $\{S_i := \sum_{j \in C_i} \|\hat{\mathbf{U}}^\top \mathbf{x}_j - \hat{\mathbf{U}}^\top \sum_{k \in C_i} \mathbf{x}_k\|_2^2\}_{i=1}^{2}$ denote the within cluster scatter values, respectively. Here C_i represent data vectors set that belongs to cluster i. Table 3 shows the clustering errors and scatter ratios of dPCA and PCA under different d values. Obviously, dPCA exhibits a lower clustering error and a higher scatter ratio.

In real-world situations, dPCA was also evaluated on real protein expression data [26] to demonstrate its ability to discover subgroups. For example, in the experimental example of Down syndrome mice, the target data $\{\mathring{\mathbf{x}}_i \in \mathbb{R}^{77}\}_{i=1}^{267}$ is composed of 267 data points and each of them consists of protein expression measurements of mice developed Down Syndrome [26]. Particularly, the entire dataset was divided into two parts. The first part $\{\mathring{\mathbf{x}}_i\}_{i=1}^{135}$ which consists of 135 data points contains protein expression measurements of mice exposed to drug treatment, yet the remaining $\{\mathring{\mathbf{x}}_i\}_{i=136}^{267}$ collects measurements of 134 mice without such treatment. On the other hand, background data $\{\mathring{\mathbf{y}}_j \in \mathbb{R}^{77}\}_{j=1}^{135}$ contains the exactly same protein expression from mice in a healthy state, which may show similar natural variation like age and sex difference as the mice with disease but won't reveal any significant differences relative to Down syndrome.

To set up the discriminative hyperparameters while applying cPCA on our datasets, the spectral clustering method presented in [1] was utilized to obtain four

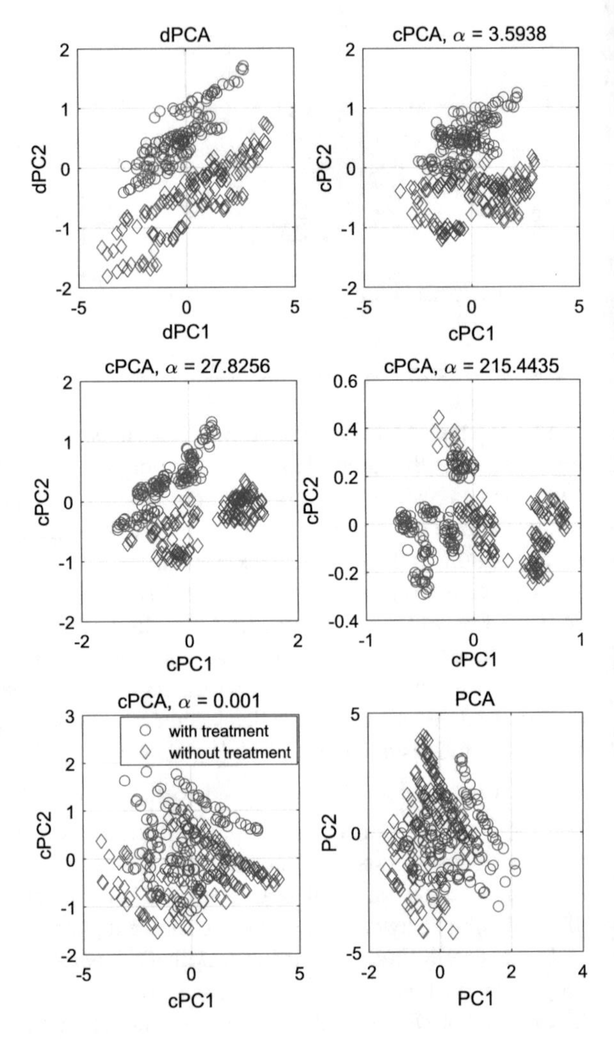

Fig. 5 Discovering subgroups in mice protein expression data

α values, which were selected from 15 logarithmically-spaced values between 10^{-3} and 10^3.

Figure 5 presents the experimental results, in which the mice exposed to treatment are represented by red circles while those not exposed to treatment are represented by black diamonds. Evidently, the low-dimensional representation results obtained by using the PCA method are distributed similarly. In contrast, the low-dimensional representations cluster two groups of mice successfully when dPCA is employed. At the price of runtime (about 15 times more than dPCA), cPCA with well tuned parameters (specifically, $\alpha = 3.5938$ and 27.8256) can also discover the two subgroups.

4 Concluding Remarks

In this chapter, we have presented two classes of methods for discovering knowledge from multiview datasets. The first class of methods including graph (G) MCCA and graph kernel (GK) MCCA can identify the most prominent patterns that are shared by all views. On the other hand, the second class of methods, termed discriminative PCA, is able to discover the most discriminative patterns between a target and a background view. We have also provided concrete mathematical formulations for both classes of methods, as well as some theoretical and experimental analyses.

References

1. Abid, A., Zhang, M.J., Bagaria, V.K., Zou, J.: Exploring patterns enriched in a dataset with contrastive principal component analysis. Nat. Commun. **9**(1), 1–7 (2018)
2. Andrew, G., Arora, R., Bilmes, J., Livescu, K.: Deep canonical correlation analysis. In: Proceedings of the Conference on Machine Learning. Atlanta, USA, June 16–21 (2013)
3. Bartlett, P.L., Mendelson, S.: Rademacher and Gaussian complexities: risk bounds and structural results. J. Mach. Learn. Res. **3**, 463–482 (2002)
4. Benton, A., Khayrallah, H., Gujral, B., Reisinger, D.A., Zhang, S., Arora, R.: Deep generalized canonical correlation analysis. arXiv preprint arXiv:1702.02519 (2017)
5. Blaschko, M.B., Shelton, J.A., Bartels, A., Lampert, C.H., Gretton, A.: Semi-supervised kernel canonical correlation analysis with application to human fMRI. Pattern Recognit. Lett. **32**(11), 1572–1583 (2011)
6. Breiman, L., Friedman, J.H.: Estimating optimal transformations for multiple regression and correlation. J. Am. Stat. Assoc. **80**(391), 580–598 (1985)
7. Carroll, J.D.: Generalization of canonical correlation analysis to three or more sets of variables. In: Proceedings of the 76th Annual Convention of the American Psychological Association, vol. 3, pp. 227–228. Washington, DC (1968)
8. Chen, J., Schizas, I.D.: Distributed efficient multimodal data clustering. In: Proceedings of the European Signal Processing Conference, pp. 2304–2308. Kos Island, Greece, Aug 28–Sep 2 (2017)
9. Chen, J., Wang, G., Giannakis, G.B.: Graph multiview canonical correlation analysis. IEEE Trans. Signal Process. **67**(11), 2826–2838 (2019)
10. Chen, J., Wang, G., Giannakis, G.B.: Nonlinear dimensionality reduction for discriminative analytics of multiple datasets. IEEE Trans. Signal Process. **67**(3), 740–752 (2019)
11. Chen, J., Wang, G., Shen, Y., Giannakis, G.B.: Canonical correlation analysis of datasets with a common source graph. IEEE Trans. Signal Process. **66**(16), 4398–4408 (2018)
12. Chen, T., Kornblith, S., Norouzi, M., Hinton, G.: A simple framework for contrastive learning of visual representations. In: International Conference on Machine Learning, pp. 1597–1607. PMLR (2020)
13. Chen, X., Han, L., Carbonell, J.: Structured sparse canonical correlation analysis. In: Artificial Intelligence and Statistics, pp. 199–207 (2012)
14. Choi, J.H., Vishwanathan, S.: Dfacto: distributed factorization of tensors. In: Ghahramani, Z., Welling, M., Cortes, C., Lawrence, N., Weinberger, K. (eds.) Advances in Neural Information Processing Systems, vol. 27, pp. 1296–1304. Curran Associates, Inc. (2014), http://papers.nips.cc/paper/5395-dfacto-distributed-factorization-of-tensors.pdf
15. Correa, N.M., Adali, T., Li, Y.O., Calhoun, V.D.: Canonical correlation analysis for data fusion and group inferences. IEEE Signal Process. Mag. **27**(4), 39–50 (2010)

16. Farquhar, J., Hardoon, D., Meng, H., Shawe-Taylor, J.S., Szedmak, S.: Two view learning: Svm-2k, theory and practice. In: Advances in Neural Information Processing Systems, pp. 355–362 (2006)
17. Fidler, S., Skocaj, D., Leibardus, A.: Combining reconstructive and discriminative subspace methods for robust classification and regression by subsampling. IEEE Trans. Pattern Anal. Mach. Intell. **28**(3), 337–350 (2006)
18. Fisher, R.A.: The use of multiple measurements in taxonomic problems. Ann. Eugenics **7**(2), 179–188 (1936)
19. Fukunaga, K.: Introduction to Statistical Pattern Recognition, 2nd edn. Academic Press, San Diego, CA, USA (2013)
20. Garte, S.: The role of ethnicity in cancer susceptibility gene polymorphisms: the example of CYP1A1. Carcinogenesis **19**(8), 1329–1332 (1998)
21. Gönen, M., Alpaydın, E.: Multiple kernel learning algorithms. J. Mach. Learn. Res. **12**, 2211–2268 (2011)
22. Gujral, E., Papalexakis, E.E.: Smacd: semi-supervised multi-aspect community detection. In: Proceedings of the SIAM International Conference on Data Mining, pp. 702–710. SIAM (2018)
23. Guo, Y., Li, S., Yang, J., Shu, T., Wu, L.: A generalized Foley-Sammon transform based on generalized Fisher discriminant criterion and its application to face recognition. Pattern Recognit. Lett. **24**(1–3), 147–158 (2003)
24. Hardoon, D.R., Szedmak, S., Shawe-Taylor, J.: Canonical correlation analysis: an overview with application to learning methods. Neural Comput. **16**(12), 2639–2664 (2004)
25. Hassani, K., Khasahmadi, A.H.: Contrastive multi-view representation learning on graphs. In: International Conference on Machine Learning, pp. 4116–4126. PMLR (2020)
26. Higuera, C., Gardiner, K.J., Cios, K.J.: Self-organizing feature maps identify proteins critical to learning in a mouse model of down syndrome. PloS ONE **10**(6), e0129126 (2015)
27. Hjelm, R.D., Fedorov, A., Lavoie-Marchildon, S., Grewal, K., Bachman, P., Trischler, A., Bengio, Y.: Learning deep representations by mutual information estimation and maximization. arXiv preprint arXiv:1808.06670 (2018)
28. Horst, P.: Generalized canonical correlations and their application to experimental data. No. 14, J. Clin. Psychol. (1961)
29. Hotelling, H.: Relations between two sets of variates. Biometrika **28**(3/4), 321–377 (1936)
30. Hou, C., Nie, F., Tao, D.: Discriminative vanishing component analysis. In: AAAI, pp. 1666–1672. Phoenix, Arizona, USA, Feb. 12–17 (2016)
31. Ibrahim, M.S., Zamzam, A.S., Konar, A., Sidiropoulos, N.D.: Cell-edge detection via selective cooperation and generalized canonical correlation. IEEE Trans. Wireless Commun. 1–10 (2021)
32. Jaffe, A., Wax, M.: Single-site localization via maximum discrimination multipath fingerprinting. IEEE Trans. Signal Process. **62**(7), 1718–1728 (2014)
33. Jiang, B., Ding, C., Tang, J.: Graph-Laplacian PCA: closed-form solution and robustness. In: Proceedings of the International Conference on Computer Vision Pattern Recognition. Portland, USA, Jun. 25–27 (2013)
34. Karl Pearson, F.R.S.: LIII. On lines and planes of closest fit to systems of points in space. The London, Edinburgh, and Dublin Phil. Mag. J. Sci. **2**(11), 559–572 (1901)
35. Kettenring, J.R.: Canonical analysis of several sets of variables. Biometrika **58**(3), 433–451 (1971)
36. Khosla, P., Teterwak, P., Wang, C., Sarna, A., Tian, Y., Isola, P., Maschinot, A., Liu, C., Krishnan, D.: Supervised contrastive learning. arXiv preprint arXiv:2004.11362 (2020)
37. Kingma, D.P., Welling, M.: Auto-encoding variational bayes. arXiv preprint arXiv:1312.6114 (2013)
38. Krizhevsky, A.: Learning multiple layers of features from tiny images. In: Master's Thesis. Department of Computer Science, University of Toronto (2009)
39. Li, T., Dou, Y.: Representation learning on textual network with personalized PageRank. Sci. China Inf. Sci. **64**(11), 1–10 (2021)

40. Li, Z., Tang, J.: Semi-supervised local feature selection for data classification. Sci. China Inf. Sci. **64**(9), 1–12 (2021)
41. Mariappan, R., Rajan, V.: Deep collective matrix factorization for augmented multi-view learning. Mach. Learn. **108**(8–9), 1395–1420 (2019)
42. Mika, S., Ratsch, G., Weston, J., Scholkopf, B., Mullers, K.R.: Fisher discriminant analysis with kernels. In: Neural Networks for Signal Processing IX: Proceedings of the IEEE Signal Processing Society Workshop, pp. 41–48. Madison, WI, USA, Aug. 25 (1999)
43. Nickel, M., Tresp, V., Kriegel, H.P.: A three-way model for collective learning on multi-relational data. In: Proceedings of the International Conference on Machine Learning, pp. 809–816 (2011)
44. Rastogi, P., Van Durme, B., Arora, R.: Multiview LSA: representation learning via generalized CCA. In: Proceedings of the North American Chapter of the Association for Computational Linguistics: Human Language Technologies, pp. 556–566. Denver, Colorado, USA, May 31–June 5 (2015)
45. Rupnik, J., Skraba, P., Shawe-Taylor, J., Guettes, S.: A comparison of relaxations of multiset cannonical correlation analysis and applications. arXiv:1302.0974 (Feb 2013)
46. Saad, Y.: Iterative Methods for Sparse Linear Systems, 2nd edn. SIAM, Philadelphia, PA, USA (2003)
47. Shahid, N., Perraudin, N., Kalofolias, V., Puy, G., Vandergheynst, P.: Fast robust PCA on graphs. IEEE J. Sel. Topics Signal Process. **10**(4), 740–756 (2016)
48. Shang, F., Jiao, L., Wang, F.: Graph dual regularization non-negative matrix factorization for co-clustering. Pattern Recognit. **45**(6), 2237–2250 (2012)
49. Shawe-Taylor, J., Cristianini, N.: Kernel Methods for Pattern Analysis, 1st edn. Cambridge University Press, Cambridge, United Kingdom (2004)
50. Sidiropoulos, N.D., De Lathauwer, L., Fu, X., Huang, K., Papalexakis, E.E., Faloutsos, C.: Tensor Decomposition for Signal Processing and Machine Learning, vol. 65, pp. 3551–3582. IEEE (2017)
51. Sindhwani, V., Niyogi, P., Belkin, M.: A co-regularization approach to semi-supervised learning with multiple views. In: Proceedings of ICML Workshop on Learning with Multiple Views, vol. 2005, pp. 74–79. Citeseer (2005)
52. Sonnenburg, S., Rätsch, G., Schäfer, C., Schölkopf, B.: Large scale multiple kernel learning. J. Mach. Learn. Res. **7**(Jul), 1531–1565 (2006)
53. Sun, S.: A survey of multi-view machine learning. Neural Comput. App. **23**(7–8), 2031–2038 (2013)
54. Tian, Y., Sun, C., Poole, B., Krishnan, D., Schmid, C., Isola, P.: What makes for good views for contrastive learning. arXiv preprint arXiv:2005.10243 (2020)
55. Wang, G., Giannakis, G.B., Chen, J.: Learning ReLU networks on linearly separable data: algorithm, optimality, and generalization. IEEE Trans. Signal Process. **67**(9), 2357–2370 (2019)
56. Wang, W., Arora, R., Livescu, K., Bilmes, J.: On deep multi-view representation learning. In: The International Conference on Machine Learning, pp. 1083–1092. Lille, France, July 6–11 (2015)
57. Witten, D.M., Tibshirani, R., Hastie, T.: A penalized matrix decomposition, with applications to sparse principal components and canonical correlation analysis. Biostatistics **10**(3), 515–534 (2009)
58. Xie, X., Sun, S.: Multi-view twin support vector machines. Intell. Data Anal. **19**(4), 701–712 (2015)
59. Yamanishi, Y., Vert, J.P., Nakaya, A., Kanehisa, M.: Extraction of correlated gene clusters from multiple genomic data by generalized kernel canonical correlation analysis. Bioinformatics **19**(1), i323–i330 (2003)
60. Yang, B.: Projection approximation subspace tracking. IEEE Trans. Signal Process. **43**(1), 95–107 (1995)
61. Yuan, Y., Sun, Q.: Graph regularized multiset canonical correlations with applications to joint feature extraction. Pattern Recognit. **47**(12), 3907–3919 (2014)

62. Zhang, L., Wang, G., Giannakis, G.B.: Going beyond linear dependencies to unveil connectivity of meshed grids. In: Proceedings of the IEEE Workshop on Computational Advances in Multi-Sensor Adaptive Processing. Curacao, Dutch Antilles, Dec. 2017
63. Zhang, L., Wang, G., Romero, D., Giannakis, G.B.: Randomized block Frank-Wolfe for convergent large-scale learning. IEEE Trans. Signal Process. **65**(24), 6448–6461 (2019)
64. Zhao, P., Jiang, Y., Zhou, Z.H.: Multi-view matrix completion for clustering with side information. In: Pacific-Asia Conference on Knowledge Discovery and Data Mining, pp. 403–415. Springer (2017)
65. Zhao, W., Xu, C., Guan, Z., Liu, Y.: Multiview concept learning via deep matrix factorization. IEEE Trans. Neural Netw. Learn. Syst. (2020)

A Supervised Ensemble Subspace Learning Model Based on Multi-view Feature Fusion Employing Multi-template EMG Signals

Aditya Saikia, Anil Hazarika, Bikram Patir, and Amarprit Singh

Abstract Multi-domain data or feature fusion models gain wide attention in various real-world applications. It is due to fact that it enables interaction of multi-modalities or large sets of single modality data and extracts low order comprehensive attributes that effectively represents inherent characteristics of multiple sets sensor measurement associated with given physiological phenomena. This paper presents a multi-view feature fusion based model that encodes the learned subspaces by generalizing Canonical correlation analysis (CCA) for diagnosis of neuromuscular disorders (i.e., amyotrophic lateral sclerosis (ALS), myopathy) using single channel intramuscular EMG signals. We first segregated each studied subject group into a number of subgroups from where multi-view features are generated by dint of two formulation pursuits, followed by subspace transformation. The extracted low dimensional features are concatenated using feature level fusion and transformed to find discriminate features which are embedded to the decision model. The method is then investigated using real-time EMG datasets. Statistical one-way analysis of variance (ANOVA) is also applied to validate the analysis, which shows that the method is significant ($p < 0.05$). The overall classification accuracy obtained is 99.4% with a specificity of 100% and sensitivities of 99.5% and 97.7%. For fair comparison, many state-of-art methods are implemented over our study datasets using common decision model and results are compared with our methods, which ensure the effectiveness and reliability of multi-view feature model for real-time applications.

Keywords Data-driven technique · Feature fusion · Electromyography · Discriminant feature

---

A. Saikia · A. Hazarika (✉)
Department of Physics, Cotton University, Assam, India

B. Patir
Department of Computer Science, Cotton University, Assam, India

A. Singh
Department of Electronics and Communication Engineering, Tezpur University, Assam, India

W. Pedrycz and S. Chen (eds.), *Recent Advancements in Multi-View Data Analytics*, Studies in Big Data 106, https://doi.org/10.1007/978-3-030-95239-6_10

1 Introduction

Multi-view learning models (MvL) are gaining attention in the field of computer vision and pattern recognition [1, 2]. Real-world problems often involve multi-view data, which form feature subsets of same or different statistical nature [3, 4]. Term multi-view indicates various forms of modalities data of a given object [1]. MvL utilizes high dimensional views in order to extract best feature set for improvement of classification performance and reliability of decision model. Uni-model inference system, such as electromyogram (EMG) analysis, often requires evidences of multiple signals of a given disorder. Signals that provides proper information of disease class is not pre-known and difficult to identify. Furthermore, uncertainty, confidence level, reliability and quality of information of each signal is different [5]. In addition, the nature of signals depends on disease duration, age of subject and experimental setting [2]. Therefore, utilization of multiple signals of a given disorder to extract features and embedding them to classifier, are essential to enhance model interpretability and robustness.

MvL employs optimization technique for multiple input features interaction in order to find a single form of discriminant vector. This features have much discriminating ability especially, when best feature sets are unknown. However, MvL often suffers dimensionality issue. Therefore, proper level of information fusion is highly essential. Two commonly used fusions are data fusion or feature level fusion and decision level fusion. In multimodal fusion, e.g., audio-visual fusion [8], the modalities data are different and they convey complementary information. In that case, decision level fusion or modalities fusion at final stage combines each modality output to make an appropriate sense. For single modality data whereas model inference highly depends on appropriate integration of data, feature level fusion is considered to be effective [5, 6]. Classical data integration methods directly integrate data [9] result in high order feature space. In contrast to that, feature level fusion enables interaction of multi-view input features using transformation and optimization, and extracts more discriminant feature that comprehensively carry information of underlying phenomena. Such applications include multimodal fusion, multisensor data fusion, structured data fusion and multiview data as mentioned in [5].

Recently Canonical correlation analysis (CCA) [10] has attracted interest in classification problems [11]. Unlike principal components analysis (PCA) [12], linear discriminant analysis (LDA) and self-organizing feature map (SOFM) [14], CCA provides a unified framework that transforms input features to a well-defined subspace and fuses the optimized features to derive discriminant vectors for classification. Xi et al. [13] adopted CCA-feature-level fusion and evaluated Type-I and II for EMG classification. Despite potential benefit and massive work in various domains, use of recently proposed multi-view feature fusion using CCA for nonlinear pattern recognition tasks is still in preliminary stage. This study introduces a feature level fusion based multi-view-CCA (mCCA) model for classification of EMG signals to diagnose neuromuscular disorders and healthy control subjects.

2 State of the Art of EMG Diagnosis

Quantitative methods [15–18] employed various EMG features (i.e., time, frequency, time-frequency etc.) to diagnose neuromuscular disorders such as-*amyotrophic lateral sclerosis* (ALS) [19]. Yousefi et al. [20] comprehensively reviewed various methods and their steady improvement in classification performance. Various methods employed motor unit action potential (MUAP) features [21], wavelet-feature, autoregressive coefficient (AR) [22], time-domain (TD), AR+RMS [23], multiscale-PCA features [24] and intrinsic mode functions [25] extracting from specific pre-diagnosed signals. In [26], bagging ensemble classifier was adopted. First DWT features were extracted from EMG signals and then statistical measures were evaluated for diagnosis of neuromuscular disorders and reported promising results. Bozkurt et al. [27] introduced ANN and combined neural network (CNN) based classification models by using autoregressive (AR) parametric and subspace-based parameters to diagnose healthy, myopathic and neurogenic disorders. Hasni et al. [28] introduced two stage cascaded SVM based algorithm which provided good classification performance with an accuracy of 95.7 %. Elamvazuthi et al. [29] introduced Multilayer Perceptron model that employed AR, RMS, zero crossing, waveform length and mean absolute Value are used to extract features. Lahmiri et al. [30] Computer-aided-diagnosis system that uses the estimated parameters of an auto-regressive moving average (ARMA) model and the variance of its disturbances to fully capture the linear dynamics and variability in the EMG signal. Recently various new classification models are also introduced in [31–34]. Subasi et al. [35] first employed multiscale principal component analysis to remove the impulsive noise from the signals. Then, dual-tree complex wavelet transform (DT-CWT) is used to extract features which were later embedded to the rotation forest ensemble classifier for classification of EMG signals. Furthermore, performances of various other models were also studied and reported good results. In [36], DWT-based statistical features were elicited from dominated MUAP to diagnose myopathy and ALS. However, it requires guidance information of MUAP morphology to determine dominated MUAPs. Furthermore, such methods often suffer difficulty in matching of wavelet coefficient from each level with the specific MUAP. In addition, use of specific coefficients from each level may not be feasible for all MUAPs [25]. It is worth noting that the performance of DWT-methods relies on the proper choice of wavelet function (i.e., mother wavelet) and level selection, which can be avoided by using methodology [37] and ground subjective knowledge. Recent popularity of DWT-method in analyzing the nonstationary signal is mainly due to its good time and frequency resolutions. It is seen that main focus of most earlier methods is on performance enhancement by adopting different feature extraction techniques (i.e., PCA, LDA, SOFM etc.) or decision models, and accordingly they reported promising results [18, 25, 36]. However, many research queries such why performance get enhanced ?, how they considered feature searched space ?, reliability of methods etc., remain unanswered or unexplored. Further, associated constraint and complexities of many methods create the bridge between in and out profession, and also make them unsuitable for real-world implementations. In that context, for signal-based inference models, challenges lie on the initial framework of learning since the performance highly depends on appropriate utilization of available signal information rather than the choice of classifier [38]. Failure occurs when a learning strategy is unable to cope the possible information within the framework. Therefore, more attention is needed on initial framework.

3 Focus of Our Algorithm

This work aims at establishing a feature fusion assumption free model to give precise and reliable prediction of various EMG signals as shown in Fig. 1. It aims is to find an efficient learning strategy that could employ multiple signals associated with a particular pre-diagnose process in order to extract lower order features. Therefore, our proposal is to use a set of signals associated with a study group for finding multi-view features (MV). Each signal contains different degree of uncertainty and information. Our rationale is, therefore, that MV that cover large-volume signals, could have more physiologically relevant information and energy contents as compared to single-signal based features. Consequently, derived features employing MVs could presumably avoid the feature biasing in the learning framework. It has the potential to be the basis of the mCCA.

In the mCCA, we adopt statistically independent MV using DWT in conjunction with directly evaluated MV. Unlike DWT methods, here DWT is employed only to transform input space and to evaluate DWT-MV. It is due to fact that synchronization of multi-domain features improves the generalization ability of feature space [8]. To the best of author knowledge, it has not been used earlier in the context of quality improvement via feature fusion technique. Experimental results and comparative

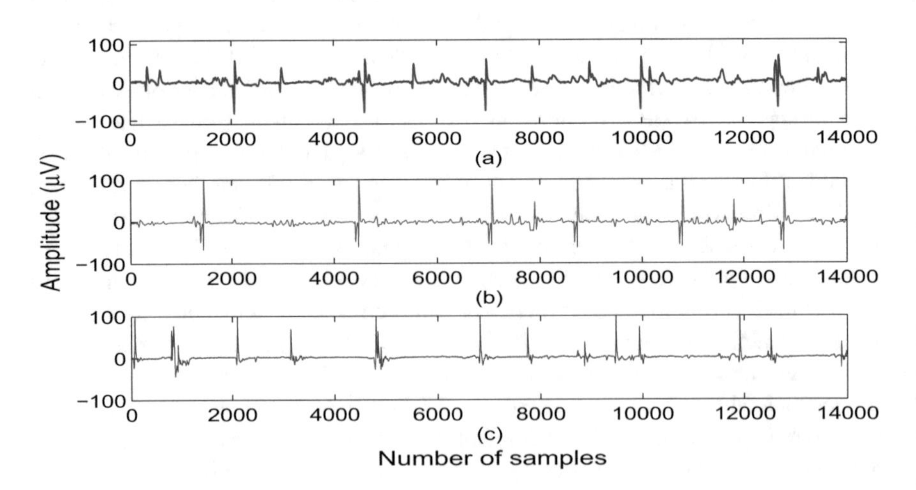

Fig. 1 Three EMG patterns—**a** Normal, **b** ALS, and **c** Myopathy

analysis reveal the effectiveness and reliable of our method. The rest of the article is organized as: Section IV describes the methodology. Results and discussion, comparative analysis and conclusion are presented in section V, VI and VII respectively.

4 Methodology

We first introduce two MV formulation strategies S-I and S-II which will be used independently to find low order features.

4.1 Strategy I (S-I)

A given dataset contains C subject groups and each group is divided into c subgroups. Each subgroup consists of n subjects and employing q signals from each subgroup an MV X_c is evaluated as shown in Fig. 2a. In this way, we obtain a set X_C for each C. In addition, DWT is performed over each signal of X_c and using DWT low-frequency components independent feature matrix $X_{c\omega}$ and set $X_{C\Omega}$ are evaluated.

$$\left.\begin{aligned} X_c &= [x_1, \ldots, x_q]^T \in \mathbb{R}^{q \times p} \\ X_{c\omega} &= [\tilde{x_1}, \ldots, \tilde{x_q}]^T \in \mathbb{R}^{q \times \tilde{p}} \end{aligned}\right\} \tag{1}$$

$$X_C = \{X_1, \ldots, X_c\};\; X_{C\Omega} = \{X_{1\omega}, \ldots, X_{c\omega}\} \tag{2}$$

where x_q in (1) is the one-dimensional p samples and $\tilde{x_q}$ is the corresponding low frequency wavelet components. Two MV sets in (2) are statistically independent and

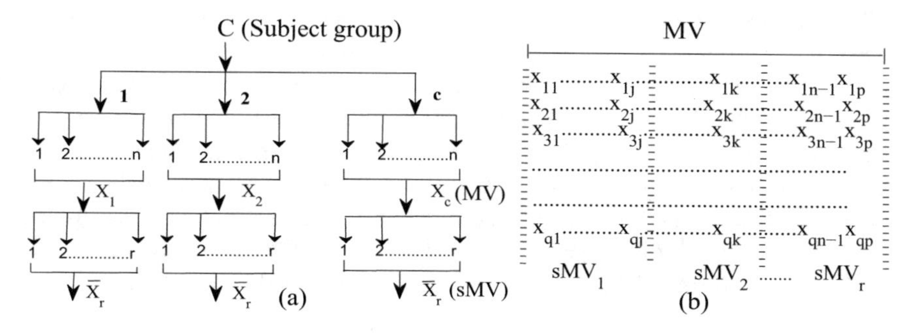

Fig. 2 **a** Generation of MVs and sMV from c subgroups of C subject groups according to S-I and S-II respectively. Here n indicates number of subject in each subgroup and r indicates the number of sMV from each MV, and **b** evaluation of sMV from MV as shown in Fig. 2 (**a**) using uniform decomposition strategy. Each row in MV represents 1D EMG signal

they correspond to same study group C, which is extended for all the subject groups (i.e., $C = 3$). Each MV and its sample delayed version is highly correlated [39]. Therefore, this pair-wise strategy is used for all possible pairs of MV in CCA transformation.

4.2 Strategy II (S-II)

In S-II, MVs in (1) are decomposed uniformly. Each signal is segmented into number of sequences with equal samples. Signal sequence and corresponding sequences of other signals are embedded in template sub-MV (sMV) as shown in Fig. 2b. Similar to S-I, two independent sets of sMV are formulated as-

$$\tilde{X}_c = \{\tilde{X}_1, \ldots, \tilde{X}_r\};\ \tilde{X}_{c\omega} = \{\tilde{X}_{1\omega}, \ldots, \tilde{X}_{r\omega}\} \tag{3}$$

Here $\tilde{X}_r$ and $\tilde{X}_{r\omega}$ indicate sMV obtained from the MV. Unlike (2), two new sets of sMV are derived from c. It is seen that the consecutive sMV retain their morphological symmetry owing to same subgroup of unimodal data as evident from visual examination of signals with EMGLAB [40] and MATLAB, despite of variations within the signal. In such case, mCCA can extract the relevant feature subsets from the pairs of consecutive sMV based on correlation [41]. In S-I, low order features are evaluated from the common subspace of MV and delay version, while in S-II these features are extracted from the common subspace of consecutive sMV. So, the fusion of such features leads to an efficient model.

4.3 Preprocessing, Signal Selection and Orientation

Although the frequency range of EMG signal is 0–1 kHz, excluding isoelectric components, its dominant energy concentrates in the range of 20–500 Hz [14]. So, signals are filtered using a twenty-order Kaiser window based filter with pass band frequency of 20–500 Hz. Also, a notch filter of 50/60 Hz is used to remove power line interference. Filtered signal are re-sampled to even number for ease of analysis. As mentioned in [2], signal varies with age of subject. Accordingly, study subject groups are divided into the subgroup. ALS, myopathy and normal has three groups of age 35–52, 53–61, 62–67 years, 19–26, 28–41, 44–63 years and 21–26, 27–29, 29–37 years. However, details of study subjects are provided in Sect. 5.1.

4.4 Multi-view Learning

CCA finds two sets of vectors for two input vectors in common subspace by measuring the correlation between transformed vectors [11]. Given two MVs X and Y (S-I) are first subjected to the PCA to remove redundancy and irrelevant information. Mean of each row from the PCA reduced matrices are removed to make centered data matrices (i.e., mean=0). CCA finds two linear transformations, also known as variates.

$$\left.\begin{aligned} u &= A_{x_1}x_1 + \cdots + A_{x_k}x_k = A_x^T X \\ v &= B_{y_1}y_1 + \cdots + B_{y_k}y_k = B_y^T Y \end{aligned}\right\} \tag{4}$$

The generalized mCCA finds weight vectors $A_x \in \mathbb{R}^d$ and $B_y \in \mathbb{R}^d$ by maximizing the correlation ρ between the variate u and v as:

$$\rho(u, v) = \max_{A_x, B_y} A_x^T C_{xy} B_y \tag{5}$$

with $A_x^T C_{xx} A_x = B_y^T C_{yy} B_y = 1$. Here C_{xx} and C_{yy} are autocovariance matrices and C_{xy} is crosscovariance matrix of X and Y. As defined in [41], optimization problem (5) can be solved by using singular value decomposition (SVD) technique. Assuming that X and Y form unitary orthogonal bases for two linear subspaces, let the SVD of $X^T Y \in \mathbb{R}^{d\times d}$ *be*[1]-

$$X^T Y = U \Lambda V^T = [U_1, U_2] D [V_1, V_2]^T = U_1 \Lambda_d V_1^T, \tag{6a}$$

$$A_x = X U_1 = [A_1, ..A_d],\ B_y = Y V_1 = [B_1, .., B_d] \tag{6b}$$

where U and V are two left and right singular orthogonal matrices of C_{xy} and C_{yx} respectively. Also U_2 falls in the null space of X, i.e., uncorrelated component [42]. The diagonal elements in $D = diag(\Lambda_d, 0)$ are in descending order and measure the correlations between the transformed vectors (e.g., A_1 and B_1). These correlations indicate the closeness of transformed vectors in orthogonal subspaces (Fig. 3). The higher value of diagonal elements indicates close proximity of vectors. Thus, low order features corresponding to the higher diagonal elements well preserves most of the energy contents of high dimensional input features. In order to avoid overfitting of model and singularity of both scatter matrices, regularized method with suitable parameters ($\alpha = 0.5$, $\beta = 0.5$) as mentioned in [42] is adopted. Nonetheless, it does not transform the projections A_x and B_y.

[1] $d \leq q = rank(X, Y)$.

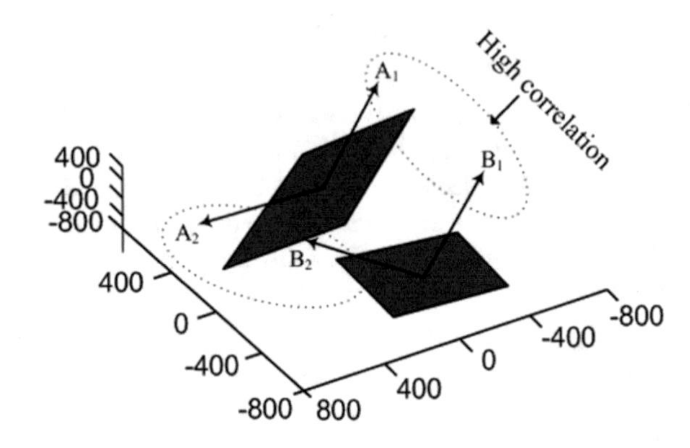

Fig. 3 Common orthogonal CCA-linear subspace of two sets of vectors. Correlation exists only between same indices pairs vector, i.e., A_1,B_1, while it is zero for different indices pairs, i.e., A_1, B_2 due to orthogonality [41]

4.5 *Feature Extraction and Reduction*

Table 2 shows the obtained MVs and sMVs for three study groups using S-I and S-II. Each subject group is divided into 3 subgroups. For each group, 3 MVs are evaluated, one for each subgroup using S-I. In S-II, each MV is divided into 8 sMVs (i.e., $r = 8$) and 24 sMV are evaluated for each study group. In the same way, DWT-MV and sMV are evaluated. In finding the MVs, various values of q *signals*[2] are selected and the optimal value is estimated.

MVs and its corresponding delay versions are pair-wisely projected to the subspace and the features are estimated using (5). Subsequently, Fig. 4a shows the correlation over three pairs of MVs in ALS group that reveals the existence of high correlation MV and delayed versions which also affirms the evidence of [39]. In order to improve the generalized ability of feature obtained through synchronized mode, the mCCA finds another three feature sets using DWT-MV. Thus, for each study group, six feature sets ($= 2 \times 3$) are evaluated in S-I.

DWT decomposes the signal into a set of basis wavelet vectors. It analyses the signal in multi-scale bands by decomposing the signal into approximation (low-frequency components, An) and detail coefficients (high-frequency components, Dn). As shown in [36], Daubechies wavelet function (db2) with second level of decomposition is efficient for EMG analysis. The obtained A2 components are considered to create wavelet MVs similar to the Table 2. The D2 components are not considered since most energy contents of biomedical signals usually fall in the lower frequency range.

In S-II, the correlations are evaluated for consecutive pairs of sMV for three MVs. This way, correlation is evaluated for twelve pairs of sMVs for each group. Figure 4b shows the mean within-group sMV-correlation which reveals the suitability of first few transformed features to capture intrinsic information from input features. Thus, highly correlated features by imposing threshold at 7 are considered for analysis. This analysis is reported for optimal choice $q = 8$ which will be discussed later

[2] The value of $q = 4, 6, 8$ and 10 are chosen in this study.

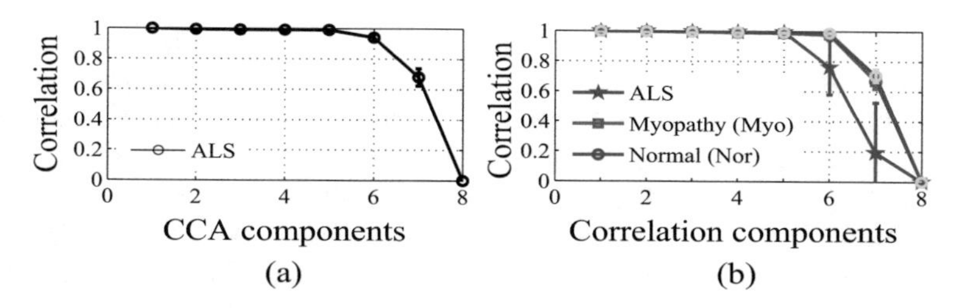

Fig. 4 Mean correlation (with $\pm\sigma$) between **a** MV and its delayed version (ALS), and **b** Consecutive sMV pairs of three groups as stated in Table 2

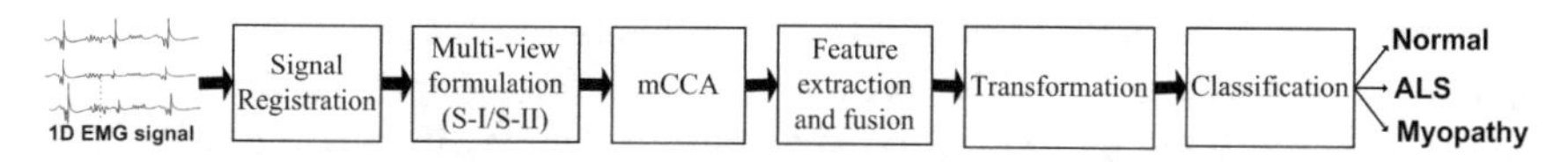

Fig. 5 Proposed multi-view feature fusion based scheme for classifying EMG signals to diagnose neuromuscular disorders-ALS and myopathy

and thus, the dimension of MV and sMV are 8×258000 (S-I) and 8×32250 (S-II) respectively. Each signal has a sample of 258000. Thus, the proposed strategy significantly reduces the feature dimensionality by incorporating both optimized-SVD and optimal choice of correlations. The proposed classification algorithm is presented through well-defined framework as shown in Fig. 5, where each step is explained comprehensively in this chapter.

4.6 Feature Fusion, Transformation and Classification

mCCA generalizes feature level fusion using summation technique for single set transformation [11] in order to fuse two independent feature sets of each pair of inputs as-

$$Z_{ij} = \sum_{t=1}^{4} T_t(u, v) = A_x^T X + B_y^T Y + C_{x1}^T X_1 + D_{y1}^T Y_1 \tag{7}$$

where T_t is canonical variate and Z_{ij} is known as generalized canonical correlation discriminant features (gCCDF) of $\{ij\}th\ pair$.[3] This could be regarded as a multi-modal fusion and it is inspired by the success of application such as in [11]. Three and twelve Z_{ij}s are evaluated for each study group using S-I and S-II respectively. Mean Z_{ij}s are used for analysis. In data mining feature transformation plays important role in extracting quality information. The Z_{ij}s are transformed using the proposed model $\Phi_i = Z_{ij} + \gamma I$, where γ and I are class indicator parameter and unitary

[3] X_1 and Y_1 are transformed DWT MVs of input X and Y; C and D are corresponding weight vectors.

matrix respectively. The dimension of Z_{ij} is matched with I by padding zeros. We empirically set $\gamma = 0$, 10, and 20 for normal, myopathy and ALS respectively.

In multi-task learning, class-structure information into the feature space is essential. Although mCCA does not include class-structure information, two-stage mCCA+LDA provides an effective solution [43], since LDA preserves the properties achieved by the mCCA. It is worth to be mentioned herein that Discriminant correlation analysis (DCA) [44] solve this issue by using inherent LDA discriminant function. However, it requires more number of steps including transformation, eigen-decomposition and optimization as compared to the mCCA+LDA. In that context, our method is much simpler and it handles the issue since LDA simultaneously maximizes and minimizes of within-scatter and between-scatter matrices to attain optimal discrimination [25]. The extracted features through proposed model steps as shown in Fig. 5 are embedded to the simplest k-nn classifier. Use of LDA with k-nn avoids the iterative training and under or overfitting which leads to higher generalization ability of model [25]. k-nn finds the euclidian distance function between the feature in the test set and neighboring healthy and pathological patterns in the training set. The pattern from the test set is categorized based on the class labels of closer patterns. The algorithm is implemented in MATLAB (The MathWorks, Inc., Natick, USA) on an Intel (R) Xeon (R) machine (Precision T3500) with processor 2.8 GHz and 8 GB of RAM.

4.7 Performance Evaluation Markers

To assess the performance, each dataset in datasets is partitioned into three subsets, 50% for training, 25% for validation and 25% for testing. The classifier has been supplied with statistical features to be assigned any one of three classes-ALS, myopathy and normal. To evaluate the model performance, we estimate the sensitivity (Sn), specificity (Sp) and overall accuracy (OA). The OA is given as:

$$OA = \frac{NC}{TN} \times 100\% \tag{8}$$

where NC and TN indicate number of correct classification and total number of classification respectively. Sn and Sp measure the degree of positive cases (ALS, myopathy) and negative case (normal) correctly the classified.

5 Results and Discussion

5.1 Database Description

Two datasets-online EMG$_{N2001}$ [40] and Guwahati Neurological Research Centre, Assam, India EMG$_{GNRC}$ datasets are used in this analysis. Datasets include three subject groups-ALS, myopathy and normal. However, none of the healthy subjects had signs or history of neuromuscular disorders. EMG_{N2001} includes 8 ALS (4 females, 4 males, age: 35–67, mean: 52.8±11.8 year (yr)), 7 myopathy (2 females, 5 males, age: 19–63, mean: 36.3±14.6 yr) and 10 normal (4 females, 6 males, age: 21–37, mean: 27.7±4.5 yr). EMG_{GNRC} includes 4 ALS (1 female, 3 males, age: 38–52, mean: 43.5±7 yr), 4 myopathy (2 females, 2 males, age: 42–59, mean: 47.5± 7.8 yr) and 4 normal (2 females, 2 males, age: 26–34, mean: 29.3±3.4 yr). EMG_{N2001} contains unequal number of signals (150 ALS, 50 myopathy and 50 normal) while the collected dataset EMG_{GNRC} has equal number of signals (20 ALS, 20 myopathy and 20 normal) of three subjects. The entire dataset includes 310 signals.

Concentric needle electrode with a leading-off area of 0.07 mm^2 was used to record the signals by inserting into the muscle sites with a surface ground electrode placed on the limb position. To eliminate the unwanted needle movement, the amplifier cable was fixed to the muscle with a piece of tape. Signals were collected under supervision on the application of a slight and constant contraction by the subjects. Afterwards, signals were examined visually in computers and audibly with the aid of a speaker. The MUAPs that satisfied the triggering criteria were frozen on the computer screen for the measurement of amplitude as well as duration. The MUAP potentials in repetitive measurements were examined to support whether a single motor unit (MU) gives rise to a MUAP or not. At a single recording site, a maximum of three different MUAPs were collected at an interval of 50 ms. Concurrently, the other monitor continuously displayed signal for the last half a second duration. When the tip of the needle was near to muscle fibers a characteristic crispy repetitive sound was heard. The recording started for 11.2 s after confirmation of signal quality which was not too noisy and complex. To explore the whole muscle, care was taken to avoid recording the same MU. This way, nearly twenty signals were recorded for analysis. Signals in EMG_{N2001} were filtered using in-built system with filter setting of 2–10 kHz and amplified with custom built-in-DISA15C01 of a gain of 4000 and sampled at the rate of 23437.5 Hz with DSP56ADC16 16 bit ADC. Signals in EMG_{GNRC} collected from author institute were acquired using in-built setup at sweep [ms/D] = 20, sen [V/D] = 50 μ and filter bandwidth 20 Hz–10 kHz with same experimental procedure. Signals of EMG_{N2001} and EMG_{GNRC} are then filtered using our filter configurations discussed in the Sect. 4.3 for analysis.

5.2 *Performance of the mCCA on Real-Time Dataset*

Figure 6 shows the correlation of intra-subgroups MV pairs as outlined in Table 1. As is evident, the low correlations indicate the significant dissimilarities of signals used to formulate the MVs. It is due to the fact that MUAP of EMG signal does not occur at same instances in different time-frames of signals. Furthermore, subgroups MVs are differed by age. So, it is necessary to employs all subgroup-MVs to extract the best features and thus, remain in favor of the proposed method.

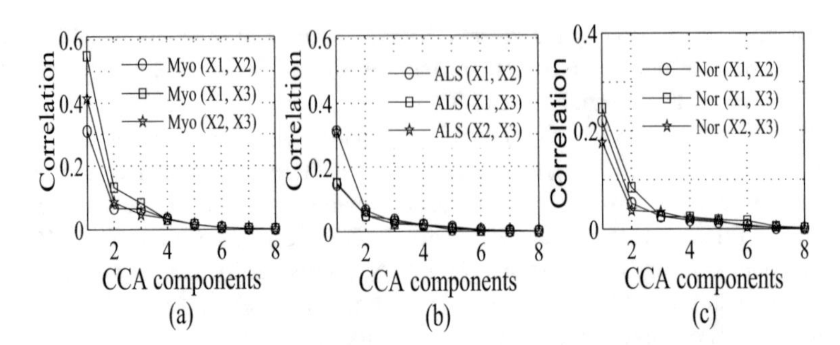

Fig. 6 Correlations are evaluated in-between three possible pairs of intra-subgroup MVs X_1, X_2 and X_3 as outlined in Table 1 for three subject groups—**a** Myopathy (Myo), **b** ALS, and **c** Normal (Nor)

Table 1 Formulation of MV and sMV using S-I and S-II

		S-I	S-II
C	c (MV)	sets of MV	Total sMV (Pair)
ALS	X_1, X_2, X_3 (3)	X_C (1)	3×8 = 24 (12)
Myopathy	X_1, X_2, X_3 (3)	X_C (1)	3×8 = 24 (12)
Normal	X_1, X_2, X_3 (3)	X_C (1)	3×8 = 24 (12)
C= 3	Total c= 3+3+3	Total= 3	Total = 72 (36)

Figure 7 shows wavelet components for DWT-MVs for optimal $q = 8$ which is obtained from performance and statistical test. This number of signals are taken from each subgroup-two from each subject (=2×4) and discriminant features are evaluated (see, Sect. 4.6). One-way analysis of variance (ANOVA) is carried out at 95% confidence level and $p = 0.05$ to assess the quality of features. Feature with $p > 0.05$ is discarded from feature matrices. Moreover, two-way ANOVA indicates that there is no benefit in adding more number of the transformed feature while deriving discriminant vectors. A minimal drop of accuracy, significant at $p = 0.05$ is observed while increasing feature dimension from 7 to 8. Features are subjected to the LDA that finds optimum decision surface among the features. In S-I, overlapping of a few features is seen. Table 2 shows p-values of features and and Fig. 8 shows the scatter distribution for S-II which show higher discrimination ability of features in comparison to the S-I features.

In order to obtain the best feature combinations for S-I and S-II, the classification performance is investigated on the training dataset of EMG_{N2001} with various combinations. For this, $k = 1$, i.e., a minimum distance classifier is chosen. Table 3 shows the comprehensive markers over three repeated measurements. As is evident, the feature combination D, E (S-II) and E (S-I) (not shown) show optimal performance and no further improvement in the performance is seen. The D (S-II) and E (S-I) are used for full-scale performance evaluation over three datasets independently using 3-fold cross-validations and results are reported graphical forms. It is seen that the

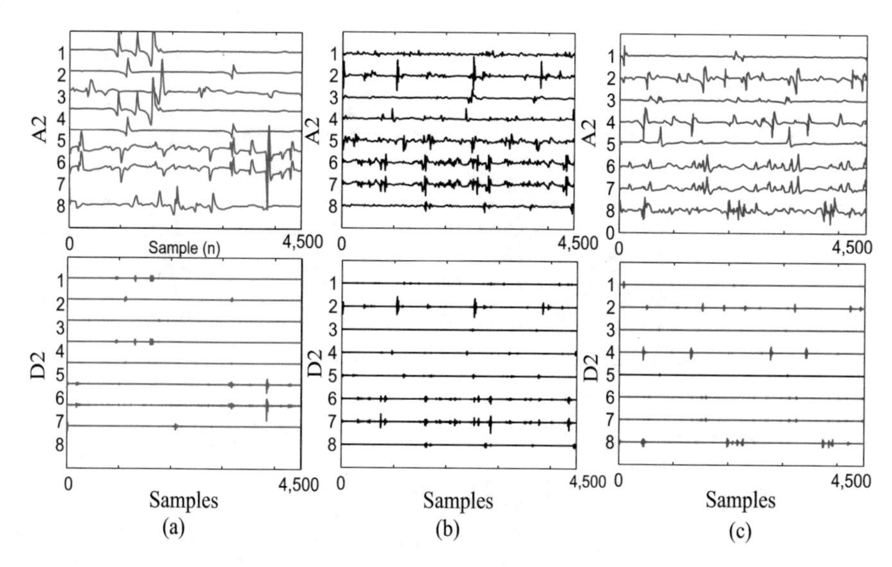

Fig. 7 Wavelet coefficients-A2s and D2s obtained from the MVs of **a** ALS, **b** myopathy, and **c** normal subject

Table 2 *p*-values of selected features

Feature	*p*-value (S-I)	*p*-value (S-II)
f_1	$<3 \times 10^{-3}$	$<1 \times 10^{-3}$
f_2	$<2 \times 10^{-3}$	$<1 \times 10^{-3}$
f_3	1×10^{-3}	$<2 \times 10^{-3}$
f_4	0	$<2 \times 10^{-3}$
f_5	$<3 \times 10^{-4}$	$<2 \times 10^{-3}$
f_6	$<4 \times 10^{-5}$	8×10^{-3}

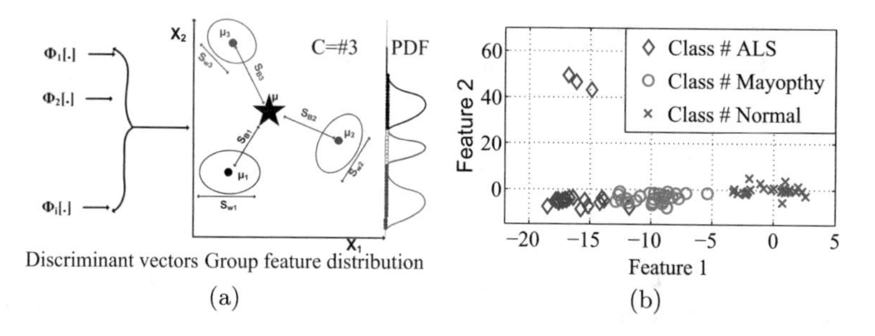

Fig. 8 LDA for C-class problem (=3) with C−1 dimension, and **b** feature distri-bution for S-II

Table 3 Classification performance of training dataset of the EMG_{N2001} for various combination of features in S-II

Combination	Feature	SnA (SnM)	Sp	OA
A	f_1	88.0 (78.6)	94.8	87.9
B	f_1, f_2	88.0 (86.6)	96.9	93.1
C	f_1, f_2, f_3	93.3 (96.6)	96.6	95.5
D	$\mathbf{f_1, f_2, f_3, f_4}$	**97.3 (97.3)**	**98.6**	**98.6**
E	f_1, f_2, f_3, f_4, f_5	**97.3 (97.3)**	**98.6**	**98.6**
F	$f_1, f_2, f_3, f_4, f_5, f_6$	93.3 (92.0)	98.2	95.5

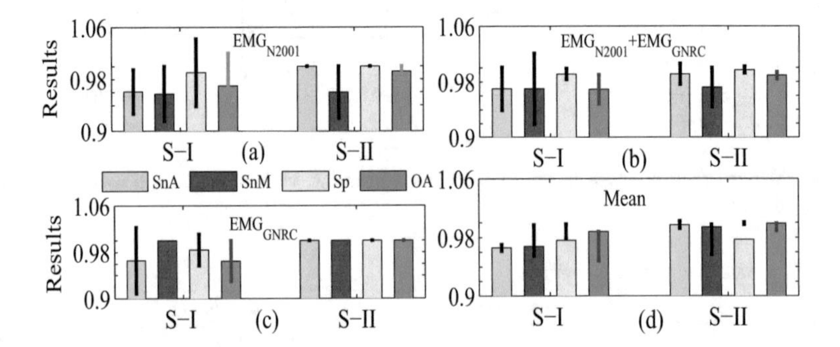

Fig. 9 Classification results of the proposed model with optimal combinations D (S-II) and E (S-I) of EMG_{N2001} over three datasets (**a**)–(**c**) and (**d**) indicates mean results of (**a**), (**b**) and (**c**)

Table 4 Mean confusion matrix formulated from the results of the classifier with S-II. The boldface diagnoal elements indicate the correct classification

Actual	Predicted class-S-II					
	ALS		Myopathy		Normal	
	μ	σ	μ	σ	μ	σ
ALS	**69**	1	0	0	1	0
Myopathy	1	1	**68**	1	1	1
Normal	0	0	0	0	**170**	0

method achieves promising results specifically with S-II over two as well as combined datasets as shown in Fig. 9a–c. Furthermore, mean results are shown in Fig. 9d. However, results with S-I are also remarkable and comparatively better, especially, in EMG_{N2001}. SnA and SnM indicate the sensitivities for ALS and myopathy. With S-II, the algorithm shows the highest performance over the EMG_{GNRC}. However, the mean OA is 99.4%, and SnA, SnM and Sp are 99.5%, 97.7% and 100% respectively (see, Fig. 9d S-II). In all the cases, the classifier with S-II achieves promising results than that of S-I. Moreover, the model provides good individual accuracy in

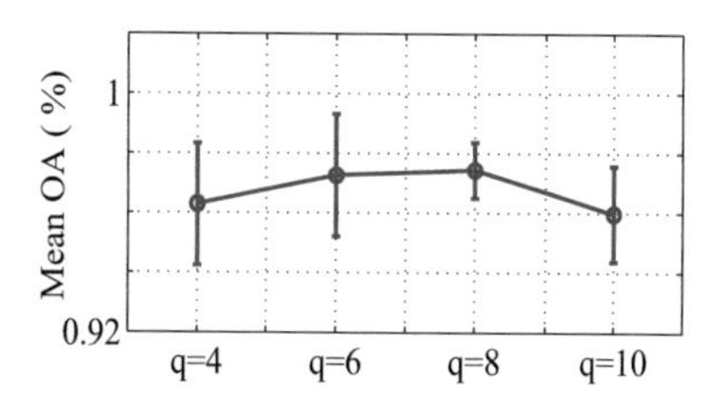

Fig. 10 Mean OA of S-I and S-II indicates the effectiveness of classification model for various choice of signals ($p = 0.0025, 0.074, 0.04$ and 0.065)

categorizing normal subjects. The confusion matrix over EMG_{N2001}+EMG_{GNRC} in Table 4 shows individual class predicting ability of model with S-II. Our model fails in categorizing only three cases (one ALS, two myopathies) while none of the control subjects is misclassified. However, in EMG_{GNRC}, the model accurately predicts the subjects (see Fig. 9c S-II). Thus, the promising results and low variance in outcomes apparently indicate the reliability of the mCCA.

The results reported for $q = 8$ is ascertained from the performance assessment and statistical test. Figure 10 shows variation of OA with q. Two-way-ANOVA test at $p < 0.001$ suggests not to add more than eight signals to avoid the computational burden. The effectiveness of our method is due to use of MVs through data reduction technique and feature level fusion. It uses large-volume inputs and provides comprehensive statistics that have high generalization ability and it also minimize the feature biasing. Ease of implementation and feature selection which enable adopting various classification models, are another vital consequences. This study ascertains the benefits of using multi-view features to extract relevant information from intramuscular EMG for pattern classification. However, upon increasing the signals beyond eight there is a drop of accuracy due to loss of useful data during the PCA. In such cases, the use of more number of principal components might provide high recognition rate. However, it will increase the dimensionality of input spaces which in turn increases the complexity of feature space. PCA provides good feature clustering assuming variance in data [45]. However, large input space alters the feature clustering to a complex pattern which will degrade the performance of the PCA. In this case, training the system with large data with an alternative data reduction technique could provide an efficient solution.

5.3 *Reliability and Scalability of the mCCA*

To ensure the reliability of the mCCA simple model discriminant analysis (DA) model with linear and quadratic discriminant functions are adopted. Main reason is that promising results of specific model (i.e., k-nn) may not ensure the quality of the feature. Extensive analysis and thorough investigations, followed by the promising results (Figs. 9 and 11) ascertain the reliability of the proposed scheme.

For scalability measure, the features obtained from the training of the model with one population are applied to classify new dataset. In other words, the features

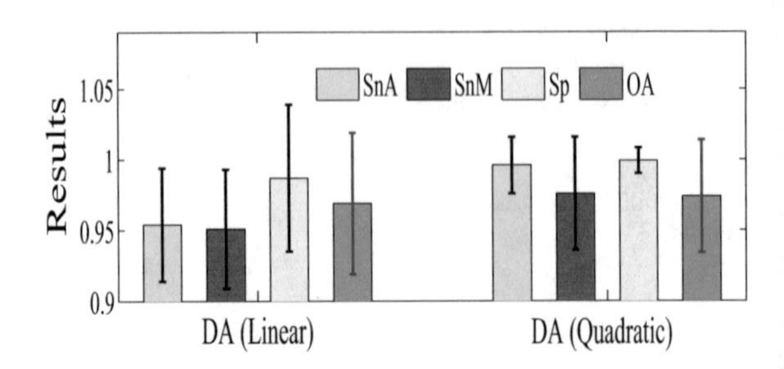

Fig. 11 Mean results with DA (linear) and DA (Quadratic) model

extracted using the datasets of EMG_{N2001} is used to estimate the performance over the EMG_{GNRC} (Fig. 9c [S-II]), wherein OA, SnA, SnM and Sp are of 100%. Figure 11 shows the similar assessment over the datasets. It is further ascertained in the context of two data sets obtained from the EMG_{GNRC} itself. Before analysis, it is ensured that there are no common samples in the datasets. Thus, promising results as shown in Figs. 9 and 11 reveal that the proposed algorithm is robust and reliable for diagnosis of neuromuscular disorders. Lower order features and ease of algorithm implementation further promote portable device based diagnosis in home care environment.

6 Comparative Analysis

For fair comparison, many prior methods are implemented over our dataset and performances are investigated using common classifier k-nn. Method [15, 22] extracted wavelet coefficients of signal and evaluated statistical measures for fuzzy-support vector machine (SVM) and particle swarm optimization (PSO)-SVM. They employed the dataset of 27 subjects with multiple signals over each subject collected in Neurology Department of University of Gaziantep. In SVM [24], optimal choice of radial basis kernel parameters-C and γ are essential for accurate classification. It also requires to ensure whether training samples of each class are uniform or not. However, in real-world, the effect of training samples are different. Use of statistical features although reduces feature dimensionality, it may not suitable for highly nonlinear data. The SVM and PSO-SVM were derived from evolutionary algorithms which deal with binary problems. The computational burden and lack of theoretical guarantees are two characteristic limitations of these algorithms. In evolutionary-SVM [16], kernel parameters are selected using GA-algorithm which involves a number of steps and have high computational time for larger datasets. RF [17] is an efficient classifier. However, it takes large memory and slow down the process in case of a large number of trees, lack of explanation and tends to overfit for noisy data. Many features-TD, AR, TDAR, AR+RMS and AR+RMS+TD [23] are capable to extract inherent input information, however, the combination of AR+RMS+TD results in a large feature dimension and additional processing steps. It was also reported that large mixture number incur very high computational expense but do not yield to give

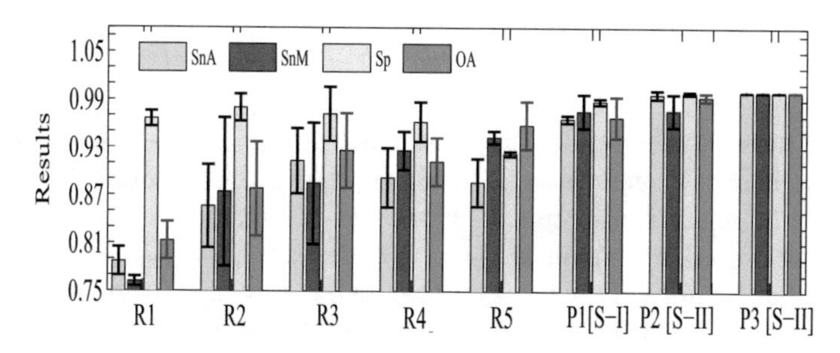

Fig. 12 Comparison of results in terms of SnA, SnM, Sp and OA. Herein R1:-Fuzzy-SVM [22], R2:- PSO-SVM [15], R3:-Evolutionary-SVM [16], R4:-RF [17] and R5: small-scale MV [7] with our schemes as indicated by P1 [S-I] and P2 [S-II] for S-I and S-II respectively. Here P3 [S-II] indicates the performance of the proposed method with S-II over EMG_{GNRC}

good result for small training data. Moreover, AR+RMS and AR+RMS+TD give 28 and 44-dimensional feature respectively. Unlike these methods, our study advocated the benefit of discriminant features employing large data for more reliable inferences.

Figure 12 shows the results of our (i.e., P1, P2, P3) and previous methods implemented over our datasets. Method [7] addresses the preliminary results over small dataset using multi-view features. Methods [15, 16, 22] and [17] employed 15, 23, 23 and 27 features (i.e., mean, average power, SD and ratio) evaluated using study signals. It is seen that the proposed method is far superior to these methods. Reported methods show promising results over small dataset. However, in EMG_{N2001}+EMG_{GNRC}, they show degraded performance. Our model improves recognition rate up to 3.6–4.2% over the best value of reported method and it also maintains the consistency in performance. It is worth noting that the nature as well as the number of MUAPs in EMG are unknown. Consequently, the supervised problem modeling with MVs leads to more reliable and accurate categorization of subjects suffering from neuromuscular disorders. Thus, we believe that it help rendering the clinically relevant facts related to the structural and functional aspect of motor units of a muscle.

Our method with S-II misclassified three cases which is presumably due to the inherent similarity of signals, order selection or dissimilarity of extracted features from the trained patterns. The integrity of advocated method is due to well-defined feature extraction and fusion strategy. Besides, it captured unique feature space that decreases the model learning complexity which in turn improves the recognition rate. The proposed method requires average of 23.78 and 36.46 s for feature fusion of includes S-I and S-II respectively. For classification task, it takes 2.1 and 3.4 s. It is reasonable due to the incorporation of large-scale information space to have better diagnosis value. Thus, there is trading-off between the quality of feature and computation cost.

7 Conclusion

The article presents subspace learning based feature fusion model which uses multi-view information extracted from EMG signal for classification of subjects suffering from neuromuscular disorders. In this model, two multi-view feature extraction strategies were proposed and investigated over datasets-EMG_{N2001} and EMG_{GNRC}. The algorithm significantly reduces the feature dimensionality with an aid of optimization technique and the results suggest that information extracted from multi-view inputs yield an excellent mean recognition accuracy up to 99.4% with a specificity of 100% and sensitivities of 99.5 and 97.7%. Furthermore, the model achieved 100% performance over EMG_{GNRC}. Admittedly, the classification efficiency relies on multi-view features extracted from multiple signals belonging to various age-based subgroups of study groups. Furthermore, optimal signal selection for multiple views creation was investigated and validated with statistical analysis which revealed that no significant improvement in classification resulted from adding more than eight signals. The outcomes are encouraging and it envisages developing a graphical user interface which would be simple, accurate and reliable enough for clinical usage.

Acknowledgements The authors would like to thank Department of Physics, Cotton University and Department of Electronics and Communication Engg for their laboratory supports.

References

1. Sun, S.: A survey of multi-view machine learning. Neural Comput. Appl. **3**(20312038) (2013)
2. Liu, M., Zhang, D., Adeli, E., Shen, D.: Inherent structure based multi-view learning with multi-template feature representation for Alzheimer's disease diagnosis. IEEE Trans. Biomed. Eng. **63**(7), 1473–1482 (2015)
3. Rav, D., Wong, C., Deligianni, F., Berthelot, M., Andreu-Perez, J., Lo, B., Yang, G.Z.: Deep learning for health informatics. IEEE J. Biomed. Health Inform. **21**, 4–21 (2017)
4. O'Leary, D.E.: Artificial intelligence and big data. IEEE Intell. Syst. **28**, 96–99 (2013)
5. Lahat, D., Adali, T., Jutten, C.: Multimodal data fusion: an overview of methods, challenges, and prospects. Proc. IEEE **103**(9), 1449–1477 (2015)
6. Hazarika, A., Dutta, L., Barthakur, M., Bhuyan, M.: Fusion of projected feature for classification of EMG patterns. In: Proceedings of the IEEE Conference on Accessibility to Digital World, pp. 69–74 (2016)
7. Hazarika, A., Bhuyan, M., Barthakur, M., Dutta, L.: Multi-view learning for classification of EMG template. In: Proceedings of IEEE Conference on Signal Processing Communication, pp. 467–471 (2017)
8. Sargin, M.E., Yemez, Y., Erzin, E., Tekalp, A.M.: Audiovisual synchronization and fusion using canonical correlation analysis. IEEE Trans. Multimedia **9**(7), 1396–1403 (2007)
9. Sun, B.Y., Zhang, X.M., Li, J., Mao, X.M.: Feature fusion using locally linear embedding for classification. IEEE Trans. Neural Netw. **21**(1), 163–168 (2010)
10. Anderson, T.W., Mathmaticien, E.W.: An Introduction to Multivariate Statistical Analysis. Wiley, New York (1958)
11. Yuan, Y.H., Sun, Q.S., Zhou, Q., Xia, D.S.: A novel multiset integrated canonical correlation analysis framework and its application in feature fusion. Pattern Recog. **44**(5), 1031–1040 (2011)

12. Dutta, L., Talukdar, C., Hazarika, A., Bhuyan, M.: A novel low cost hand-held tea flavor estimation system. IEEE Trans. Ind. Electron. **65**(6), 4983–4990 (2017)
13. Xi, X., Tang, M., Luo, Z.: Feature-level fusion of surface electromyography for activity monitoring. Sensors **18**(2), 614 (2018)
14. Chu, J.U., Moon, I., Lee, Y.J., Kim, S.K., Mun, M.S.: A supervised feature-projection-based real-time EMG pattern recognition for multifunction myoelectric hand control. IEEE/ASME Trans. Mechatron. **12**(3), 282–290 (2007)
15. Subasi, A.: Classification of EMG signals using PSO optimized SVM for diagnosis of neuromuscular disorders. Comput. Biol. Med. **43**(5), 576–586 (2013)
16. Subasi, A.: A decision support system for diagnosis of neuromuscular disorders using evolutionary support vector machines. Signal, Image Video Process. **9**(2), 399–408 (2015)
17. Gokgoz, E., Subasi, A.: Comparison of decision tree algorithms for EMG signal classification using DWT. Biomed. Signal Process. Control **18**, 138–144 (2015)
18. Kamali, T., Boostani, R., Parsaei, H.: A multi-classifier approach to MUAP classification for diagnosis of neuromuscular disorders. IEEE Trans. Neural Syst. Rehabil. Eng. **22**(1), 192–200 (2014)
19. Waclawik, A.J.: Neurodegenerative disorders: amyotrophic lateral sclerosis and inclusion body myositis, neurology board review manual. Neurology **8**(3) (2004)
20. Yousefi, J., Wright, A.H.: Characterizing EMG data using machine-learning tools. Comput. Biol. Med. **51**(1), 1–13 (2014)
21. Katsis, C.D., Exarchos, T.P., Papaloukas, C., Goletsis, Y., Fotiadis, D.I., Sarmas, I.: A two-stage method for MUAP classification based on EMG decomposition. Comput. Biol. Med. **31**(9), 12321240 (2007)
22. Subasi, A.: Medical decision support system for diagnosis of neuromuscular disorders using DWT and fuzzy support vector machines. Comput. Biol. Med. **42**(8), 806–815 (2012)
23. Huang, Y., Englehart, K.B., Hudgins, B., Chan, A.D.: A Gaussian mixture model based classification scheme for myoelectric control of powered upper limb prostheses. IEEE Trans. Biomed Eng. **52**(11), 1801–1811 (2005)
24. Gokgoz, E., Subasi, A.: Effect of multiscale PCA de-noising on EMG signal classification for diagnosis of neuromuscular disorders. J. Med. Syst. **38**(4), 1–10 (2014)
25. Naik, G., Selvan, S., Nguyen, H.: Single-channel EMG classification with ensemble-empirical-mode-decomposition-based ICA for diagnosing neuromuscular disorders. IEEE Trans. Neural Syst. Rehabil. Eng. **24**(7), 734–743 (2016)
26. Subasi, A., Yaman, E., Somaily, Y., Alynabawi, H.A., Alobaidi, F., Altheibani, S.J.P.C.S.: Automated EMG signal classification for diagnosis of neuromuscular disorders using DWT and bagging. Procedia Comput. Sci. **140**, 230–237 (2018)
27. Bozkurt, M.R., Subasi, A., Koklukaya, E., Yilmaz, M.: Comparison of AR parametric methods with subspace-based methods for EMG signal classification using stand-alone and merged neural network models. Turk. J. Electr. Eng. Comput. Sci. **24**(3), 15471559 (2016)
28. Hasni, H., Yahva, N., Asirvadam, V.S., Jatoi, M.A.: Analysis of electromyogram (EMG) for detection of neuromuscular disorder. In: 2018 International Conference on Intelligent and Advanced System (ICIAS), IEEE, pp. 1–6 (2018)
29. Elamvazuthi, I., Duy, N.H.X., Ali, Z., Su, S.W., Khan, M.K.A.A., Parasuraman, S.: Electromyography (EMG) based classification of neuromuscular disorders using multi-layer perceptron. Procedia Comput. Sci. **76**, 223–228 (2015)
30. Lahmiri, S.: Improved electromyography signal modeling for myopathy detection. In: IEEE International Symposium on Circuits and Systems, pp. 03 (2018)
31. Lo, Y.L., Najjar, R.P., Teo, K.Y.: A reappraisal of diagnostic tests for myasthenia gravis in a large Asian cohort. J. Neurol. Sci. **376**, 153158 (2017)
32. Jose, S., George, S.T., Subathra, M.S.P., Handiru, V.S., Jeevanandam, P.K., Amato, U., Suviseshamuthu, E.S.: Robust classification of intramuscular EMG signals to aid the diagnosis of neuromuscular disorders. IEEE Open J. Eng. Med. Biol. 235–242 (2020)
33. Hazarika, A., Dutta, L., Barthakur, M., Bhuyan, M.: A multiview discriminant feature fusion-based nonlinear process assessment and diagnosis: application to medical diagnosis. IEEE Trans. Instrument. Mcas. **68**(7), 2498–2506 (2018)

34. Hazarika, A., Barman, P., Dutta, L., Talukdar, C., Subasi, A., Bhuyan, M.: Real-time implementation of a multi-domain feature fusion model using inherently available large-volume sensor data. IEEE Trans. Ind. Informat. **15**(12), 30–38 (2019)
35. Subasi, A.: Diagnosis of neuromuscular disorders using DT-CWT and rotation forest ensemble classifier. IEEE Trans. Instrument. Meas. **69**, 1940–1947 (2019)
36. Doulah, A.S.U., Fattah, S.A., Zhu, W.P., Ahmad, M.O.: Wavelet domain feature extraction scheme based on dominant motor unit action potential of EMG signal for neuromuscular disease classification. IEEE Trans. Biomed. Circuits Syst. **8**(2), 155–164 (2014)
37. Maitrot, A., Lucas, M.F., Doncarli, C., Farina, D.: Signal-dependent wavelets for electromyogram classification. Med. Biol. Eng. Comput. **43**(4), 487–492 (2005)
38. Hargrove, L.J., Englehart, K., Hudgins, B.: A comparison of surface and intramuscular myoelectric signal classification. IEEE Trans. Biomed. Eng. **54**(5), 847–853 (2007)
39. De Clercq, W., Vergult, A., Vanrumste, B., Van Paesschen, W., Van Huffel, S.: Canonical correlation analysis applied to remove muscle artifacts from the electroencephalogram. IEEE Trans. Biomed. Eng. **53**(12), 2583–2587 (2006)
40. Nikolic, M.: Detailed Analysis of Clinical Electromyography Signals: EMG Decomposition, Findings and Firing Pattern Analysis in Controls and Patients with Myopathy and Amytrophic Lateral Sclerosis, Ph.D. dissertation, University of Copenhagen, Kbenhavn, Denmark, Aug. 2001
41. Kim, T.K., Kittler, J., Cipolla, R.: Discriminative learning and recognition of image set classes using canonical correlations. IEEE Trans. Pattern Anal. Mach. Intell. **29**(6), 1005–1018 (2007)
42. Sun, L., Ji, S., Ye, J.: Canonical correlation analysis for multilabel classification: a least-squares formulation, extensions, and analysis. IEEE Trans. Pattern Anal. Mach. Intell. **33**(1), 194–200 (2011)
43. Sharma, A., Kumar, A., Daume, H., Jacobs, D.W.: Generalized multiview analysis: a discriminative latent space. In: Proceedings of the IEEE Conference on Computer Vision and Pattern Recognition, pp. 2160–2167 (2012)
44. Haghighat, M., Abdel-Mottaleb, M., Alhalabi, W.: Discriminant correlation analysis: real-time feature level fusion for multimodal biometric recognition. IEEE Trans. Inform. Forensics Secur. **11**, 1984–1996 (2016)
45. Hyvarinen, A., Oja, E., Karhunen, J.: Independent Component Analysis. Wiley, New York (2001)

Performance Profiling of Operating Modes via Multi-view Analysis Using Non-negative Matrix Factorisation

Michiel Dhont, Elena Tsiporkova, and Veselka Boeva

Abstract In industrial settings, continuous monitoring of the operation of assets generates a vast amount of data originating from a multitude of very diverse sources. This data allows to study and understand asset performance in real operating conditions, paving the way for failure prediction, machine setting optimisation and many other industrial applications. However, it is not always feasible and neither wise to approach data analytics for such applications by merging all the available data into a single data set, which often leads to information loss. The literature lacks methods to inspect asset performance based on splitting the data in different views corresponding to different types of monitored parameters. The multi-view data analysis method proposed in this work allows to extract operating modes for an industrial asset and subsequently, profile their performance. In this two-step approach, the endogeneous (internal working) data view is first exploited to detect and characterise distinct operating modes, while an exogeneous (operating context) data representation (disjoint with the endogeneous view) of these operating modes is subsequently used to derive prototypical performance profiles via non-negative matrix factorisation. The application potential and validity of the proposed method is illustrated based on real-world data from a wind turbine.

Keywords Non-negative matrix factorisation · Performance profiling · Multi-view data · Multi-dimensional binning

---

M. Dhont (✉) · E. Tsiporkova
EluciDATA Lab of Sirris, Bd A. Reyerslaan 80, 1030 Brussels, Belgium
e-mail: michiel.dhont@sirris.be

E. Tsiporkova
e-mail: elena.tsiporkova@sirris.be

M. Dhont
Department of Electronics and Information Processing (ETRO), VUB, Pleinlaan 2, 1050 Brussels, Belgium

V. Boeva
Blekinge Institute of Technology, Blekinge Tekniska Högskola, Biblioteksgatan 4, 371 79 Karlskrona, Sweden
e-mail: veselka.boeva@bth.se

W. Pedrycz and S. Chen (eds.), *Recent Advancements in Multi-View Data Analytics*, Studies in Big Data 106, https://doi.org/10.1007/978-3-030-95239-6_11

1 Introduction

Operating modes are typically referring to different phases or states of the operation of an industrial asset, machinery or production facility including startup, shutdown, normal operations, maintenance, etc. [1]. Besides for startups and shutdowns, it is not always easy to discriminate between the different operating modes since even the normal operations may involve different operational settings depending on all kinds of contextual factors, e.g., product requirements, environmental conditions, human interventions, etc. Moreover, transitions from one to another operating mode are often occurring in a continuous fashion and difficult to detect.

Operating modes may impact substantially performance and operational reliability. For instance, operating at sub-optimal setup in each context leads to performance degradation and lower productivity. Therefore, it is of essential importance to reveal and characterise the relationship between operating mode and performance. This is a challenging task since the operational behaviour and the associated performance of many industrial assets (vehicles, wind turbines, compressors, milling machines, …) are impacted by a multitude of factors of very diverse nature, e.g., originating from the internal working of the asset (oil temperature, pressure, speed, vibration, …) or measured in the environment where the asset is operating (ambient temperature, wind speed, humidity, …) or set by an operator (machine configuration, alarm thresholds, …). Moreover, those influencing factors are often also highly interdependent, which makes it extremely difficult to trace back certain performance to distinct operating modes. The present state-of-practice during O&M (operation and maintenance) is to inspect asset performance via operative curves. The latter are typically used to elucidate the relation between the output parameter versus the most prominent causal parameter (e.g., wind speed in case of wind turbines). Although these charts can capture the most prominent causal relation with the output, they cannot take full advantage of the rich data collected by the asset, and are not able to reveal more complex relations or distinct operating modes. Therefore, new more elaborated performance profiling and analysis methods are needed.

In this work, we propose a multi-view analysis approach empowering the establishment of a direct link between operating mode and prototypical performance. The proposed workflow is partially building upon the layered integration approach introduced in [2] and extending further towards context-aware performance profiling via Non-negative Matrix Factorisation (NMF).

The inherent multi-source nature of the real-world data makes it impossible to directly integrate different data sets without information loss. Moreover, it is essential to first identify distinct operating modes solely relying on the internal working of the asset of interest, detached from the other influencing factors. This is done by considering only the internal data (endogeneous) view, i.e., only the parameters capturing the internal working of an asset. Subsequently, the corresponding multi-dimensional time series data is clustered, grouping together timestamps for which the values of the considered internal parameters relate to each other in a similar way.

Their min-max ranges define and characterise each of the asset-specific operating modes.

Next, an alternative representation of each operating mode in terms of expected performance is derived by considering the exogeneous data view, i.e., the environmental factors impacting performance. Namely, for each operating mode, a dedicated data set is constructed, composed of the corresponding values of the parameters capturing the operating context of the asset, and subjected to intelligent multi-dimensional binning. The goal is to derive relatively homogeneous multi-dimensional windows of operating conditions from which performance indicator(s) can be extracted. Finally, the resulting bins per operating mode are transformed into a performance indicator matrix, to be factorised by NMF. In this way, distinctive performance profiles are derived, exploiting the inherent capability of NMF to extract feature profiles that represent typical performance behaviour in function of the contextual parameters. An explicit link can be established between those performance profiles and the operating modes characterised during the internal data view analysis.

The validity and the application potential of the proposed approach is demonstrated on a real-world data set of wind turbines.

More concrete, the contributions of this work can be summarised as follows:

- a multi-view data analysis workflow enabling the establishment of a direct link between operating mode and prototypical performance;
- a multi-dimensional binning approach allowing to capture diverse operating contexts;
- a context-aware performance profiling methodology exploiting the inherent capability of NMF to extract feature profiles that represent typical performance behaviour;
- validation of the proposed multi-view profiling workflow on real-world wind turbine data.

2 Related Work

2.1 *Profiling with NMF*

Non-negative matrix factorisation (NMF) [3] approximates a given matrix $\boldsymbol{X}$ with exclusively positive values into two factors (matrices) $\boldsymbol{S}$ and $\boldsymbol{W}$, in such a way that

$$\underset{M\times N}{\boldsymbol{X}} \simeq \underset{M\times R}{\boldsymbol{S}} \times \underset{R\times N}{\boldsymbol{W}}.$$

Taking into account that the approach requires that both the input and output matrices have exclusively positive values, the interpretation of the decomposition is in many application contexts much more intuitive. Lee and Seung [4] have illustrated this by comparing the decomposition of multiple factorisation methods on a database of facial images. In contrast to the eigenfaces of the principal component analysis (PCA), the prototypical profiles of NMF are really parts-based (i.e., isolated parts of a face) rather than shady contours of a face as with PCA. The latter can be explained by the fact that both output matrices ($\boldsymbol{S}$ and $\boldsymbol{W}$) are constrained to be positive, leading to only additive combinations for reconstructing the original matrix $\boldsymbol{X}$. In the case of PCA, parts of the eigenfaces are neutralised by negative components, making the eigenfaces less intuitive to interpret.

These natural representation capabilities of NMF to extract profiles which can be represented in a matrix format are exploited in several studies over a wide range of domains. Xu et al. [5] have used NMF to extract biologically meaningful patterns out of ^{1}H nuclear magnetic resonance data for both healthy and people with diabetes II. The results allow to accurately discriminate the metabolic status of healthy and diabetes II patients. Qin et al. [6] have used NMF for unsupervised profiling of network traffic. The latter allows to optimise the network management and do intrusion detection.

Mandache et al. [7] have demonstrated how NMF, applied on data from a non-invasive dynamic cell imaging (DCI) technique, can be used to detect cancerous breast tissue. In their approach, NMF reveals components that correspond to cells, fibres, motion artefacts, baseline signal and noise. In a follow up step, the identified components are used as features for discriminating between normal versus cancerous breast tissue.

2.2 Multi-view Data Analysis Approaches

Multi-view data analysis approaches have received a lot of attention recently due to the availability of data sets with a large number of diverse parameters. Industrial assets are typically equipped with a multitude of sensors, which often enables the collection of data from the operational state, environmental state, machine settings, etc. However, in order to conduct meaningful analytics on such complex data one should be able to consider and analyse separately different data representations and additionally, benefit optimally from the extracted knowledge by applying an appropriate integration analysis. Multi-view data analysis algorithms aim to solve these challenges by respecting the latent structure of the different data views [8].

In addition to the heterogeneous nature of the data generated today in real-world industrial applications, the available data sets are often not labelled. Well-known tools that are broadly used for analysis and extraction of interesting patterns from non-labelled data are clustering techniques. Traditional clustering algorithms however, are not suitable to deal with the multi-source characteristic of the data. New multi-view clustering algorithms that are able to address this challenge have been developed recently, e.g., [9, 10] provide an overview and comparison of different published multi-view clustering algorithms. The challenges of incomplete views are addressed in series of papers [11–13]. Some studies treat multi-view clustering as a multi-objective optimisation problem, e.g., [14]. Although there is a wide variety of algorithms, each algorithm can typically be assigned to one of the three following approaches [15]:

- The multi-view data set is first transformed into a common lower dimensional subspace. Next, well-known clustering approaches (e.g., k-means) are applied on the reshaped data set [16, 17].
- Cluster algorithms incorporate the multi-view structure by adapting the loss function to the latent architecture of the views [18, 19].
- Each view of the data points is clustered separately. Afterwards, the final cluster is chosen by majority voting. This is often referred to as *late fusion* or *late integration* [20, 21].

2.3 Multi-view Data Analysis Techniques Based on NMF

Several studies based on NMF for conducting multi-view data analysis have already appeared in data science literature [12, 15, 22–26]. An algorithm based on NMF for large scale incomplete distributed data sets is introduced in [12]. Liu et al. [15] have proposed a way to exploit the well-know (single-view) clustering capabilities of NMF for multi-view clustering. This is achieved by developing an optimisation algorithm which is able to regularise view-specific coefficient matrices into a common consensus. Greene and Cunningham [22] use the *late fusion* approach, discussed in Sect. 2.2, to cluster a multi-view text data set (news articles from 3 different papers). In this context, NMF is used to reconcile the view-specific clusters into overarching clusters. Kalayeh et al. [23] use NMF for image annotation on partially labelled data (where labels represent an additional view, in the form of binary values). As a result of this joint factorisation, a consensus for weight and profile matrices is obtained, naturally solving the problem of feature fusion. To anticipate for data set imbalance issues, weight matrices are proposed. A co-regularised multi-view NMF approach with correlation constraint for non-negative representation learning is developed in [24]. The developed approach imposes correlation constraint on the low-dimensional space to learn a common latent representation shared by different views.

A very different use case of applying NMF for multi-view data scenarios is discussed by Zhang et al. in [27]. In their research, the issue of missing links between distinct views of the same instances is tackled. The latter is achieved by minimising both the disagreement between each pair of views, as well as the loss function of NMF within each view. A variant of this use case has been considered by Gong et al. [26], which propose a model to impute missing values in multi-view data. By use of NMF, regional latent similarities, geographic positions as well as the correlations between multiple views are taken into account.

This paper proposes an NMF-based profiling approach, which exploits the multi-view context in a multi-layered incremental fashion. The underlying framework is based on the (generic) layered integration approach introduced by Dhont et al. in [2]. The approach identifies first operating modes based on an internal (endogeneous) view, followed by the extraction of prototypical profiles based on an alternative (exogeneous) view, and thus enables comparison of prototypical behaviour across multiple operating modes.

3 Materials and Methods

3.1 Data and Computer Code

The proposed method is illustrated on supervisory control and data acquisition (SCADA) data from an onshore wind turbine. The selected wind turbine is located in La Haute Borne, France and managed by Engie.[1] The data set[2] covers a period between 07/2011 and 03/2017. Within this time period, 137 parameters (e.g., torque, rotor bearing temperature, generator speed and grid frequency) are stored with a frequency of 10 min.

The research was conducted using the Python[3] programming language. The code can be provided on request.

3.2 Endogeneous Versus Exogeneous Data Views

As already discussed in the introduction, the operational behaviour and the associated performance of many industrial assets (e.g., vehicles, wind turbines, compressors, milling machines, ...) are impacted by a multitude of factors of very diverse nature. In general, it can be discriminated between two main subsets of factors: (1) endogeneous, i.e., originating from the internal working of the asset (e.g., oil temperature, pressure, speed, …) and therefore characterising the operational behaviour and (2) exogeneous, i.e., originating from the environment where the asset is operating (e.g., ambient temperature, wind speed, humidity, …) and thus directly correlated to the observed performance. Subsequently, the continuous monitoring of any asset will generate complex time series data sets, which can be decomposed into two different data views, *endogeneous or operation-related view* and *exogeneous or performance-related view*, to be considered explicitly during analysis in order to exploit fully the richness of the data.

Formally, let us assume that an industrial asset is monitored via N endogeneous and M exogeneous parameters. Thus, for any timestamp t, a column vector can be constructed of the respective endogeneous parameter values:

[1] Engie is a multinational French energy company. Url: www.engie.com.

[2] The SCADA data set can be found here: www.opendata-renewables.engie.com/explore.

[3] More info on Python can be found here: https://www.python.org/.

$$\boldsymbol{p}(t) = \begin{bmatrix} p(1,t) \\ p(2,t) \\ \vdots \\ p(N,t) \end{bmatrix},$$

where $p(i,t)$ is the recorded value of an endogeneous parameter i ($i = 1, 2, \ldots, N$) at time t. Analogously, for the same timestamp t, another column vector can be constructed of the respective exogeneous parameter values:

$$\boldsymbol{q}(t) = \begin{bmatrix} q(1,t) \\ q(2,t) \\ \vdots \\ q(M,t) \end{bmatrix},$$

where $q(j,t)$ is the recorded value of an exogeneous parameter j ($j = 1, 2, \ldots, M$) at time t. Thus, for a given time period of a length T, two data matrices

$$\boldsymbol{P} = [\boldsymbol{p}(1), \boldsymbol{p}(2), \ldots, \boldsymbol{p}(T)] \qquad \boldsymbol{Q} = [\boldsymbol{q}(1), \boldsymbol{q}(2), \ldots, \boldsymbol{q}(T)]$$

are then available for analysis with $\boldsymbol{P} \in \mathbb{R}^{N \times T}$ representing the endogeneous data view and $\boldsymbol{Q} \in \mathbb{R}^{M \times T}$ the exogeneous data view.

3.3 Endogeneous Data View: Operating Mode Identification

In [2], we proposed a novel approach, which performs an integration of heterogeneous real-world data sets originating from multiple sources (fleet of assets) in several incremental steps (layers). In this work, we build upon the initial analysis layer, which identifies different operating modes by clustering the multivariate endogeneous time series data set $\boldsymbol{P}$ along the time axis. This approach was already exploited in the work of Iverson [28] for the purpose of real-time inductive health monitoring. Practically, the method groups/clusters timestamps for which the values of the operational parameters relate to each other in a similar way. Each timestamp cluster will define a range of allowable values for each operational parameter and thus generate parametric characterisation of the respective operating mode.

Formally, the column vectors of the endogeneous data matrix $\boldsymbol{P}$ are subjected to a clustering algorithm in order to arrive at $K \in \mathbb{N}_+$ timestamp partitions $\rho_1, \rho_2, \ldots, \rho_K$. Concretely for any $k = 1, 2, \ldots, K$, partition $\rho_k = \{t_1^k, t_2^k, \ldots, t_{T_k}^k\}$ contains the indices of the column vectors of $\boldsymbol{P}$ belonging to the same cluster with $t_i^k \in \{1, 2, \ldots, T\}, i = 1, 2, \ldots, T_k$ and $T = T_1 + T_2 + \ldots + T_K$. Thus the endogeneous data view $\boldsymbol{P}$ can be decomposed into K sub-views $\boldsymbol{P} = [\boldsymbol{P}_1, \boldsymbol{P}_2, \ldots, \boldsymbol{P}_K]$:

$$\boldsymbol{P}_k = [\boldsymbol{p}(t_1^k), \boldsymbol{p}(t_2^k), \ldots, \boldsymbol{p}(t_{T_k}^k)],$$

identifying supposedly different operating modes $\boldsymbol{OM}_k$ of the asset (for $k = 1, 2, \ldots, K$). The operating modes can be further characterised by deriving properties/features from the matrix values. The nature of those properties will depend on the potential application in mind. In case of Iverson [28], so-called bounding boxes (the min-max ranges of the corresponding compositions of the endogeneous parameters per cluster) were used to define each of the operating modes, i.e.,

$$\boldsymbol{OM}_k = [\min(\boldsymbol{P}_k), \max(\boldsymbol{P}_k)],$$

where $\boldsymbol{OM}_k \in \mathbb{R}^{N\times 2}$ and $k = 1, 2, \ldots, K$.

Subsequently, detecting any composition of parameters during real-time monitoring not fitting within any of the bounding boxes of the operating modes (i.e., within the min-max ranges) is then considered as a deviation from normal operation. However, for other applications it can be more relevant to consider mean and standard deviation, to derive probability distributions or even to extract other more elaborated features from the time series data. Thus, it can be generalised that from each endogeneous data sub-view $\boldsymbol{P}_k$ ($k = 1, 2, \ldots, K$), a set of (endogeneous) features can be derived:

$$F^k = \{f_i^k | i = 1, 2, \ldots, N\},$$

identifying and characterising the corresponding operating mode of the asset $\boldsymbol{OM}_k$, i.e.,

$$F^k \rightarrow \boldsymbol{OM}_k.$$

3.4 Exogeneous Data View: Performance Profiling

Understanding how internal operating modes of an industrial asset impact its performance requires to investigate how asset performance changes along with different operating contexts. Therefore, the data associated with the factors (e.g., exogeneous) mostly impacting performance needs to be explored with the aim to derive relatively homogeneous multi-dimensional windows/bins of operating conditions. Subsequently, diverse performance indicators can be extracted from those bins. The quality and robustness of those performance indicators will depend on how well the binning solution captures the underlying data structure, i.e., reflecting data distribution and density.

In general, multi-dimensional binning approaches aim at dividing the parameter space into multi-dimensional bins often referred to hypercubes [29]. A hypercube is defined as a cube of N dimensions as illustrated in Fig. 1 and properties of interest can be derived from each cube. The assumption is that the data points characterised with similar parameter values (so they end up in the same hypercube), exhibit similar properties. However, partitioning a multi-dimensional space into relatively homogeneous hypercubes can be very computationally expensive when a large number of dimensions is considered. For the sake of generalisation, the methodology in this

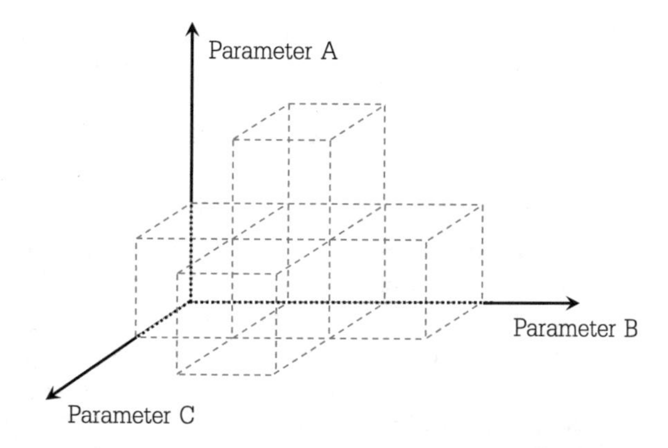

Fig. 1 Three-dimensional hypercubes

section is built around a general hypercube binning, while an elegant and scalable 3D binning solution is realised in the results section in the context of our use case.

Formally, hypercube binning allows to extract different operational contexts from the exogeneous data view $\boldsymbol{Q} \in \mathbb{R}^{M \times T}$ and subsequently, derive baseline performance per operating context by considering performance-related measurements (e.g., production rate, energy consumption, etc.). Let us assume that the number of partitions L_j (for $j = 1, 2, \ldots, M$) are defined in advance for each of the exogeneous time series of $\boldsymbol{Q}$. The column vectors of $\boldsymbol{Q}$ will be then partitioned into $L = L_1 \times L_2 \times \ldots \times L_M$ hypercubes (bins) in total. Each combination of $l = (l_1, l_2, \ldots, l_M)$ (for $l_j = 1, 2, \ldots, L_j$ and $j = 1, 2, \ldots, M$) defines a unique hypercube $\chi(l)$ capturing a specific operational context.

Our aim is to study and characterise performance behaviour as a function of different operational contexts. For this purpose, let us further assume that for each hypercube $\chi(l)$ some performance indicator $g(l)$ is extracted. Next, for any exogeneous parameter j ($j = 1, 2, \ldots, M$), a column vector $\boldsymbol{v}^j(l_j)$ can be composed by stacking the values of $g(l_1, l_2, \ldots, l_j, \ldots, l_M)$, where all possible combinations of $(l_1, l_2, \ldots, l_{j-1}, l_{j+1}, \ldots, l_M)$ are considered for fixed $l_j = 1, 2, \ldots, L_j$. More concretely:

$$\boldsymbol{v}^j(l_j) = \begin{bmatrix} v^j(1, l_j) \\ v^j(2, l_j) \\ \vdots \\ v^j(L_1, l_j) \\ v^j(L_1+1, l_j) \\ \vdots \\ v^j(2L_1, l_j) \\ v^j(2L_1+1, l_j) \\ \vdots \\ v^j(L^j - L_M + 1, l_j) \\ \vdots \\ v^j(L^j, l_j) \end{bmatrix} = \begin{bmatrix} g(1, 1, \ldots, 1, l_j, 1, \ldots, 1) \\ g(2, 1, \ldots, 1, l_j, 1, \ldots, 1) \\ \vdots \\ g(L_1, 1, \ldots, 1, l_j, 1, \ldots, 1) \\ g(1, 2, \ldots, 1, l_j, 1, \ldots, 1) \\ \vdots \\ g(L_1, 2, \ldots, 1, l_j, 1, \ldots, 1) \\ g(1, 3, \ldots, 1, l_j, 1, \ldots, 1) \\ \vdots \\ g(L_1, L_2, \ldots, L_{j-1}, l_j, L_{j+1}, \ldots, L_{M-1}, 1) \\ \vdots \\ g(L_1, L_2, \ldots, L_{j-1}, l_j, L_{j+1}, \ldots, L_{M-1}, L_M) \end{bmatrix},$$

where $L^j = L_1 \times L_2 \times \ldots \times L_{j-1} \times L_{j+1} \times \ldots \times L_M$. Subsequently, for each exogeneous parameter j ($j = 1, 2, \ldots, M$), matrix

$$\boldsymbol{V}^j = [\boldsymbol{v}^j(1), \boldsymbol{v}^j(2), \ldots, \boldsymbol{v}^j(L_j)] \tag{1}$$

with $\boldsymbol{V}^j \in \mathbb{R}_+^{L^j \times L_j}$ is constructed. Matrix $\boldsymbol{V}^j$ is representing the performance indicators arranged as rows along the L_j binning partitions of exogeneous parameter j and can be further decomposed, using Non-Negative Matrix Factorisation (NMF), into the product of two non-negative matrices.

NMF is a method to approximate a non-negative matrix by two factors. It is often used for dimensionality reduction, but can also be very powerful methodology for data analysis due to its inherent clustering property. In this study, NMF is used to discover latent prototypical components of the data. Formally, each matrix $\boldsymbol{V}^j$ ($j = 1, 2, \ldots, M$) can be approximated using NMF into the product of two non-negative matrices $\boldsymbol{W}^j$ and $\boldsymbol{S}^j$, exploiting the inherent capability of NMF to derive R^j feature profiles that represent the typical performance behaviour in function of exogeneous parameter j across all different operational contexts:

$$\boldsymbol{V}^j \simeq \boldsymbol{W}^j \boldsymbol{S}^j,$$

where $\boldsymbol{W}^j \in \mathbb{R}_+^{L^j \times R^j}$ are the weights, $\boldsymbol{S}^j \in \mathbb{R}_+^{R^j \times L_j}$ are the typical performance profiles, and R^j ($0 < R^j < \min(L^j, L_j)$) is the number of components (to be chosen, trade-off between accuracy and interpretability, as demonstrated later on in Sect. 4.3). Algorithm 1 summarises the sequential steps to obtain the performance profiles in pseudo code, given the exogeneous data view $\boldsymbol{Q}$ and the timestamps of the operating mode of interest.

In this way, the exogeneous data view $\boldsymbol{Q}$ can be associated with a set of matrices:

$$\boldsymbol{S} = \{\boldsymbol{S}^j | j = 1, 2, \ldots, M\},$$

each of them capturing the prototypical performance behaviour of the asset as a function of different exogeneous parameters. $\boldsymbol{S}$ can be considered as a multi-dimensional performance profile of the asset.

Note that, each row vector $\boldsymbol{v}^j(i)$ ($j = 1, 2, \ldots, M$ and $i = 1, 2, \ldots, L^j$) can be approximated as shown in Eq. (2), where $\boldsymbol{w}^j(i)$ represents the ith row vector of $\boldsymbol{W}^j$. Conceptually, one could interpret the rows of $\boldsymbol{S}^j$ as the available building blocks to reconstruct $\boldsymbol{v}^j(i)$ based on the weights from the ith row of $\boldsymbol{W}^j$ [30].

$$\boldsymbol{v}^j(i) \simeq \boldsymbol{w}^j(i)\boldsymbol{S}^j, \quad 1 \leq i \leq L^j, 1 \leq j \leq M. \tag{2}$$

Algorithm 1 Performance Profiling Algorithm

Require: extraction of prototypical profiles from an operating mode
Input: operating mode k timestamps $[t_1^k, t_2^k, \ldots, t_{T_k}^k]$ and exogeneous data view $\boldsymbol{Q}$
Output: weight matrices $\boldsymbol{W}^m$ and prototypical profile matrices $\boldsymbol{S}^m$ (for $m = 1, 2, \ldots, M$)

1: $\boldsymbol{Q}_k \leftarrow [\boldsymbol{q}(t_1^k), \boldsymbol{q}(t_2^k), \ldots, \boldsymbol{q}(t_{T_k}^k)]$
2: **for** $j \leftarrow 1$ to M **do**
3: \\Reshape $\boldsymbol{Q}_k$:
4: $\boldsymbol{V}^j \leftarrow [\boldsymbol{v}^j(1), \boldsymbol{v}^j(2), \ldots, \boldsymbol{v}^j(L_j)]$
5:
6: \\Apply non-negative matrix factorisation:
7: $\boldsymbol{W}^j, \boldsymbol{S}^j \leftarrow NMF(\boldsymbol{V}_j)$
8: **end for**
9: **return** $([\boldsymbol{W}^1, \boldsymbol{W}^2, \ldots, \boldsymbol{W}^M], [\boldsymbol{S}^1, \boldsymbol{S}^2, \ldots, \boldsymbol{S}^M])$

3.5 Linking the Data Views: Operating Modes Versus Performance Profiles

This section is concerned with leveraging the results obtained in the foregoing sections. The ultimate goal is to derive an explicit link between the results generated in the different data views, i.e., to link operating modes to performance profiles. For this purpose, the following workflow is considered (see Fig. 2):

- Divide the asset data set into endogeneous $\boldsymbol{P} \in \mathbb{R}^{N \times T}$ and exogeneous $\boldsymbol{Q} \in \mathbb{R}^{M \times T}$ data views as explained in Sect. 3.2.
- Following the approach described in Sect. 3.3, derive K partitions $\rho_1, \rho_2, \ldots, \rho_K$ from the endogeneous data view $\boldsymbol{P}$. Note that $K \in \mathbb{N}_+$ is either predefined or determined experimentally and

$$\rho_k = \{t_1^k, t_2^k, \ldots, t_{T_k}^k\},$$

  for $k = 1, 2, \ldots, K$ and $T = T_1 + T_2 + \ldots + T_K$.
- Subsequently, the following operations are conducted for each partition ρ_k ($k = 1, 2, \ldots, K$):

  - An endogeneous data sub-view is defined

$$\boldsymbol{P}_k = [\boldsymbol{p}(t_1^k), \boldsymbol{p}(t_2^k), \ldots, \boldsymbol{p}(t_{T_k}^k)],$$

    which, as described in Sect. 3.3, induces a set of features $F^k = \{f_i^k | i = 1, 2, \ldots, N\}$ characterising an operating mode $\boldsymbol{OM}_k$, i.e., $F^k \to \boldsymbol{OM}_k$.
  - A respective exogeneous data sub-view is defined

$$\boldsymbol{Q}_k = [\boldsymbol{q}(t_1^k), \boldsymbol{q}(t_2^k), \ldots, \boldsymbol{q}(t_{T_k}^k)],$$

which, following the approach described in Sect. 3.4, can be associated with a multi-dimensional performance profile $\boldsymbol{S}_k = \{\boldsymbol{S}_k^j | j = 1, 2, \ldots, M\}$. The latter can be traced back to (induced by) operating mode $\boldsymbol{OM}_k$, i.e., $\boldsymbol{OM}_k \rightarrow \boldsymbol{S}^k$.

- The *endogeneous feature characterisation* of each operating mode $\boldsymbol{OM}_k$ is linked directly to a certain performance behaviour expressed via an *exogeneous multi-dimensional prototypical profile*, i.e.,

$$\{f_i^k | i = 1, 2, \ldots, N\} \rightarrow \{\boldsymbol{S}_k^j | j = 1, 2, \ldots, M\}.$$

4 Results

In this section, the proposed methods summarised in Fig. 2 are validated and illustrated on data originating from a real-world wind turbine as described in Sect. 3.1.

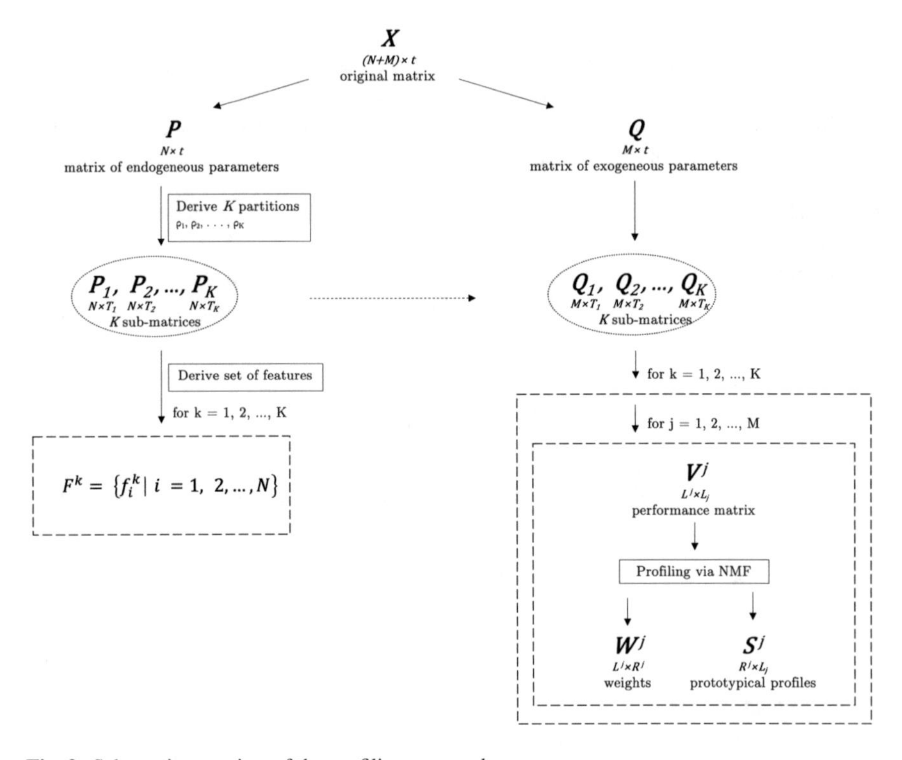

Fig. 2 Schematic overview of the profiling approach

Table 1 Selected endogeneous parameters

P 1	Generator bearing 1 temperature	**P 7**	Gearbox oil sump temperature
P 2	Generator bearing 2 temperature	**P 8**	Gearbox inlet temperature
P 3	Pitch angle (sine)	**P 9**	Gearbox bearing 1 temperature
P 4	Pitch angle (cosine)	**P 10**	Gearbox bearing 2 temperature
P 5	Torque	**P 11**	Generator stator temperature
P 6	Rotor bearing temperature	**P 12**	Generator speed

4.1 Identification of Endogeneous and Exogeneous Data Views

The raw data (see Sect. 3.1) contains 137 different parameters. Prior to the application of the proposed above methodology, a separate subset of endogeneous, related to the internal working of the machinery, and exogeneous, related to the environment in which the asset is being exploited, need to be selected in order to obtain a distinct endogeneous and exogeneous data view as explained in Sect. 3.2.

During the data preparation phase, a substantial number of the 137 parameters have been pruned away due to redundancy (e.g., "wind speed sensor 1", "wind speed sensor 2" and "average wind speed"), high correlation (e.g., "generator speed", "generator converter speed" and "rotor speed") or due to having too much missing data points. From the selected parameters, 12 provide information of the operational state of the asset, and represent thus the endogeneous view of the wind turbine data (see Table 1). Considering we have data for each 10 min during a period between 07/2011 and 03/2017, more than 300000 time measurements are at our disposal. Consequently, the resulting *endogeneous data set* $\boldsymbol{P}$ has a dimension of 12 (parameters) by 300000 (time stamps).

For the exogeneous view 3 qualitative parameters which are typically considered as directly having impact on the production performance have been selected. These parameters are "wind speed", "wind direction" and "ambient temperature". As a result, the dimension of the *exogeneous data set* $\boldsymbol{Q}$ is 3 (parameters) by 300000 (time stamps). The advantage of such a limited amount of exogeneous parameters is the ability to apply an optimised three-dimensional binning approach in order to achieve context-aware bins without suffering the curse of dimensionality [31]. For more details on the parameter selection procedure, we refer to our previous work [2].

4.2 Endogeneous Data View: Operating Mode Identification

As discussed in Sect. 3.3, we identify K operating modes by use of the 12 selected parameters from the endogeneous data view (see Table 1). To arrive at these operating modes, k-means clustering [32] has been used. Considering that there is no prior

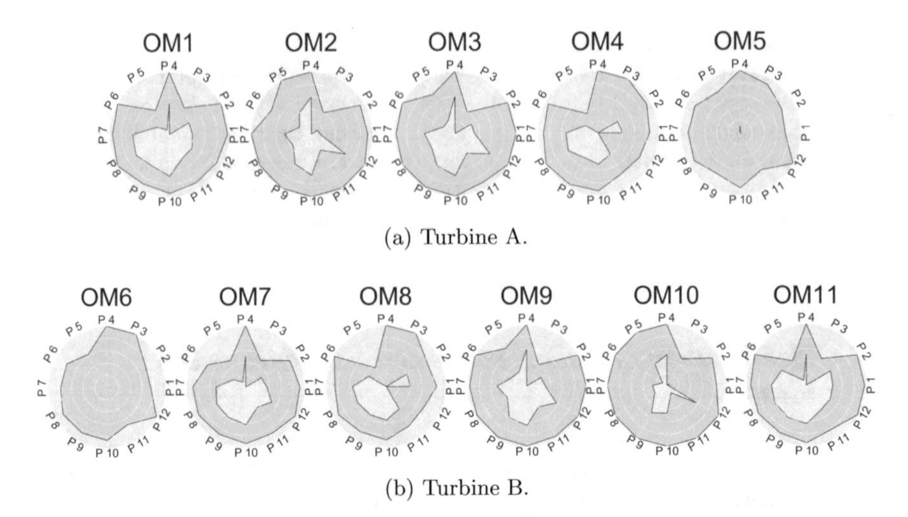

(a) Turbine A.

(b) Turbine B.

Fig. 3 Spider plots of the normalised endogeneous parameter ranges per operating mode (OM) for two different turbines co-located in the same fleet

knowledge of the underlying data structure, the optimal number of clusters K was determined by applying a majority voting approach on a range of cluster validation measures. Clustering solutions have been generated via a brute force approach for a range of k values, allowing to elect the optimal clustering scheme supported by the pool of validation measures. In this pool of validation measures, two are assessing separation and compactness properties—*Davis-Bouldin Index* [33] and *Silhouette Index* [34], two measures are assessing the within-cluster variability—*within cluster sum of squares* [35] and *Calinski Harabasz Index* [36] and one is assessing the connectedness—*Connectivity* [37]. The latter approach resulted in 5 clusters, each representing a distinct operating mode. Since this step is identical as the first layer in [2], we refer to there for more specific details.

Each of the 5 operating modes can be characterised by the value ranges of the endogeneous parameters. One can clearly observe distinct parameter ranges for each operating mode by use of spider plots in Fig. 3a. Such spider plots can be interpreted as fingerprints for the operating modes and allow for visual comparison across multiple operating modes. The same approach has been applied on a neighbouring wind turbine for which a similar data set was available. The associated spider plots are shown in Fig. 3b. As one can see, a different number of operating modes (6 in total) has been detected for the second turbine. However, some operating modes seem to be present in both wind turbines (e.g., OM1—OM11, OM3—OM9 and OM4—OM8).

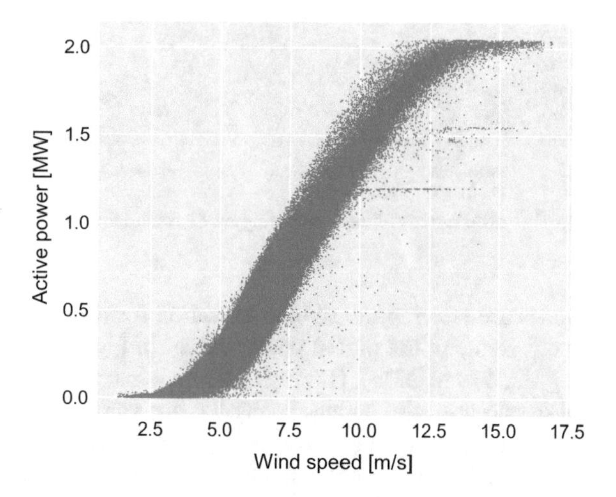

Fig. 4 Power curve for the full time period of the Engie wind turbine considered in this study

4.3 Exogeneous Data View: Performance Profiling

In Sect. 4.2 operating modes were extracted based on the endogeneous view. Next, the operating modes need to be mapped into the exogeneous view. In this subsection the operating modes are profiled in terms of the influence of the exogeneous parameters on the performance (energy production).

Power Curve In the wind domain, production performance is typically represented by use of a scatter plot, indicating the relation between the active power and the wind speed. This is mostly due to contractual issues guaranteeing certain production volume for given wind speed, which is the exogeneous parameter with the highest influence on the produced active power. The scatter plot is called a power curve and allows to have compact and intuitive representation of the turbine performance as a function of the wind speed. However, it fails to capture the influence of other exogeneous parameters when more elaborate insights or subtle phenomena need to be investigated. Figure 4 depicts the power curve for the full time period of the wind turbine considered in this study. One can observe the typical S-shape in this image, characterised by a minimum wind speed which is necessary to activate the blades (cut in) and a maximum wind speed from where a turbine is shut down to avoid damage (cut out).

Exogeneous Views In this subsection, the respective exogeneous data sub-views are considered for each of the five operating modes derived in Sect. 4.2. In Fig. 5, the power curves for each of the 5 operating modes are depicted. It can be observed that each operating mode is inducing only a part of the overall power curve in Fig. 4. This is a clear indication that the different value compositions of the internal parameters defined by the corresponding operating modes are tightly linked to the changing contexts (wind speed) in which the turbine is operating, i.e., distinctive operating

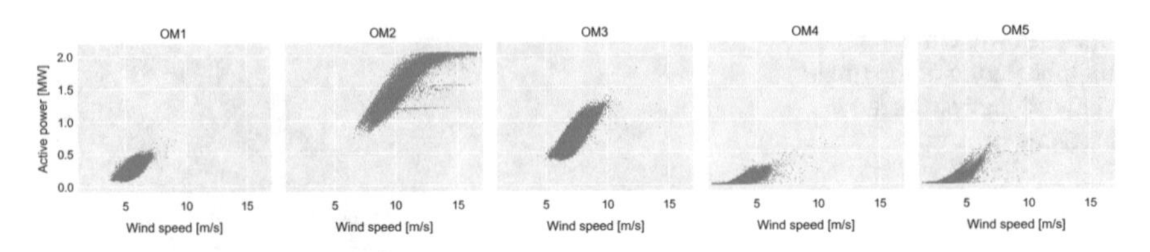

Fig. 5 Power curves of each operating mode separately

modes are steering the turbine operation for different external conditions, respectively. Some of the partial power curves in Fig. 5 appear quite similar at first sight (e.g., OM4 and OM5). However, zooming into the endogeneous characterisations of those two operating modes in Fig. 3a, one can observe that a substantial difference between the value ranges for most of the 12 parameters is present. The latter indicates that the turbine might exhibit rather similar production performances in terms of power curves, but the underlying internal states inducing these performances can be quite different in terms of load and wear of the different turbine components. This is relevant for maintenance and remaining useful life time applications.

Further, using the kernel density estimation (KDE) methodology [38], we have generated the probability distributions of the three exogeneous parameters per operating mode. Those can be consulted in Fig. 6. The largest difference between the operating modes can be observed for the wind speed parameter (see Fig. 6a), i.e., the different operating modes (with exception of OM4 and OM5) exhibit different coverage of the wind speed values with some overlap. The distribution of the wind direction values indicates two peaks around 40° and 220°, which have a difference of exact 180°. The latter indicates a dominant wind axis present at the turbine location. Note that OM4 and OM5 deviate here (again) from the general trend, as they seem to have a rather balanced occurrence of all possible wind direction values. Figure 6c illustrates the coverage of the ambient temperatures per operating mode. Although the operating modes mainly overlap in the temperature dimension, it is interesting to observe that OM4 and OM5 are the most diverging operating modes. It appears that the differentiation between these two operating modes is mostly based on turbine operations for rather similar wind speed and wind direction ranges, but different ambient temperatures. Thus, although the detection of the operating modes was based on the endogeneous view, it appears that each operating mode covers a unique, composition of (overlapping) value ranges, subspace in the exogeneous view.

Cylindrical Binning Following the approach described in Sect. 3.4, in order to enable a more elaborate performance profiling based on the three exogeneous parameters, data points are first binned into relatively homogeneous three-dimensional bins, each representing a window of similar external conditions. Since the exogeneous parameters contain one angular parameter (wind direction), a more intuitive variant of the general hypercube approach can be used. More concretely the two-dimensional binning approach which we introduced in [39] as *circular binning* is extended to 3D. The circular binning is applicable for any pair of parameters as far as one of them is of a circular nature, e.g., wind direction. Practically, the data points are mapped into a polar coordinate system, where the circular variable is the angular coordinate and the other variable is the radial coordinate. Many variables besides wind direction can be considered to be circular in certain application context, e.g., the hours of the days or the days of the week when studying seasonality aspects. Apart from the more intuitive interpretation, this approach allows bins to extend bin borders across the artificial/theoretical minimum and maximum borders (e.g., $0^\circ \rightarrow 360^\circ$).

In order to incorporate an additional exogeneous parameter (e.g., ambient temperature) into the operating context profiling, we extend our circular binning approach to *cylindrical binning* by mapping the data vectors of any three variables to cylindrical coordinates radial, angular and height, respectively, as visualised in Fig. 7. Again, the cylindrical binning is applicable to any three-dimensional vector space with the only requirement that one of them is circular. The binning is executed in two subsequent steps:

1. The circular binning is executed as specified in [39];
2. The data cylinder is sliced vertically along the height coordinate into a predefined number of equal-density circles.

Next, the cylindrical binning is applied to all the data vectors of the exogeneous view (no division in separate operating modes). The bin edges per dimension (wind speed, wind direction and ambient temperature), have been chosen in such a way to guarantee an average bin size close to 200 data points, while aiming at twice as much edges in the wind direction compared to the other dimensions. The motivation for this arrangement is that the distribution of the wind direction contains two clear peaks, resulting in a more complex distribution compared to those of the wind speed and ambient temperature (see Fig. 6). This resulted in 1458 ($= 19 \times 9 \times 9$) unique bins, where 19 refers to the wind direction and the two nines to the wind speed and ambient temperature, respectively. Thanks to this intuitive binning strategy, we can visualise a heat map of each slice as it is done in Fig. 8. In contrast to a traditional power curve (see Fig. 4), the slices of Fig. 8 allow to inspect the influence of all three exogeneous parameters. Note that we constructed the slices in 3D to facilitate the interpretation

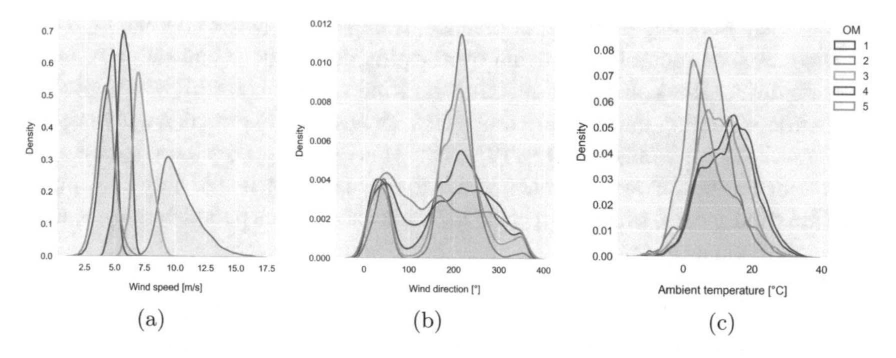

Fig. 6 Kernel density estimation (KDE) for the exogeneous parameters per OM

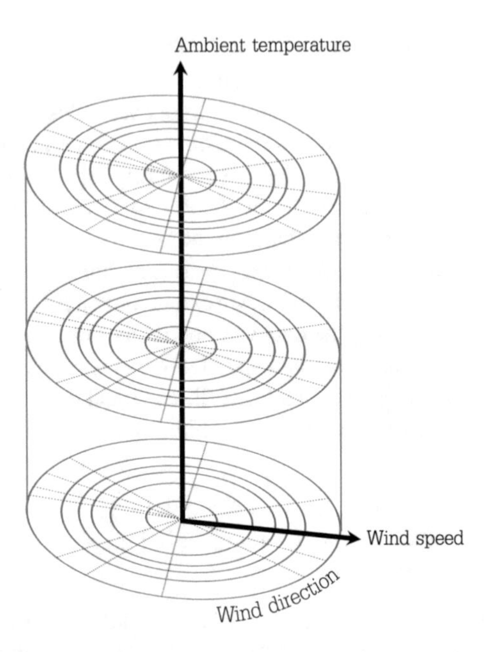

Fig. 7 Cylindrical binning

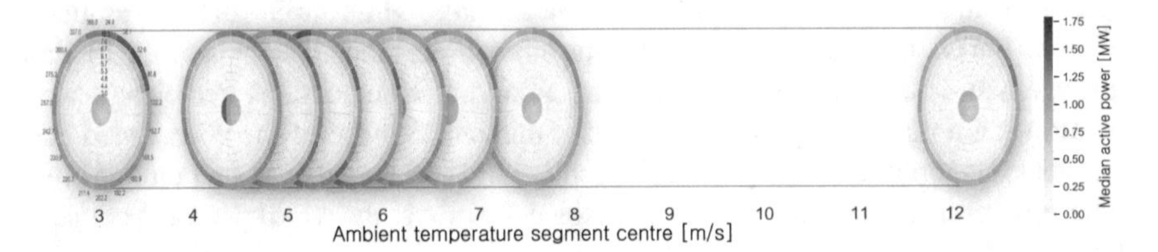

Fig. 8 Median active power per cylindrical bin. (The radial dimension represents the wind speed segment centres, the angular dimension represents the wind direction segment centres and the length represents the ambient temperature segment centres.)

of cylindrical binning. However, in practice, it might be a better alternative to plot the slices next to each-other to avoid overlapping parts. The visualisation in Fig. 8 supports further the well-known fact that the wind speed has the highest impact on production. Zooming into the different slices, shows a higher median active power in the northeastern wind directions (0°–90°). This effect appears to be stronger for lower temperatures. However, note that due to the high influence of only one dimension (the wind speed), smaller effects of the other exogeneous parameters are blurred in this representation.

Overall Profiling In this subsection, the NMF profiling methodology proposed in Sect. 3.4 is applied to the cylindrical bins generated in the foregoing subsection for the whole exogeneous view without discriminating between different operating modes. In this study, we opted for expressing the profiles as a function of the most influential parameter, the wind speed. The latter allows to reveal the influence of the other exogeneous parameters (wind direction and ambient temperature) in terms of prototypical performance profiles. As explained in Sect. 3.4, a performance matrix $\boldsymbol{V}^{\text{wind speed}}$ needs to be constructed (see Eq. (1)). In this matrix each column represents a different wind speed bin range, while rows have an entry for all possible combinations of the wind direction and ambient temperature bins. As performance indicator the median active power per bin is used, normalised by the overall median active power. The latter improves the interpretability of the data.

As mentioned in Sect. 3.4, it is not trivial to determine the appropriate number of latent components R of the factorisation of matrix $\boldsymbol{V}^{\text{wind speed}}$ in a case where no prior knowledge is available of the latent data structure. The higher this number, the more accurate one can represent the original matrix $\boldsymbol{V}^{\text{wind speed}}$. However, this is at the expense of less compressed and interpretable prototypical performance representations since determining the R value is a question of trade-off between accuracy and representative profiling. For this purpose, we inspect the values of two factorisation validation measures for a range of all possible R values. The first one is measuring the reconstruction error by means of the Frobenius norm [40] as shown in Eq. (3). In addition, the explained variance [41] is calculated as shown in Eq. (4) for each of the possible R values, providing the second measure for the factorisation quality.

$$\|\boldsymbol{E}\|_{Fro} = \sqrt{\sum_{ij} \boldsymbol{E}_{ij}^2} \qquad \boldsymbol{E} = \boldsymbol{P} - \boldsymbol{W}\boldsymbol{S} \tag{3}$$

$$\sigma_{\text{explained}}^2 = \frac{\sigma_{\boldsymbol{WS}}^2}{\sigma_{\boldsymbol{V}}^2} \tag{4}$$

The factorisation validation results are shown in Fig. 9. It is natural to expect that both measures improve by increasing R. However, the criteria we apply for discovering the optimal R is to focus on a sudden stagnation in one or both measures. In this way, R value of 4 was selected as most appropriate since increasing R to 5 results in hardly any improvement in terms of both the Frobenius norm (see Fig. 9a) and the explained variance (see Fig. 9b).

Subsequently, by applying NMF for $R = 4$, $\underset{171\times 9}{\boldsymbol{V}^{\text{wind speed}}}$ is decomposed in two non-negative matrices: (1) a matrix $\underset{4\times 9}{\boldsymbol{S}}$, where each row represents the prototypical performance profiles; (2) a matrix $\underset{171\times 4}{\boldsymbol{W}}$, where each row represents the weights needed to reconstruct the active power production of that specific combination of wind direction and ambient temperature.

Figure 10a depicts each prototypical performance profile. The vertical dotted lines indicate the edges of the bins in the wind speed dimension. To enhance the comparability, each prototypical performance profile has been scaled linearly between 0 and 1. It is interesting to observe that profiles 1 and 3 represent a continuous increas-

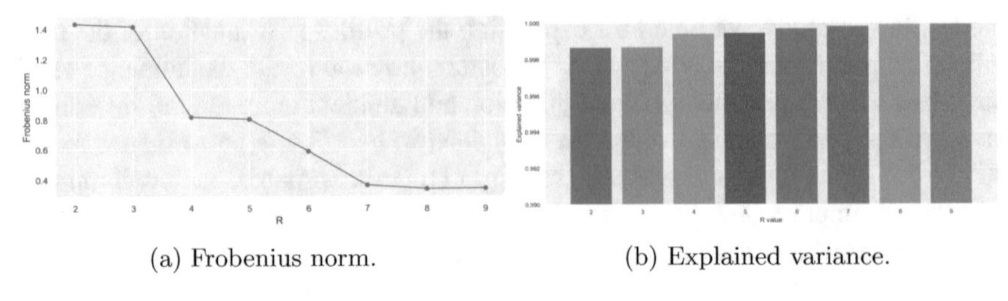

(a) Frobenius norm. (b) Explained variance.

Fig. 9 NMF validation results for multiple R values

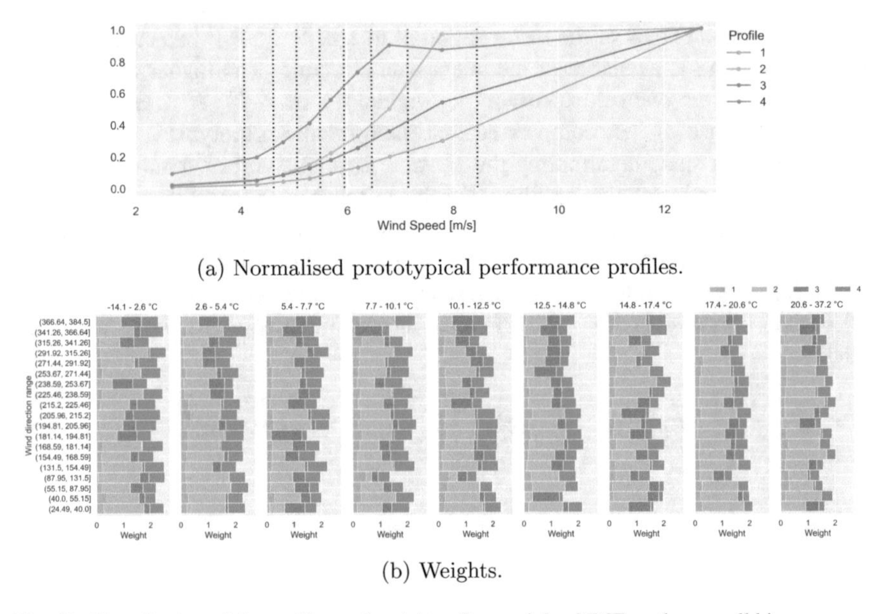

(a) Normalised prototypical performance profiles.

(b) Weights.

Fig. 10 Visualisation of the profiles and weights after applying NMF on the overall bins

ing pattern, while profiles 2 and 4 tend towards an S-shape as typically represented by a power curve (e.g., Fig. 4). In Fig. 10b, stacked bar charts, split per ambient temperature bin, have been constructed from the weights of each prototypical performance profile per wind direction bin. Note that the two S-shape like profiles (2 and 4) have in general the highest weights. In particular, profile 2 seems to be the most dominating based on its weights, which is not surprising since this profile seems to express the most typical performance behaviour as a function of increasing wind speed. Further, comparing the weights in terms of ambient temperature (see Fig. 10b), it can be observed that the importance of prototypical profiles 3 and 4 decreases for the higher temperature ranges. Furthermore, no other additional insights about how the derived performance profiles link to the internal working of the turbine can be derived from this representation.

Operating Mode Profiling By extracting prototypical performance profiles from all available data, there is a risk to overlook operating mode specific characteristics. For this reason, the same approach as above was repeated for each of the 5 operating modes. The amount of bin edges per dimension were determined separately per operating mode by the same 2 criteria (a mean bin size of around 200 points, and twice as many edges along the wind direction compared to the ambient temperature). This operating mode-specific binning strategy allows to zoom in on those conditions which occur the most frequent. However, this results in prototypical performance profiles which are representing different wind speed partitions, making it a bit more challenging to compare profiles amongst different operating modes.

Note that in all the figures below depicting the NMF profiling obtained per operating mode, the vertical dotted lines indicate the edges of the bins in the wind speed dimension. Moreover, in order to facilitate interpretation and comparison across the operating modes, each prototypical performance profile has been scaled linearly between 0 and 1.

Operating mode 1. As shown in Fig. 5, the first operating mode OM1 appears to be active in circumstances when fairly low wind speeds occur. This is also confirmed by the spider plot for OM1 in Fig. 3a, where it can be clearly seen that this mode is characterised by rather low torque values (parameter P5). The NMF decomposition revealed three distinctive prototypical performance profiles as shown in Fig. 11a. Examining the weights corresponding to the three profiles, depicted in Fig. 12b for the different ambient temperature bins, it is clear that profile 2 is the most dominant one. It captures the turbine behaviour when the wind is relatively low (e.g., wind gust), i.e., the turbine blades can start moving but the wind speed is not sufficient to significantly produce energy. It is interesting to observe that it requires the wind speed to increase well beyond 6 m/s in order for the turbine to fully start. This is a very relevant insight, which is further confirmed by profiles 1 and 3, which express stagnation and even drop in production for probably unstable wind speeds above 6 m/s. Figure 11b reveals further that for the coldest temperature range (-14.0 to $5.0\ ^\circ$ C) all three profiles have equal size/importance, while with the rise of the ambient temperature, the third profile is gradually disappearing.

Operating mode 2. The second operating mode OM2 is very different from the previous one. It is active in contexts when there is an optimal wind flow, resulting in a high production of active power (see Fig. 5). The NMF decomposition derives 4 different behaviour profiles as illustrated in Fig. 12a. Profile 1 expresses the linear increase of production in response to a linear increase in wind speed. Profile 3 captures some more complex behaviour. Namely, a clear stagnation in production is observed till about wind speed of 11 m/s, which then evolves to linear increase as far as the wind speed rises further. The wind speed of 11 m/s seems also essential for the remaining two profiles 2 and 4 since beyond this wind speed both profiles exhibit a linear drop in production. This might be related to curtailments, i.e., contexts when a wind turbine may need to be shut down to mitigate issues associated with turbine loading, export to the grid, or certain regulation conditions.

Note that the exogeneous conditions captured by OM2 occur less frequent, leading to a smaller amount of bins, i.e., 4:9:4 for wind speed, wind direction and ambient

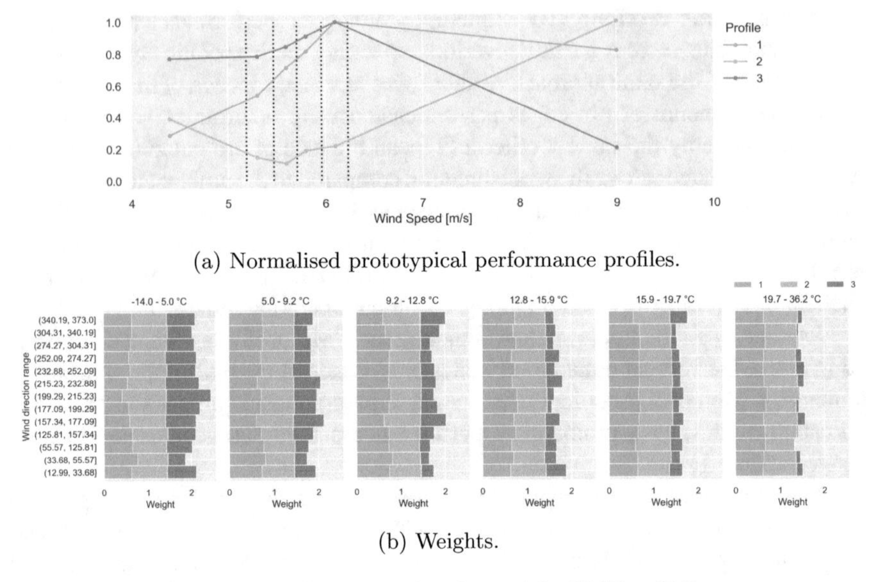

(a) Normalised prototypical performance profiles.

(b) Weights.

Fig. 11 Visualisation of the profiles and weights after applying NMF on OM1

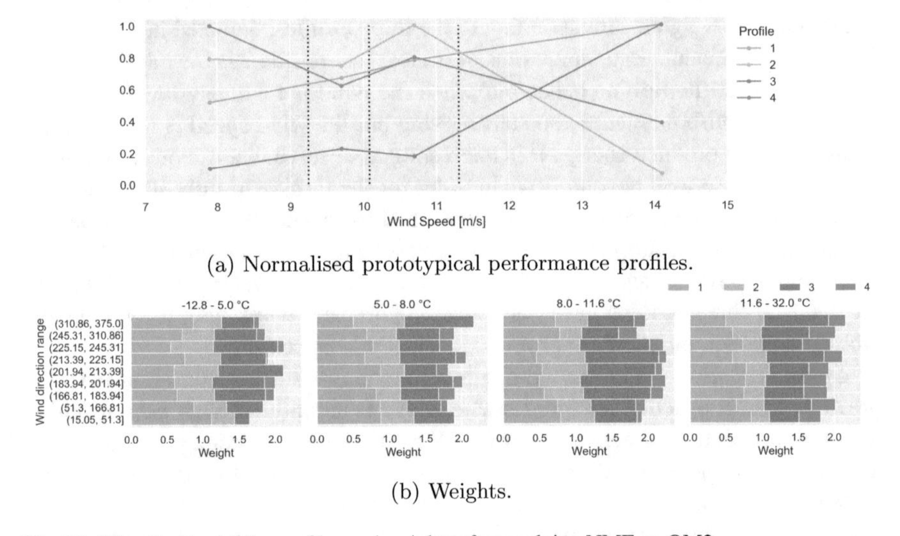

(a) Normalised prototypical performance profiles.

(b) Weights.

Fig. 12 Visualisation of the profiles and weights after applying NMF on OM2

temperature, respectively. Figure 12b depicts the corresponding weights per ambient temperature bin. It can be observed that although quite similar in shape, profiles 2 and 4 are active in different temperature ranges e.g., profile 4 is mostly present for the highest temperature bin, while profile 2 is well represented across all temperature ranges.

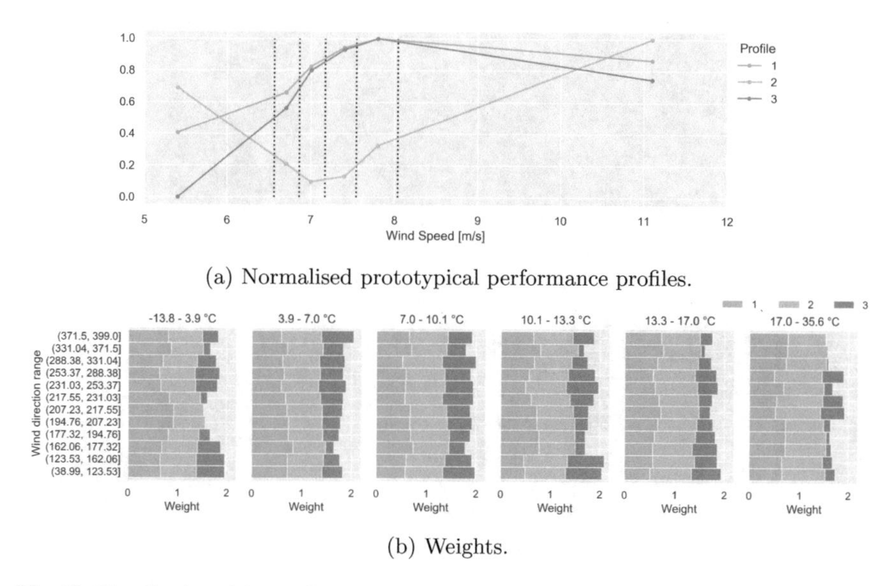

(a) Normalised prototypical performance profiles.

(b) Weights.

Fig. 13 Visualisation of the profiles and weights after applying NMF on OM3

Operating mode 3. According to Fig. 5, the third operating mode OM3 is representing the straight part of the power curve, where wind speed and active power exhibit a positive linear correlation to each-other. Examining the spider plots in Fig. 3a reveals some similarities between OM1 and OM3, e.g., the composition of the two pitch angle parameters (P3 and P4) and also most of the remaining parameters have quite comparable ranges. The only well pronounced difference is between the ranges of the torque and the generator speed (P5 and P12, respectively) parameters, which are composed of relatively higher values for OM3 in comparison to those for OM1.

As already discussed above, OM1 is capturing contexts when the turbine is struggling to fully start due to insufficient wind speed. OM3 appears to express contexts when the wind speed is of moderate power and most probably it is also rather unstable. The three prototypical performance profiles in Fig. 13a, extracted for this operating mode via NMF, support this hypothesis. In particular, profile 2 captures the poor production performance for wind speed ranges 6–8 m/s, which is probably due to unstable wind behaviour. It needs for the wind to gain in power beyond 8 m/s in order for the energy production to respond with linear increase. Further, profiles 1 and 3 seem to capture the opposite phenomenon, when rather stable wind speeds are observed till about 8 m/s, but further increase in wind speed does not induce increase of production. Further, zooming into the profiles' weights depicted in Fig. 13b, it can be observed that profiles 1 and 2 seem to be present across all temperature bins, while profile 3 has clearly a lower overall presence in the warmest temperature bin.

Operating mode 4. As shown in Fig. 5, operating mode OM4 contains data points from the most left part of the power curve, i.e., the lowest wind speeds, even lower than the values captured by OM1. OM4 seems to capture pre-start behaviour of the turbine

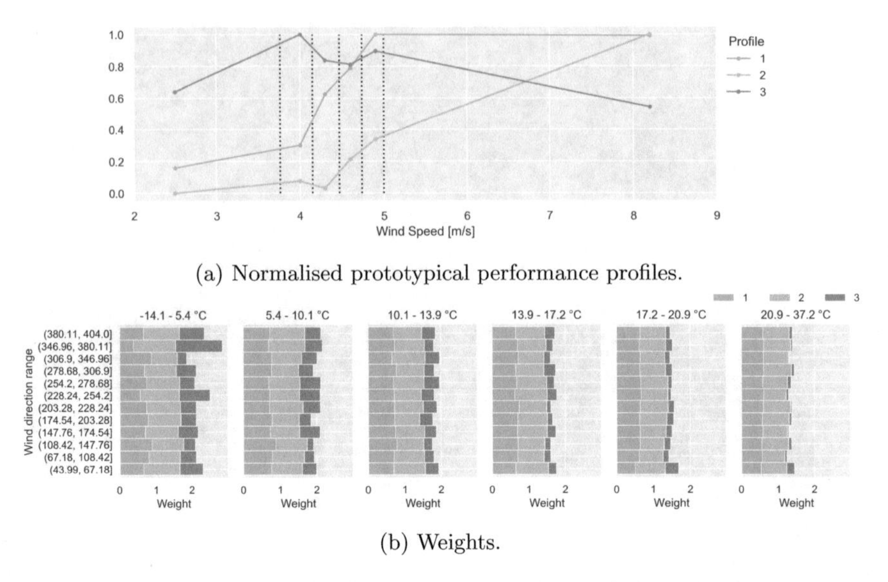

(a) Normalised prototypical performance profiles.

(b) Weights.

Fig. 14 Visualisation of the profiles and weights after applying NMF on OM4

and the main difference with OM1 is that the wind speed never excels sufficiently in power and stability in order for the turbine to reach an optimal production state. Figure 3a reveals that the pitch angle (P3 and P4) parameters have very different ranges for OM4 in comparison to the first three operating modes. The range of the pitch angle sine now also covers the highest possible values (i.e., 1), while the cosine covers also the lowest possible values (i.e., 0). This indicates that a pitch angle was sometimes further from its optimal power production settings (0° [42]). Such non-optimal production settings might again be related to contexts when a wind turbine may need to be shut down to mitigate issues associated with turbine loading, electricity oversupply, or certain regulation conditions (e.g., limitation of rotation speed to reduce noise).

The extracted via NMF profiles are depicted in Fig. 14a. Profile 1 appears to mimic a power curve behaviour for very narrow wind speed range of values between 4 and 5 m/s. Profile 2 captures some relatively moderate production activity when the wind speed gains in power beyond 5 m/s. Profile 3 is just an expression of the wind blades struggling to shift to continuous movement because of low and unstable wind speeds. Moreover, according to Fig. 14b, profile 3 is mostly present for lower ambient temperatures, while profiles 1 and 2 seem to capture well prototypical behaviours for this operating mode across all temperature bins.

Operating mode 5. Inspecting Fig. 5, it seems as if the last operating mode is a kind of transition state between OM4 and OM1. The spider plots of Fig. 3a reveal that OM4 and OM5 exhibit similar ranges for the pitch angle (P3 and P4) parameters, i.e., non optimal pitch angle settings are also encountered in OM5. However, the remaining parameters have very different ranges across all three modes (OM1, OM4

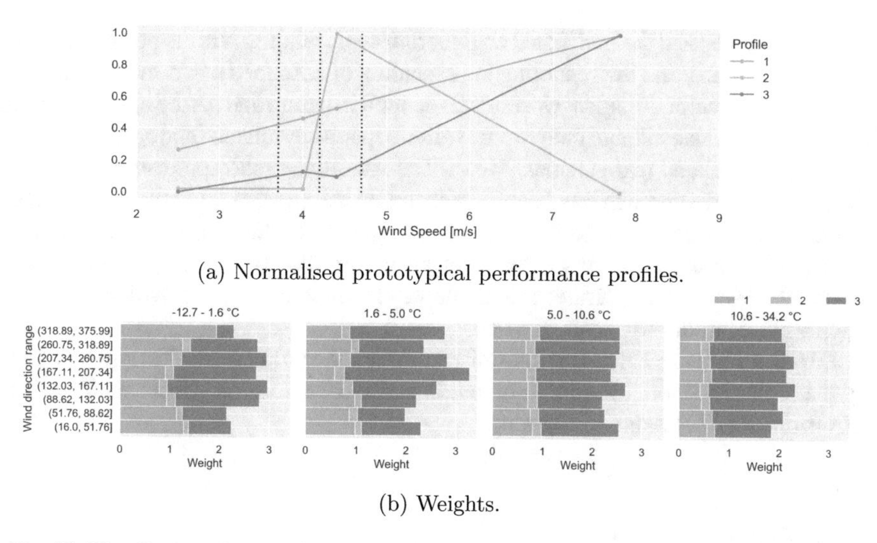

Fig. 15 Visualisation of the profiles and weights after applying NMF on OM5

and OM5). For instance, OM5 seems to differ a lot from OM1 and OM4 in its torque values (P5) and also the values for the generator bearing temperatures (P1 and P2) have quite different ranges in comparison to those of OM1.

The prototypical performance profiles derived via NMF for OM5 are depicted in Fig. 15a. Profile 1, and to some extend profile 3, represents a moderate production performance, which is linearly correlated to relatively limited increase in wind speed. Profile 2 captures the response of the turbine blades to a sudden gust in wind, which does not sustain in time. The corresponding profile weights in Fig. 15b confirm that profile 2 occurs very rare.

The foregoing analysis of the prototypical profiles derived via NMF per operating mode demonstrates that the initial segmentation into operating modes based on the internal behaviour of an asset (endogeneous view) allows to identify and discriminate between different asset operating contexts (exogeneous views), which reveal very different performance behaviour. Linking internal operating modes of the asset to operating contexts facilitates the better understanding of the overall asset behaviour not only in terms of performance, but also in terms of load and wear. This is essential to gain a better understanding of the asset functioning and health in general.

5 Conclusion

In this work, we proposed a multi-view analysis approach empowering the establishment of a direct link between operating mode and prototypical performance. The extracted knowledge reveals valuable insights about the operation-performance dynamics far beyond the traditional engineering analyses (e.g., via the power curve in the wind domain), and thus enabling the execution of advanced performance studies. The proposed methodological workflow is general in nature and can be applied in any context where the available data can be split in operation-related (endogeneous) and performance-related (exogeneous) views. The validity and the application potential of the proposed approach has been demonstrated on a real-world data set of wind turbines. The obtained results are very encouraging about the ability of the approach to detect latent components of the performance profiles for each operating mode separately. We intend to further extend the validation to other assets and to consider also its applicability for fleets of assets.

Funding: This research was subsidised through the projects MISTic and ReWind by the Brussels-Capital Region—Innoviris and received funding from the Flemish Government (AI Research Program).

References

1. Mackey, R., Iverson, D., Pisanich, G., Toberman, M., Hicks, K.: Integrated system health management (ISHM) technology demonstration project final report. In: IEEE Sensors Applications Symposium (2006)
2. Dhont, M., Tsiporkova, E., Boeva, V.: Layered integration approach for multi-view analysis of temporal data. In: International Workshop on Advanced Analytics and Learning on Temporal Data, pp. 138–154. Springer (2020)
3. Févotte, C., Idier, J.: Algorithms for nonnegative matrix factorization with the β-divergence. Neural Comput. **23**(9), 2421–2456 (2011)
4. Lee, D.D., Seung, H.S.: Learning the parts of objects by non-negative matrix factorization. Nature **401**(6755), 788–791 (1999)
5. Xu, L., Dong, J., Chen, Z., Dai, X.: Non-negative matrix factorization for diabetes ii metabolic profiling analysis. In: 2007 1st International Conference on Bioinformatics and Biomedical Engineering, pp. 641–643. IEEE (2007)
6. Qin, M., Lei, K., Bai, B., Zhang, G.: Towards a profiling view for unsupervised traffic classification by exploring the statistic features and link patterns. In: Proceedings of the 2019 Workshop on Network Meets AI & ML, pp. 50–56 (2019)
7. Mandache, D., á la Guillaume, E.B., Olivo-Marin, J.-C., Meas-Yedid, V.: Blind source separation in dynamic cell imaging using non-negative matrix factorization applied to breast cancer biopsies. In: IEEE 18th International Symposium on Biomedical Imaging (ISBI), vol. 2021, pp. 1605–1608. IEEE (2021)
8. Liu, L., Zhang, Z., Huang, Z.: Flexible discrete multi-view hashing with collective latent feature learning. Neural Process. Lett. **52**(3), 1765–1791 (2020)
9. Fu, L., Lin, P., Vasilakos, A.V., Wang, S.: An overview of recent multi-view clustering. Neurocomputing **402**, 148–161 (2020)
10. Yang, Y., Wang, H.: Multi-view clustering: a survey. Big Data Min. Anal. **1**(2), 83–107 (2018)
11. Liu, X., et al.: Late fusion incomplete multi-view clustering. IEEE Trans. Pattern Anal. Mach. Intell. **41**(10), 2410–2423 (2019)
12. Shao, W.: Online multi-view clustering with incomplete views. In: IEEE International Conference on Big Data (Big Data), vol. 2016, pp. 1012–1017 (2016)
13. Ye, Y., et al.: Incomplete multiview clustering via late fusion. Comput. Intell. Neurosci. 1–11 (2018)
14. Jiang, B., et al.: Evolutionary multi-objective optimization for multi-view clustering. In: 2016 IEEE CEC 2016, pp. 3308–3315 (2016)

15. Liu, J., Wang, C., Gao, J., Han, J.: Multi-view clustering via joint nonnegative matrix factorization. In: Proceedings of the 2013 SIAM International Conference on Data Mining. SIAM (2013), pp. 252–260
16. Chaudhuri, K., Kakade, S.M., Livescu, K., Sridharan, K.: Multi-view clustering via canonical correlation analysis. In: Proceedings of the 26th Annual International Conference on Machine Learning, pp. 129–136 (2009)
17. Chen, M.-S., Huang, L., Wang, C.-D., Huang, D.: Multi-view clustering in latent embedding space. In: Proceedings of the AAAI Conference on Artificial Intelligence, vol. 34(04), pp. 3513–3520 (2020)
18. Kumar, A., Rai, P., Daume, H.: Co-regularized multi-view spectral clustering. Adv. Neural Inf. Process. Syst. **24**, 1413–1421 (2011)
19. Wang, X., Guo, X., Lei, Z., Zhang, C., Li, S.Z.: Exclusivity-consistency regularized multi-view subspace clustering. In: Proceedings of the IEEE Conference on Computer Vision and Pattern Recognition, pp. 923–931 (2017)
20. Bruno, E., Marchand-Maillet, S.: Multiview clustering: a late fusion approach using latent models. In: Proceedings of the 32nd International ACM SIGIR Conference on Research and Development in Information Retrieval, pp. 736–737 (2009)
21. Xu, J., Ren, Y., Li, G., Pan, L., Zhu, C., Xu, Z.: Deep embedded multi-view clustering with collaborative training. Inf. Sci. **573**, 279–290 (2021)
22. Greene, D., Cunningham, P.: A matrix factorization approach for integrating multiple data views. In: Joint European Conference on Machine Learning and Knowledge Discovery in Databases, pp. 423–438. Springer (2009)
23. Kalayeh, M.M., Idrees, H., Shah, M.: Nmf-knn: image annotation using weighted multi-view non-negative matrix factorization. In: Proceedings of the IEEE Conference on Computer Vision and Pattern Recognition, pp. 184–191 (2014)
24. Ou, W., Long, F., Tan, Y., Yu, S., Wang, P.: Co-regularized multiview nonnegative matrix factorization with correlation constraint for representation learning. Multimedia Tools Appl. **77**(10), 12 955–12 978 (2018)
25. Wang, J., Tian, F., Yu, H., Liu, C.H., Zhan, K., Wang, X.: Diverse non-negative matrix factorization for multi-view data representation. IEEE Trans. Cybern. **48**(9), 2620–2632 (2017)
26. Gong, Y., Li, Z., Zhang, J., Liu, W., Yin, Y., Zheng, Y.: Missing value imputation for multi-view urban statistical data via spatial correlation learning. IEEE Trans. Knowl. Data Eng. (2021)
27. Zhang, X., Zong, L., Liu, X., Yu, H.: Constrained NMF-based multi-view clustering on unmapped data. In: Proceedings of the AAAI Conference on Artificial Intelligence, vol. 29(1) (2015)
28. Iverson, D.L.: Inductive system health monitoring. NASA (2004)
29. Murgia, A., Tsiporkova, E., Verbeke, M., Tourwé, T.: Context-aware performance benchmarking of a fleet of industrial assets. Archives of Data Science, Series A, vol. 5 (2020)
30. Dhillon, I.S., Sra, S.: Generalized nonnegative matrix approximations with Bregman divergences. In: NIPS, vol. 18. Citeseer (2005)
31. Bellman, R., Corporation, R., Collection, K.M.R.: Dynamic Programming, ser. Rand Corporation research study. Princeton University Press (1957). [Online]. https://books.google.be/books?id=wdtoPwAACAAJ
32. MacQueen, J., et al.: Some methods for classification and analysis of multivariate observations. In: Proceedings of 5th Berkeley Symposium on Statistics and Probability (1967)
33. Davies, D.L., Bouldin, D.W.: A cluster separation measure. IEEE Trans. Pattern Anal. Mach. Intell. **2**, 224–227 (1979)
34. Rousseeuw, P.J.: Silhouettes: a graphical aid to the interpretation and validation of cluster analysis. J. Comput. Appl. Math. **20** (1987)
35. Krzanowski, W.J., Lai, Y.: A criterion for determining the number of groups in a data set using sum-of-squares clustering. Biometrics 23–34 (1988)
36. Caliński, T., Harabasz, J.: A dendrite method for cluster analysis. Commun. Stat. Theory Methods **3**(1), 1–27 (1974)

37. Handl, J., Knowles, J., Kell, D.B.: Computational cluster validation in post-genomic data analysis. Bioinformatics **21**(15), 3201–3212 (2005)
38. Sheather, S.J.: Density estimation. Stat. Sci. 588–597 (2004)
39. Dhont, M., Tsiporkova, E., Boeva, V.: Advanced discretisation and visualisation methods for performance profiling of wind turbines. Energies (2021), submitted
40. Bhatia, R.: Matrix Analysis, vol. 169. Springer (2013)
41. O'Grady, K.E.: Measures of explained variance: cautions and limitations. Psychol. Bull. **92**(3), 766 (1982)
42. Gumilar, L., Afandi, A.N., Sias, Q.A., Nugroho, W.S., Sholeh, M., Gunawan, A.: Comparative study: pitch angle variation for making power curve and search maximum power of horizontal axis wind turbine. In: AIP Conference Proceedings, vol. 2228, no. 1, p. 030005. AIP Publishing LLC (2020)

A Methodology Review on Multi-view Pedestrian Detection

Rui Qiu, Ming Xu, Yuyao Yan, and Jeremy S. Smith

Abstract Although surprisingly good progress has been made in monocular pedestrian detection with emerging deep learning techniques, the existing algorithms still suffer from heavy occlusion between people. To cope with this problem, multiple cameras are often used to provide complimentary information. This chapter presents a comprehensive review on the algorithms for multi-view pedestrian detection. The existing methods are sorted into low-level fusion, intermediate-level fusion, high-level fusion and deep-learning based fusion, in terms of the degree of information fusion. The challenges in developing deep multi-view algorithms are also described in this chapter.

Keywords Multi-view · Data fusion · Pedestrian detection · Video surveillance

1 Introduction

Pedestrian detection is a research area which attracts great attention in the computer vision community. It has a variety of applications in video surveillance, traffic monitoring and autonomous driving. At the same time, pedestrian detection is also a fundamental step in many related research fields, such as pedestrian tracking, person re-identification, pose recognition and event classification. Although surprisingly good progress has been made in monocular pedestrian detection with deep learning techniques, the existing algorithms still suffer from the occlusion between people.

R. Qiu · M. Xu (✉) · Y. Yan
Department of Electrical and Electronic Engineering, Xi'an Jiaotong-Liverpool University, Suzhou 215123, China
e-mail: ming.xu@xjtlu.edu.cn

R. Qiu
e-mail: rui.qiu@liverpool.ac.uk

R. Qiu · M. Xu · J. S. Smith
Department of Electrical Engineering and Electronics, University of Liverpool, Liverpool L69 3BX, UK
e-mail: j.s.smith@liverpool.ac.uk

W. Pedrycz and S. Chen (eds.), *Recent Advancements in Multi-View Data Analytics*, Studies in Big Data 106, https://doi.org/10.1007/978-3-030-95239-6_12

They tend to miss heavily occluded pedestrians in the detection and cannot localise partially occluded pedestrians on the ground plane. To cope with this problem, multiple cameras are often used to provide complimentary information. In contrast with the booming deep monocular pedestrian detection, there are few successful works in multi-camera people detection by using deep learning methods, which is due to the shortage of large-scale annotated multi-view video datasets and effective multi-view information fusion methods.

In this chapter a review on multi-view pedestrian detection is presented according to the information fusion degrees. The existing multi-view algorithms are categorized into low-level fusion, intermediate-level fusion, high-level fusion and deep-learning based fusion. Although there exist some excellent reviews on pedestrian detection before, they are focused on monocular pedestrian detection [1, 2]. The novelty of this chapter lies in that it is the first review of multi-view pedestrian detection and it presents the current challenges in deep multi-view pedestrian detection.

The rest of this chapter is organized as follows: In Sect. 2, a historical review on monocular pedestrian detection is presented; In Sect. 3, the low-level information fusion methods based on camera handovers are introduced; In Sect. 4, the intermediate-level information fusion methods, which associate and fuse single-view features, are reviewed; In Sect. 5, the high-level information fusion methods, which fuse the foreground bitmaps in multiple views, are introduced; The emerging deep-learning based multi-view methods, as well as the challenges in developing such algorithms, are described in Sect. 6; The benchmark video datasets and frequently used metrics for the performance evaluation of multi-view pedestrian detection algorithms are introduced in Sect. 7; Finally conclusions are presented in Sect. 8.

2 Monocular Pedestrian Detection

Early single-camera video surveillance was mainly based on the analysis and tracking of foreground blobs in moving object detection. The foregrounds were extracted by using either background subtraction or image differencing methods. The former is to compare each frame against a fixed or adaptive background image, whilst the latter is to compare each frame against a preceding frame. The pixels of significant intensity variation are thought of as foreground pixels and are connected into foreground regions in connected component analysis. Since the predominant targets in video surveillance scenarios are pedestrians and vehicles, it is trivial to differentiate these two types of moving targets by considering their sizes and height/width ratios. Using some uninterpreted low-level features, e.g. the bounding box, centroid, colour histogram or colour template of each foreground region, combined with tracking information can solve the detection problem for pedestrians in low density, single-camera video surveillance [3–8]. To cope with the partial occlusion of pedestrians, some researchers modelled the foreground blobs as joint connected body parts [9] or parameterised shapes [10]. In this case, even if pedestrians are partly occluded, they can still be recognised from their body parts. These methods require the classification of moving objects before the tracking stage so as to assign a specific object model to that type of objects. However, there exist un-modelled objects or non-rigid objects which are difficult to model, e.g. cyclists. Since these early methods were based on background subtraction or image differencing, they are sensitive to illumination variations and pedestrians moving slowly or standing still, which causes false positives and false negatives in the foreground detection. In addition, each foreground region detected as above may contain multiple grouped pedestrians and the low-level features of such a region may not reflect any individual member.

With the development of machine learning techniques, the methods using artificially extracted features combined with classifiers have been widely used in single-camera object detection. These methods use a sliding window to scan the image, compare the enclosed sub-image with an image pyramid, and use a classifier to determine whether each window encloses a specific class of objects. By using Haar-like features [11, 12], Edgelet features [13], Shapelet features [14] or Histograms of Oriented Gradients (HOG) features [15], the feature-based methods obtain good results in pedestrian detection. At this stage, since no single feature can outperform the HOG features, many researchers focus on the improvement of the HOG features [16] and combine it with other features [17]. This approach has good detection results for the objects which have minor occlusion or deformation, but it often fails in severe occlusion. In order to cope with such deformation and occlusion, part-based detection methods have been proposed. Each part of an object is detected separately so that the object can still be detected even if the deformation or occlusion occurs. Based on [18], Felzenszwalb et al. [19, 20] proposed a Deformable Part Model (DPM), in which the HOG features of each part of pedestrians are trained separately and the corresponding position of each part is trained to fit a distribution. This model is widely used to detect occluded pedestrians [21].

In recent years, deep learning has achieved excellent performances in single-camera object detection [22, 23]. In particular, there exist the detection methods based on object proposals [24], such as R-CNN [25, 26], Fast R-CNN [27] and Faster R-CNN [28]. In these methods, object proposals are used to replace the image pyramid to improve efficiency. On the basis of these works, deep instance segmentation algorithms, such as Mask R-CNN [29] and Mask Scoring R-CNN [30], were also proposed, which is effective to segment each individual in a group of objects. At the same time, end-to-end detection methods have been proposed, which can significantly improve the detection speed with limited loss of accuracy. This family of approaches includes YOLO [31], YOLO2 [32] and SSD [33], etc. On the other hand, when these deep-learning based methods are used in pedestrian detection, they still have problems in the detection of heavily occluded pedestrians and the localisation of partially occluded pedestrians on the ground. When cameras are placed at eye-level, severe occlusion cannot be avoided in high density crowds. The solution to this problem lies in multi-camera video surveillance.

Placing multiple cameras at different locations to monitor a scene can provide a broader field of view and collect complementary observations in video analysis. For example, when the dynamic occlusion of moving objects occurs in a camera view, the involved objects may be visible in other camera views. Using multiple cameras can improve the accuracy and robustness in information fusion. In addition, the 3D position of an object in the air can be determined from the intersection of a pair of image rays, each of which comes from a different camera view. This is impossible using just a single camera. According to the degree of information fusion, the current multi-view surveillance systems can be categorised as low-level fusion, intermediate-level fusion, high-level fusion and deep learning based fusion.

3 Low-Level Information Fusion

The early multi-camera video surveillance systems used the methods which switch the tracking of objects across camera views [34–36]. In these methods, each camera detects and tracks moving objects separately. When an object moves out of the field of view (FOV) of the current camera, it is switched to another camera. Therefore, multiple cameras are used to enlarge the limited field of view of a single camera. The existing researches are focused on when to switch the camera, which camera is optimal for an object, and how to find corresponding objects between cameras.

Cai and Aggarwal [34] tracked each object in a camera view and switched it to another camera view when it was predicted that the current camera would not have a good view. The good view is defined as the camera view which has a high-confidence match, between the object in this camera view and the object passed from the original camera, and will observe the object over the greatest number of frames in the future. The matching confidence is evaluated by extracting features from the upper human bodies.

Javed et al. [35] proposed a method to calculate the FOV borders, of each camera, in other camera views, for a set of uncalibrated cameras. The borders are called field of view lines which are used to trigger the handoff between two camera views. When an object is close to a field of view line, the correspondence of this object between two camera views is established. The advantage of this method is that there is no feature matching step which is difficult in widely separated cameras.

Quaritsch et al. [36] designed a migration region in the overlapping FOV of two camera views. Before an object moves out of the FOV, it enters the migration region, which triggers the handover of the tracker to the next camera view where the tracker can continue to track that object.

In these methods, the same object is only detected and tracked by one camera at the same time. During the switch of the cameras, the exchange of information is very limited. Only the parameters of the tracker or the features of the tracked object are passed. These methods cannot use multiple cameras to solve the problems caused by the lack of information from a single camera view, e.g. dynamic occlusion. Therefore, this method of camera switching is classified as low-level information fusion.

4 Intermediate-Level Information Fusion

When the spatial correspondence in the overlapping FOV of multiple cameras is obtained, the extracted features or tracking trajectories of the same object, in different camera views, can be associated and fused to obtain a global estimate of the object's location. The extracted features include, but are not limited to, foot points [37, 38], bounding boxes [39], centroids [40, 41], principal axes [42–44] and colour histograms [45, 46]. Usually the region-based features, such as centriods and principal axes, are more robust than the extreme-point features such as foot points and bounding boxes, since the former are extracted over all the pixels of a foreground region. The correspondence of the objects between camera views can be obtained by using either homographies [37, 39, 43] or epipolar lines [41, 47, 48]. The homography is the relationship between the coordinates of a pair of captured images, on the same plane, from two cameras. The homography mapping is a point-to-point transformation, but it assumes that all the objects are located on a common plane, which is realistic in video surveillance due to the ground plane. The epipolar line is the ray which starts from a camera and passes through a specific point in the image captured by that camera. The epipolar line based methods require camera calibration and can be used to locate objects in the air. However, they map points to lines and need to perform a one dimensional search to associate the objects in different views. In addition to the geometric constraint approach as above, colour information has also been frequently used to match pedestrians across multiple camera views [42, 45, 46]. However, the colour cue is vulnerable to the pedestrians wearing clothes of similar colours and the occlusion between pedestrians. Therefore, the colour approach is not so robust as the geometric approach.

Khan and Shah [37] applied the ground-plane homography, from each camera view to a reference top view, and fused the detection results of multiple camera views on the top view to estimate the location of each pedestrian. Xu et al. [38] took into account the measurement uncertainty of football players in each camera view. When the measurements in a camera view are projected to a top view, the uncertainty changes according to the distances from the players to the camera. By considering the uncertainty in multi-camera information fusion, the locations of the football players are estimated more accurately.

Kang et al. [39] projected the bounding box of each detected pedestrian from one camera view to other camera views. According to the relationship between the bounding boxes detected in a camera view and the projected bounding boxes from other camera views, the occurrence that a pedestrian is partially occluded by a static obstacle or two pedestrians are merged into the same bounding box can be detected. This method has a good result when the pedestrian density is low and the occlusion relationship is relatively simple.

The principal axis is another good feature in pedestrian detection. Kim and Davis [42] used colour information to segment each detected foreground region into different pedestrians, where each pedestrian has a colour model. Then, the principal axis of each pedestrian in a camera view is obtained and projected to the top view. The location of each pedestrian is detected by finding the intersections of all the corresponding principal axes. Since the colour segmentation as above is time-consuming, Hu et al. [43] segmented each foreground silhouette using a vertical projection histogram. The principal axes are calculated from the segmented foreground regions and then projected to the top view through homography mapping. Different from [42], Hu et al. fused the principal axes on the top view without the prior knowledge of which principal axes belong to the same pedestrian. Du and Piater [44] further integrated particle filters into the principal axis-based method, which makes the detection results more robust.

Using epipolar lines to associate and fuse the observations of objects can offer measurements in 3D space. Chang and Gong [47] calculated the epipolar line of the top of each pedestrian's head in other camera views as the height observation of the pedestrian. This information is combined with colours to identify corresponding pedestrians. In Mittal and Davis's work [48], the foreground is segmented into sub-regions based on colours. The sub-regions are then matched across pairs of views by using colours. The midpoints of each matched pair in different camera views are projected on the top view as epipolar lines. By analysing the intersections of the epipolar lines, the pedestrian can be located. Black and Tim [41] calculated the epipolar line of the centroid of each foreground region in world coordinates and identified the intersections of these epipolar lines as the 3D locations of pedestrians.

Since the image features in these methods are calculated in each camera view separately, these methods are sensitive to the grouping and occlusion of pedestrians. These methods are classified into intermediate-level information fusion.

5 High-Level Information Fusion

In the last decade, high-level information fusion has been favoured by the video surveillance community, in which individual camera views no longer provide extracted features but foreground regions to the fusion centre. Therefore, this category of fusion methods is not sensitive to the occlusion and grouping of pedestrians and can be used in more crowded scenarios. The high-level fusion algorithms usually consist of two stages: candidate proposal and candidate selection. The candidates are often proposed by using either a bottom-up approach or a top-down approach. The former projects the foregrounds in individual views to the ground plane by using homography transformations and identifies the heavily overlaid regions as the locations of pedestrian candidates. The latter discretizes the ground plane into a grid of positions and finds the likelihood of pedestrian presence, at each position, from the foreground observations in all camera views. On the other hand, the methods for candidate selection are diversified.

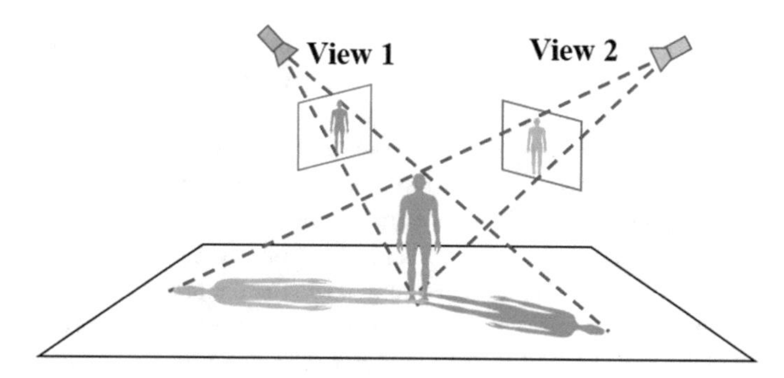

Fig. 1 The bottom-up approach: the foregrounds in multiple camera views are projected to and overlaid in a reference top view

5.1 The Bottom-Up Approach

The bottom-up approach is to project the foreground likelihood or foreground silhouettes from multiple camera views to a reference view by using homography transformations. In the reference view, the location of each object is determined by the analysis of the overlaid foreground projections, as shown in Fig. 1. This method was firstly proposed by Elfes [49] who projected the camera view of a moving robot to the reference ground plane. Otsuka and Mukawa [50] used an elliptical cylinder to approximate each pedestrian and projected the foreground region to the top view as a visual cone. The occlusion relationship is then analysed from the intersections of the visual cones in the top view.

Khan and Shah [51] projected the foreground likelihood maps of the individual camera views into a reference camera view by using ground-plane homography and calculated a joint foreground likelihood map by multiplying them together. By thresholding the joint foreground likelihood map, the locations of the pedestrians are detected and then warped back to the individual views. The advantage of this method is the foreground pixels of a front pedestrian can still support the occluded pedestrian who is hidden behind. Once the foreground pixels belonging to the hidden pedestrian are observed in other camera views, that pedestrian can be detected. However, this may lead to false positives (phantoms) in pedestrian detection, since the foreground pixels belonging to non-corresponding pedestrians in different camera views may be falsely intersected. In addition, this method is sensitive to broken foreground regions. Applying the multi-plane homography mapping [52], which projects and overlays the foreground likelihood maps of individual camera views onto the reference view, according to the homographies of a set of planes parallel to the ground plane and at different heights, can relieve the effect of the broken foregrounds.

Eshel and Moses [53] projected the foreground intensities from multiple camera views to a reference view using the head-plane homographies. Then pixel-wise correlation is calculated in these foreground intensity projections. Pedestrians can be detected at those locations where the projected intensities from multiple camera views are highly correlated. This method can detect pedestrians by using the cameras hanging above the heads of pedestrians. It may fail when eye-level cameras are used, because this method is vulnerable to occlusion. In addition, it depends too much on the assumed dissimilarity of the colours between different pedestrians.

Ge and Collins [54] fused foreground silhouettes on a top view and generated the occupancy likelihood rays in each camera view, which are based on the polar coordinates with the location of the camera as the origin. The pedestrian locations are sampled by using the occupancy likelihood rays as the proposed distributions in the Markov Chain Monte Carlo method. In this method, the pedestrians are modelled as cylinders in the world coordinate and projected to the individual camera views as a rectangle. The objective of this method is to find the optimal pedestrian locations so that the generated rectangles can interpret the foreground silhouettes well.

Utasi and Benedek [55] projected the foreground silhouettes of each camera view to the top view, by using the multi-plane homographies at different heights, to estimate the locations and heights of pedestrians. Similar to [54], they modelled pedestrians as cylinders and then extended the Bayesian Marked Point Process (MPP) [56] to 3D space to generate a finite number of the cylinders which fit the foreground silhouettes well. The advantage of this method is the usage of the height information of pedestrians based on pixel-level features, but the drawback is such features are sensitive to broken foregrounds, which may lead to missed detection.

In Liu et al.'s work [57], each foreground region in a camera view is approximated by a group of vanishing line segments which are then projected to the discretized ground plane. The positions of pedestrians are inferred by counting the number of intersections of the projected line segments at each grid location on the ground. By considering the physical shape and size of the foreground silhouette of a pedestrian, geometry-based rules are applied to filter out phantoms. This method is a faster implementation of Khan and Shah's method [52] and is at least one hundred times faster than the latter.

5.2 The Top-Down Approach

In the top-down approach, the ground plane is divided into a grid, and the occupancy probability at each location of the grid is estimated on the basis of the back-projection of some kind of generative models in each of multiple camera views [58]. The generative models, which are usually based on the average height and width of pedestrians, are compared with the foreground silhouettes to validate the estimated pedestrians' locations, as shown in Fig. 2.

Fleuret et al. [58] computed a probabilistic occupancy map (POM) on the discretized ground plane. Each pedestrian is represented by a rectangle model of the average size of pedestrians. They minimised the Kullback-Leibler divergence between the approximated occupancy probability of each location and the posterior distribution observed from foreground silhouettes. Then they updated the approximated

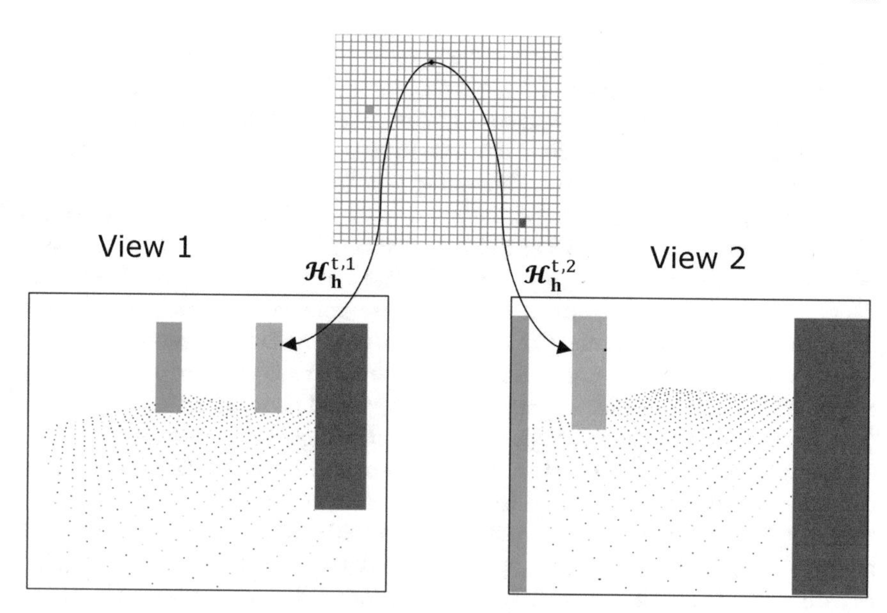

Fig. 2 The top-down approach: each location of the ground grid is projected back to individual camera views to find the foreground evidence of a pedestrian

occupancy probabilities iteratively to find the optimal set of rectangles covering more foreground pixels but fewer background pixels in all camera views.

Evans et al. [59] calculated a high-resolution probabilistic occupancy map by back-projecting each pixel of the top view to the individual camera views. At each corresponding location in these views, they estimated the foreground ratio of a vertical line segment or a narrow rectangle, of the average height of pedestrians, sitting at that location. A joint occupancy likelihood is calculated by multiplying such foreground ratios, for the same pixel of the top view, across multiple views. They realised that the resultant top-view image is equivalent to Khan and Shah's method [52], if the foreground homographic transformations are applied to an infinite number of parallel planes in the latter. They then filtered out false positive detections by using tracking information.

Alahi et al. [60] formulated pedestrian detection as a linear inverse problem regularized by assuming pedestrians' sparse locations on the discretized ground plane. Then an iterative process was carried out to find the binary occupancy vector which contains the minimum number of non-zero elements and fits the multi-camera foreground regions. They also proposed Repulsive Spatial Sparsity (RSS), which sets the minimum distance between pedestrians to reduce pedestrian candidates. After that, a real-time version based on the same framework was proposed by using a greedy optimisation algorithm based on set covering [61].

Peng et al. [62] used the Multiview Bayesian Networks (MvBN) to analyse the occlusion relationship between potential pedestrians and estimate the positions of real pedestrians. They then expanded their research by using self-adaptive heights to model different pedestrians [63], which makes this method more robust than their previous work using the average height of pedestrians. In [63], each candidate rectangle can independently shift within a small range in the individual camera views to cope with inaccurate camera calibration or synchronisation.

Yan et al. [64] proposed an algorithm based on the POM framework. They divided the ground plane into a grid of locations and calculated the joint occupancy likelihood of each location by taking into account a template matching response and head/foot likelihoods in individual camera views. Pedestrian candidates are proposed at those locations with high occupancy likelihoods. Then each candidate is represented by a rectangle of the average size of pedestrians. When such rectangles are projected back to the individual camera views, they divide foreground regions into sub-regions. In the candidate selection stage, a logic minimization method is used to find the essential candidates, each of which covers at least a sub-region alone in a camera view and is therefore thought of as a pedestrian. The remaining candidates are selected to cover the other sub-regions in a repeated process, which alternates between merging redundant candidates and finding emerging "essential" candidates. This algorithm can efficiently reduce the search space for the optimized set of candidates.

5.3 *Pros and Cons of both Approaches*

The bottom-up approach projects the foregrounds of multiple camera views to the ground plane or a set of parallel planes, which is to fuse the foregrounds at a single height or a set of different heights. Therefore, it is sensitive to broken foreground regions, at that height, which are inherent in the background subtraction. For example, if the feet of a pedestrian are lost in the foreground detection in a camera view, the foreground projection on the ground plane will have no intersection with that from the other camera view and therefore that pedestrian will be lost in the multi-view detection. In contrast, the top-down approach is to integrate the foregrounds at a range of heights, which is more robust to broken foreground regions.

The bottom-up approach warps regions of irregular or even broken shapes to a reference view, whilst the top-down approach back-projects points to individual camera view. Therefore, the top-down approach is much easier to implement. On the other hand, the bottom-up approach is only focused on the foreground regions in the individual views, whilst the top-down approach needs to project each location of the discretized ground plane and calculate the foreground ratio within each back-projected rectangle in the individual views. The top-down approach is more time consuming than the bottom-up approach. Using integral images [65] in the top-down approach can greatly relieve the burden of repeatedly counting foreground pixels.

The bottom-up approach cannot accurately localise pedestrians, when pedestrians are crowded or when more than one pedestrian is lined up between two opposite cameras. In these scenarios, the intersections of foreground projections tend to have large areas. In contrast, the top-down approach has no similar problems. Therefore, the bottom-up approach is appropriate to be used to detect pedestrians in low or medium density, whilst the top-down approach can be used for pedestrians in medium or high density.

6 Deep Learning Based Information Fusion

Deep learning based monocular detection, such as the R-CNN family, has impressed the computer vision community with its much improved performance in object detection. These networks can effectively handle single-view pedestrian detection tasks and can adapt well to different scenarios. However, there are few successful studies on deep learning based multi-view pedestrian detection, because some problems exist in model training and information fusion.

In recent years, several researchers have explored the methods to apply deep learning techniques to multi-view pedestrian detection. These methods can be divided into two classes. The first class is to apply deep learning in the front end only. That is to extract features, from individual camera views, by using some deep learning method and fuse the multi-view features to predict the locations of pedestrians. This class is in a spirit similar to the intermediate-level information fusion. The second class of deep multi-view pedestrian detectors is based on end-to-end models, which is to use deep neural networks to directly locate pedestrians from the input of multi-view images. This class of methods is in a spirit similar to the high-level information fusion.

6.1 Deep Front-End Models

Deep front-end models, combined with multi-view feature fusion, are an effective way for multi-view pedestrian detection. In this approach, convolutional neural networks (CNNs) are often used to replace background subtraction for foreground or feature extraction in individual camera views. Pedestrians or their instance segmentation masks may be also detected by using deep monocular detectors in these views. Since these monocular detectors are no longer inherent with the problems of background subtraction, they are more robust and can detect pedestrians in a single frame rather than in a video. In addition, they can separate the foreground masks of different pedestrians in a group and have good generalization capabilities across different datasets or scenarios. Therefore, the multi-view pedestrian detectors based on a deep front-end model are expected to have an improved performance.

Xu et al. [66] proposed a deep-learning based approach for pedestrian tracking. This method uses Faster R-CNN [28] in the detection stage to locate the pedestrians in each camera view. Each detected pedestrian is represented by its bounding box. The bottom points of these bounding boxes are projected to the reference ground plane. The projected points are then clustered and fused to estimate the ground location for each pedestrian. Since Faster R-CNN is not robust in the localization of partially occluded pedestrians on the ground, the performance of this method on the multi-view pedestrian detection task is mediocre [68, 71].

Lopez-Cifuentes et al. [67] proposed a pedestrian detection method based on both a semantic segmentation network and an object detection network. The former is used to segment the regions of pedestrians and scene backgrounds. The classes of regions, which are associated with the ground, are projected to the top view and combined to form an Area of Interest (AOI) region. The object detection network extracts the locations of different pedestrians in each view. The bottom point of the bounding box of each detected pedestrian is then projected to the ground plane and clustered with the projected bottom points from other views. The points are clustered into a group if they are close to each other and each of them comes from a different view. Similar to [62], the pedestrians' locations in each view are iteratively regressed to the pedestrians' foregrounds generated by the semantic segmentation network. This method requires more GPU resources than other deep learning based methods, because it uses both the object detection network and semantic segmentation network.

Deep Occlusion [68], proposed by Baque et al., is a deep learning based and POM-like [58] network. In this method, the individual views are put into a CNN network to extract features. A Gaussian network then classifies these extracted features into eight classes of body parts for a pedestrian. The information in multiple views is fused in a similar way to the POM. Each grid cell on the discretized ground plane has a corresponding bounding box of the average pedestrians' height in each view. These generated bounding boxes are matched with the eight body-part classes in each view to obtain the probability of whether each bounding box contains a pedestrian. Deep Occlusion transforms the overall occupancy problem on the ground plane into an energy minimization problem, which can be solved by Conditional Random Field (CRF). This method achieves a good performance on the EPFL Wildtrack dataset [69].

6.2 Deep End-to-End Models

For deep multi-view pedestrian detectors based on end-to-end models, they use deep neural networks to directly locate pedestrians from multi-view images. These methods can be further sub-divided into anchor-based and anchor-free methods. The anchor-based methods are the same with the top-down approach, in the category of high-level information fusion, which divides the ground plane into a grid of locations. The anchor-free methods are similar to the bottom-up approach, in the category of high-level information fusion, which projects the feature maps from multiple camera views to a top view.

The Anchor-Based Method Chavdarova and Fleuret proposed the DeepMCD [70], which is an anchor-based end-to-end multi-view detection network. This method firstly discretizes the ground into a grid of cells and considers the location of each cell as potentially occupied by a pedestrian. Then, for the information fusion of the multiple views at the same grid cell, each cell is projected back to the individual camera views using the ground-plane homographies. A bounding box with the average pedestrians' height is created, sitting at the projected position in each individual view.

The sub-image within each bounding box is cropped and used as an input to a backbone. In the training stage, negative samples with a comparable number of positive samples are also taken as inputs to the backbone. Negative samples are generated in two ways. One way is shifting the positive samples and the other is combining the positive samples of different pedestrians. Meanwhile, the training on these sub-images is enhanced with a data augmentation method, which is called "input dropout", to improve the robustness of the network for processing partially occluded pedestrians.

The authors designed six types of masks to simulate occlusion scenarios. When the sub-images as above are input to the backbone, they are randomly occluded by one of these six masks or no mask. Then the backbone extracts features from the processed positive and negative sub-images. From the feature maps after multi-view embedding, an MLP predicts which cells on the ground plane are the locations of pedestrians. Non-Maximum Suppression (NMS) is used to ensure a minimum distance between the final detected pedestrians.

The Anchor-Free Method Hou et al. proposed MVDet [71], an anchor-free end-to-end pedestrian detection network, which contains a backbone network and two classifiers. The network uses a lightweight ResNet18 [72] as the backbone to extract features from the input camera views. These feature maps are provided to two branches for different detection tasks. In the first branch, the head-foot pairs of pedestrians are detected, in the individual views, with a classifier of a two-layer convolutional network. In the second branch, the feature maps are projected to the ground plane by using ground-plane homographies. These projected feature maps are concatenated on the ground plane. Then the ground-plane classifier predicts the locations of pedestrians on these concatenated feature maps. In this way, the features at the same location in different views are aligned on the ground plane due to geometric constraints. In the final step, an NMS is also used in MVDet to ensure a minimum distance between the final detected pedestrians. MVDet achieves the state-of-the-art performance on the EPFL Wildtrack [69] and MultiviewX [71] datasets which contain pedestrians in high density.

6.3 *The Challenges*

The following issues exist in the current deep learning based methods for multi-view pedestrian detection and need to be addressed in future.

1. Lack of training samples. The difficulties in the annotation of multi-view pedestrian datasets lead to a small number of annotated frames available in existing multi-view video datasets. For example, there are only 200 frames annotated

in the EPFL Terrace dataset and 400 frames annotated in the EPFL Wildtrack dataset. This makes it rather difficult for deep learning based methods to be properly trained on these datasets. Current end-to-end methods are heavily reliant on the number of samples and the scene diversity in the training datasets. Therefore, insufficient training samples will directly affect the performance of the end-to-end networks.
2. Division of training sets and testing sets. In contrast with the monocular pedestrian detectors, the multi-view detectors based on end-to-end models have been trained and tested on the same video dataset. Therefore, the frame division for the training and testing becomes an issue that can affect the performance evaluation. Meanwhile, since the density of the pedestrians in a multi-view pedestrian dataset is not evenly distributed across all the annotated frames [69], this can also affect the training and testing results.
3. A large amount of GPU memory and computational resources are required. Compared with monocular detection networks, multi-view detection networks need to calculate the feature maps of multiple camera views. This causes a linear increase in the demand for GPU memory and processing capacity [71, 72].
4. Weak generalizability. The camera number and viewpoints, along with the scene background, vary in different datasets. When the network is trained on a multi-view pedestrian dataset with a small number of samples and of poor diversity, it tends to overfit that dataset, and its performance on other datasets will be significantly degraded. For example, the DeepMCD algorithm [70] achieves an MODA of 0.601, if trained and tested using the same dataset, but has a decreased MODA of 0.334, if trained with one dataset and tested with another [69].

To solve the problem of insufficient training samples, it is promising to synthesise large-scale multi-view video datasets by using computer graphics techniques, like the MultiviewX dataset [71]. In addition, to develop a semi-automatic tool to accelerate the annotation of the existing large-scale multi-view datasets like EPFL Wildtrack [69] is another choice. To solve the greatly increased demand for GPU memory and computational resources, the deep front-end approach may be used and followed by some optimization algorithm.

7 Performance Evaluation

The performance evaluation of multi-view pedestrian detection algorithms is usually based on some benchmark video datasets with ground-truth data publicly available. In addition, some common performance metrics are used by most of the multi-view algorithms.

7.1 Multi-view Video Datasets

PETS2009 Dataset The PETS2009 dataset [73] is a widely used multi-view pedestrian dataset. It was captured by using eight cameras at a crossroad in a campus. The original dataset is referred to as City Centre (CC) dataset, whilst that without C2 is referred to as S2L1 dataset. Within the eight cameras, C1-C4 have a resolution of 768×576 pixels and C5-C8 have a resolution of 720×576 pixels. 795 frames were captured by each camera. The ground truth data of C1 are available to public. The static occlusion in C1/C3 and the inaccurate calibration/synchronization of C5-C8 are the major challenges for multi-camera pedestrian detection methods.

EPFL Terrace Dataset EPFL Terrace dataset [74] contains 5000 frames captured by using four eye-level cameras on a terrace. The image resolution is 360×288 pixels. One frame in every 25 frames was annotated and the ground truth data are available to public. Up to 9 pedestrians appear in a small area simultaneously in this video. The challenge of this dataset is the heavy occlusion between pedestrians, due to the eye-level cameras, and illumination changes.

EPFL RLC Dataset The EPFL RLC dataset [74] contains 8001 frames shot at 60 fps in the EPFL Rolex Learning Center. Each of the three cameras has a resolution of 1920×1080 pixels. The ground truth data in frames 7700 to 8001 of View 1 are available. The static occlusion and the limited number of cameras are the major challenges for performance evaluation.

EPFL WILTRACK Dataset The EPFL WILTRACK dataset [69, 75] is a high-resolution and large-scale video dataset. It contains seven eye-level camera views with an image resolution of 1920×1080 pixels. The ground truth data of 400 annotated frames are available. An AOI was defined as a $12\,\text{m} \times 36\,\text{m}$ rectangle. Within this area, all the seven cameras can capture an average of 23.8 people per frame. The major challenges of this dataset are the severe occlusion between pedestrians and the detection of small distant targets.

MultiviewX Dataset The MultiviewX dataset [71] is a synthetic 3D multi-view pedestrian dataset created by using the Unity engine. It contains 400 annotated frames of 6 virtual camera views with a resolution of 1920×1080 pixels. The AOI is a $16\,\text{m} \times 25\,\text{m}$ rectangle and an average of 40 pedestrians appear in each frame. The synthetic scene in this dataset contains different backgrounds, buildings and trees; Shaded areas are simulated by virtual light sources. The challenges of this dataset are the shaded areas and crowded pedestrians.

APIDIS Dataset APIDIS is a video dataset of a basketball match shot in an indoor basketball court by using seven cameras with a resolution of 1600×1200 pixels. The video has a frame rate of 22 fps. Cameras C1, C2, C4, C6 and C7 were installed around the basketball court, while C3 and C5 were installed overhead in the middle of the basketball court. The difficulties of this dataset are in the detection of running/jumping players and heavy occlusions.

7.2 Performance Metrics

To evaluate the quantitative performance of a multi-view pedestrian detection algorithm, we need to compare the location (and maybe the size) of each detected pedestrian with that of each ground-truth pedestrian. If they are sufficiently close to each other, they are thought of as being matched. There are two ways to match the detections with the ground truth pedestrians. The first way depends on the ground-plane distance r and the threshold is often set as $r = 0.5$ m which is about the average width of pedestrians. The second way for finding a match is to calculate the Intersection-over-Union ratio (IoU) [76] between the bounding box of a detected pedestrian and that of a ground-truth pedestrian in each camera view. The IoU threshold is usually set to $IoU = 0.5$.

The following performance metrics have been frequently used: MDR, FDR, TER, PRECISION, RECALL, MODA and MODP. MDR is the missed detection rate, FDR is the false detection rate, and TER is the total error rate; PRECISION is the ratio of the ground-truth pedestrians over all the detections; RECALL is the ratio of the ground-truth pedestrians which are detected. MODA (Multi-Object Detection Accuracy) [76] assesses the accuracy using the normalized counts of missed detections and false detections. MODP (Multi-Object Detection Precision) [76] assesses the localization precision using the average IoU value of all the true-positive detections. A lower value in MDR, FDR or TER indicates a better performance, whilst a higher value in PRECISION, RECALL, MODA or MODP indicates a better performance. Suppose GT, TP, FP and FN are the numbers of ground-truth pedestrians, true positives, false positives and false negatives, respectively. We have:

$$MDR = FN/GT \tag{1}$$

$$FDR = FP/GT \tag{2}$$

$$TER = MDR + FDR \tag{3}$$

$$PRECISION = TP/(TP + FP) \tag{4}$$

$$RECALL = TP/(TP + FN)\ . \tag{5}$$

The performances of some benchmark algorithms for multi-view pedestrian detection have been evaluated and compared using the performance metrics as above. The video datasets used are PETS2009 CC dataset, PETS2009 S2L1 dataset and EPFL Terrace dataset. Since the codes of most of these algorithms are not publicly available, the originally published results using the same datasets and camera views are included in this comparison. The evaluation of these algorithms was based on the ground distance with the same threshold $r = 0.5$ m. In the performance comparison as shown in Table 1, "C" is the number of camera views, 'Eval.' indicates who made the evaluation, "PREC." is PRECISION and "REC." is RECALL. The bold fonts are used to indicate the best result in the same comparison. The QM algorithm [64] proposed by us was compared with some benchmark non-deep multicamera algorithms such as POM [58], 3DMPP [55], MvBN [63] and MVSampler [54]. It was also com-

Table 1 Performance comparison of multicamera detection algorithms ($r = 0.5\,\text{m}$)

C	Method	Eval.	MDR	FDR	TER	PREC.	REC.	MODA	MODP
PETS2009 CC dataset									
2	POM [58]	[55]	N/A	N/A	0.267	N/A	N/A	N/A	N/A
	3DMPP [55]	[55]	N/A	N/A	0.309	N/A	N/A	N/A	N/A
	MvBN [63]	[63]	0.10	0.03	0.13	0.97	0.90	N/A	N/A
	QM [64]	[64]	**0.045**	**0.027**	**0.072**	**0.973**	**0.955**	0.928	0.790
PETS2009 S2L1 dataset									
4	POM [58]	[63]	0.30	0.07	0.37	0.91	0.70	N/A	N/A
	MVSampler [54]	[70]	0.11	0.16	0.27	0.85	0.89	N/A	N/A
	MvBN [63]	[63]	0.05	0.06	0.11	0.94	0.95	N/A	N/A
	DeepMCD [70]	[70]	**0.04**	0.06	0.10	0.94	**0.96**	N/A	N/A
	QM [64]	[64]	0.042	**0.013**	**0.055**	**0.987**	0.958	0.945	0.838
EPFL terrace dataset									
2	POM [58]	[55]	N/A	N/A	0.845	N/A	N/A	N/A	N/A
	3DMPP [55]	[55]	N/A	N/A	0.370	N/A	N/A	N/A	N/A
	MvBN [63]	[63]	0.19	**0.05**	0.24	**0.94**	0.81	N/A	N/A
	QM [64]	[64]	**0.120**	0.098	**0.218**	0.900	**0.880**	0.782	0.764
3	POM [58]	[55]	0.331	0.355	0.686	0.653	0.669	N/A	N/A
	3DMPP [55]	[55]	0.083	**0.048**	0.131	**0.950**	0.917	N/A	N/A
	QM [64]	[64]	**0.037**	0.062	**0.099**	0.935	**0.939**	0.901	0.785
4	RCNN-2D/3D [66]	[68]	0.50	0.61	1.11	0.39	0.50	−0.11	0.28
	POM-CNN [68]	[68]	0.22	0.20	0.42	0.80	0.78	0.58	0.46
	Deep occlusion [68]	[68]	0.18	0.11	0.29	0.88	0.82	0.71	0.48
	MVDet [71]	ours	0.112	**0.016**	0.128	**0.982**	0.888	0.872	0.700
	QM [64]	[64]	0.034	0.037	0.071	0.96	0.97	0.930	0.791
	QM + DeepLab [64]	[64]	**0.020**	0.023	**0.043**	0.98	**0.98**	**0.957**	**0.792**

pared with some deep multicamera detection algorithms such as RCNN-2D/3D [66], POM-CNN [68], DeepMCD [70], Deep Occlusion [68] and MVDet [71]. The QM + DeepLab algorithm [64] in Table 1 is an alternative implemetation of the QM algorithm, in which the background subtraction for foreground detection was replaced by a deep-learning based semantic segmentation algorithm.

Of the seven performance metrics in Table 1, TER and MODP are the two dominant metrics. This is due to the following facts: MDR and FDR are encoded in TER, PRECISION and RECALL are correlated with MDR and FDR; MODA is correlated with TER. In terms of TER and MODP, the QM and QM + DeepLab algorithms have the best performance in these multi-view pedestrian detection methods. In addition, these two methods do not need pre-training on the video datasets

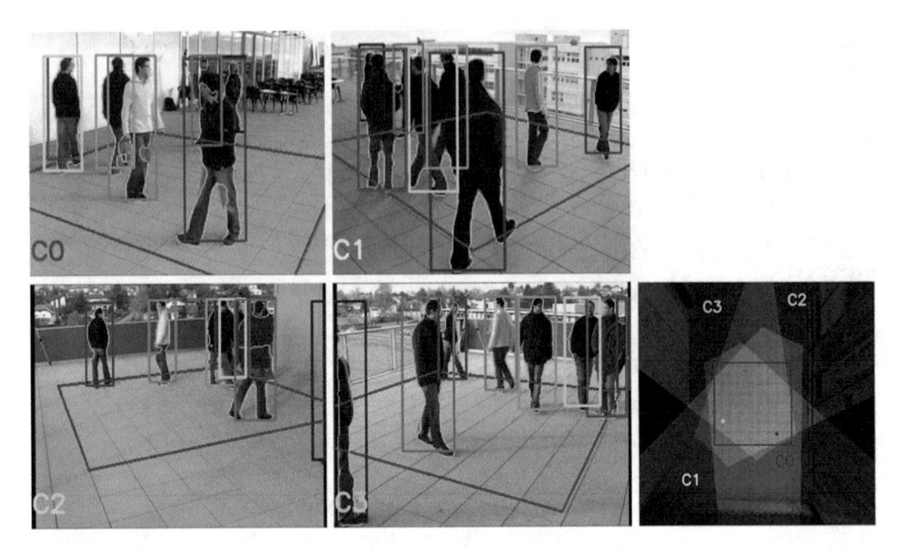

Fig. 3 The detection results of the QM algorithm at frame 1650 on the EPFL Terrace dataset: camera views C0-C3 and a synthetic top view

used for testing, which is unlike what happens in the deep end-to-end methods such as DeepMCD [70], Deep Occlusion [68] and MVDet [71] that used 90% of the annotated frames of a dataset for training and the remaining 10% for testing. At the same time, the other high-degree fusion methods, such as MvBN [63] and 3DMPP [55], also have a comparable or even better performance than the deep-learning based methods. Therefore, the great potential of the deep-learning methods has not been fully revealed in multi-view pedestrian detection.

Some qualitative results of multi-view pedestrian detection are demonstrated in Figs. 3 and 4. Figure 3 shows the detection results of the QM algorithm at frame 1650 of the EPFL Terrace dataset, which correspond to camera views C0-C3 and a synthetic top view. Figure 4 are the detection results of the MVDet algorithm at frame 1815 of the EPFL Wildtrack dataset, which correspond to camera views C1-C7 and a virtual top view. The area of interest (AOI) is represented by a red rectangle on the ground plane. Each detected pedestrian is represented in a distinguished colour which is consistent in all the camera views and the top view. In the camera views, each detected pedestrian is enclosed by a solid box. In the top view, each detected pedestrian is represented by a disk. Although the pedestrians are very crowded and/or in heavy occlusion in both examples, the QM and MVDet algorithms can still detect them correctly.

Fig. 4 The detection results of the MVDet algorithm at frame 1815 on the EPFL Wildtrack dataset: camera views C1-C7 and a virtual top view

8 Conclusions

We have presented a review on the methods for multi-view pedestrian detection, in which the existing methods are sorted in a framework in terms of the degree of information fusion among multiple camera views. In the low-level fusion methods, there are few information exchanges between camera views. In the intermediate-level fusion methods, the single-view features are integrated in a reference view, but the features are not very robust when pedestrians are grouped or occluded. The high-level fusion methods are robust in the presence of occlusion and broken foregrounds. The deep-learning based fusion methods are emerging and their potential in multi-view applications has not been fully revealed.

Acknowledgements This work was supported by National Natural Science Foundation of China (NSFC) under Grant 60975082 and Xi'an Jiaotong-Liverpool University under Grant RDF-17-01-33.

References

1. Wojek, C., Dollar, P., Schiele, B., Perona, P.: Pedestrian detection: an evaluation of the state of the art. IEEE Trans. Pattern Anal. Mach. Intell. **34**(4), 743–761 (2012)
2. Benenson, R., Omran, M., Hosang, J., Schiele, B.: Ten years of pedestrian detection, what have we learned?. In: Proceedings of European Conference on Computer Vision, pp. 613–627 (2014)
3. Bremond, F., Thonnat, M.: Tracking multiple nonrigid objects in video sequences. IEEE Trans. Circuits Syst. Video Technol. **8**(5), 585–591 (1998)
4. Dockstader, S.L., Tekalp, A.M.: Tracking multiple objects in the presence of articulated and occluded motion. In: Proceedings of the Workshop on Human Motion, pp. 88–95 (2000)
5. Javed, O., Shah, M.: Tracking and object classification for automated surveillance. In: Proceedings of European Conference on Computer Vision, pp. 343–357 (2002)
6. Rosales, R., Sclaroff, S.: Improved tracking of multiple humans with trajectory prediction and occlusion modelling. In: Proceedings of IEEE Workshop on Interpretation of Visual Motion, pp. 117–123 (1998)
7. Intille, S.S., Davis, J.W., Bobick, A.F.: Real-time closed-world tracking. In: Proceedings of IEEE Conference on Computer Vision and Pattern Recognition, pp. 697–703 (1997)
8. Xu, M., Ellis, T., Godsill, S.J., Jones, G.A.: Visual tracking of partially observable targets with suboptimal filtering. IET Comput. Vis. **5**(1), 1–13 (2011)
9. Haritaoglu, I., Harwood, D., Davis, L.S.: W4: real-time surveillance of people and their activities. IEEE Trans. Pattern Anal. Mach. Intell. **22**(8), 809–830 (2000)
10. Remagnino, P., Baumberg, A., Grove, T., et al.: An integrated traffic and pedestrian model-based vision system. In: Proceedings of IEEE International Conference on Computer Vision, pp. 380–389 (1997)
11. Lienhart, R., Maydt, J.: An extended set of Haar-like features for rapid object detection. In: Proceedings of IEEE International Conference on Image Processing, pp. 900–903 (2002)
12. Viola, P., Jones, M.J.: Robust real-time face detection. Int. J. Comput. Vis. **57**(2), 137–154 (2004)
13. Wu, B., Nevatia, R.: Detection of multiple, partially occluded humans in a single image by Bayesian combination of edgelet part detectors. In: Proceedings of IEEE International Conference on Computer Vision, pp. 90–97 (2005)
14. Sabzmeydani, P., Mori, G.: Detecting pedestrians by learning shapelet features. In: Proceedings of IEEE Conference on Computer Vision and Pattern Recognition, pp. 1–8 (2007)
15. Dalal, N., Triggs, B.: Histograms of oriented gradients for human detection. In: Proceedings of IEEE Conference on Computer Vision and Pattern Recognition, pp. 886–893 (2005)
16. Zhu, Q., Yeh, M.C., Cheng, K.T., Avidan, S.: Fast human detection using a cascade of histograms of oriented gradients. In: Proceedings of IEEE Conference on Computer Vision and Pattern Recognition, pp. 1491–1498 (2006)
17. Dollar, P., Tu, Z., Perona, P., Belongie, S.: Integral channel features. In: Proceedings of British Machine Vision Conference, pp. 1–11 (2009)
18. Fischler, M.A., Elschlager, R.A.: The representation and matching of pictorial structures. IEEE Trans. Comput. **100**(1), 67–92 (1973)

19. Felzenszwalb, P., McAllester, D., Ramanan, D.: A discriminatively trained, multiscale, deformable part model. In: Proceedings of IEEE Conference on Computer Vision and Pattern Recognition, pp. 1–8 (2008)
20. Felzenszwalb, P., Girshick, R.B., McAllester, D., Ramanan, D.: Object detection with discriminatively trained part-based models. IEEE Trans. Pattern Anal. Mach. Intell. **32**(9), 1627–1645 (2010)
21. Tang, S., Andriluka, M., Schiele, B.: Detection and tracking of occluded people. Int. J. Comput. Vis. **110**(1), 58–69 (2014)
22. Uijlings, J.R., Van De Sande, K.E., Gevers, T., Smeulders, A.W.: Selective search for object recognition. Int. J. Comput. Vis. **104**(2), 154–171 (2013)
23. Dai, J., Li, Y., He, K., Sun, J.: R-FCN: object detection via region-based fully convolutional networks. In: Proceedings of Advances in Neural Information Processing Systems, pp. 379–387 (2016)
24. Cheng, M.M., Zhang, Z., Lin, W.Y., Torr, P.: BING: binarized normed gradients for objectness estimation at 300 fps. In: Proceedings of IEEE Conference on Computer Vision and Pattern Recognition, pp. 3286–3293 (2014)
25. Girshick, R., Donahue, J., Darrell, T., Malik, J.: Rich feature hierarchies for accurate object detection and semantic segmentation. In: Proceedings of IEEE Conference on Computer Vision and Pattern Recognition, pp. 580–587 (2014)
26. Girshick, R., Donahue, J., Darrell, T., Malik, J.: Region-based convolutional networks for accurate object detection and segmentation. IEEE Trans. Pattern Anal. Mach. Intell. **38**(1), 142–158 (2016)
27. Girshick, R.: Fast R-CNN. In: Proceedings of IEEE International Conference on Computer Vision, pp. 1440–1448 (2015)
28. Ren, S., He, K., Girshick, R., Sun, J.: Faster R-CNN: towards real-time object detection with region proposal networks. IEEE Trans. Pattern Anal. Mach. Intell. **39**(6), 1137–1149 (2017)
29. He, K., Girshick, G., Dollár, P, Girshick, R.: Mask R-CNN. In: Proceedings of IEEE International Conference on Computer Vision, pp. 2980–2988 (2017)
30. Huang, Z., Huang, L., Gong, Y., Huang, C., Wang, X.: Mask scoring R-CNN. In: Proceedings of IEEE Conference on Computer Vision and Pattern Recognition (2019)
31. Redmon, J., Divvala, S., Girshick, R., Farhadi, A.: You only look once: unified, real-time object detection. In: Proceedings of IEEE Conference on Computer Vision and Pattern Recognition, pp. 779–788 (2016)
32. Redmon, J., Farhadi, A.: YOLO9000: better, faster, stronger. In: Proceedings of IEEE Conference on Computer Vision and Pattern Recognition, pp. 6517–6525 (2017)
33. Liu, W., et al.: SSD: single shot multibox detector. In: Proceedings of European Conference on Computer Vision, pp. 21–37 (2016)
34. Cai, Q., Aggarwal, J.K.: Automatic tracking of human motion in indoor scenes across multiple synchronized video streams. In: Proceedings of IEEE International Conference on Computer Vision, pp. 356–362 (1998)
35. Javed, O., Rasheed, K.Z., Shah, M.: Camera handoff: tracking in multiple uncalibrated stationary cameras. In: Proceedings of Workshop on Human Motion, pp. 113–118 (2000)
36. Quaritsch, M., Kreuzthaler, M., Rinner, B., Bischof, H., Strobl, B.: Autonomous multicamera tracking on embedded smart cameras. EURASIP J. Embed. Syst. **2007**(1), Article no. 092827 (2007)
37. Khan, S., Shah, M.: Consistent labeling of tracked objects in multiple cameras with overlapping fields of view. IEEE Trans. Pattern Anal. Mach. Intell. **25**(10), 1355–1360 (2003)
38. Xu, M., Orwell, J., Lowey, L., Thirde, D.: Architecture and algorithms for tracking football players with multiple cameras. IEE Proc.-Vis., Image Signal Process. **152**(2), 232–241 (2005)
39. Kang, J., Cohen, I., Medioni, G.: Continuous tracking within and across camera streams. In: Proceedings of IEEE Conference on Computer Vision and Pattern Recognition, pp. 267–272 (2003)
40. Stein, G.P.: Tracking from multiple view points: Self-calibration of space and time. In: Proceedings of IEEE Conference on Computer Vision and Pattern Recognition, pp. 521–527 (1999)

41. Black, J., Ellis, T.: Multi camera image tracking. Image Vis. Comput. **24**(11), 1256–1267 (2006)
42. Kim, K., Davis, L.S.: Multi-camera tracking and segmentation of occluded people on ground plane using search-guided particle filtering. In: Proceedings of European Conference on Computer Vision, pp. 98–109 (2006)
43. Hu, W., Hu, M., Zhou, X., Tan, T., Lou, J., Maybank, S.: Principal axis-based correspondence between multiple cameras for people tracking. IEEE Trans. Pattern Anal. Mach. Intell. **28**(4), 663–671 (2006)
44. Du, W., Piater, J.: Multi-camera people tracking by collaborative particle filters and principal axis-based integration. In: Proceedings of Asian Conference on Computer Vision, pp. 365–374 (2007)
45. Orwell, J., Remagnino, P., Jones, G.A.: Multiple camera color tracking. In: Proceedings of IEEE International Workshop Visual Surveillance, pp. 14–24 (1999)
46. Krumm, J., Harris, S., Meyers, B., Brumitt, B., Hale, M., Shafer, S.: Multi-camera multi-person tracking for easyliving,. In: Proceedings of IEEE International Workshop Visual Surveillance, pp. 3–10 (2000)
47. Chang, T.H., Gong, S.: Tracking multiple people with a multi-camera system. In: Proceedings of IEEE Workshop on Multi-Object Tracking, pp. 19–26 (2001)
48. Mittal, A., Davis, L.S.: M2tracker: a multi-view approach to segmenting and tracking people in a cluttered scene using region-based stereo. In: Proceedings of European Conference on Computer Vision, pp. 18–36 (2002)
49. Elfes, A.: Using occupancy grids for mobile robot perception and navigation. Computer **22**(6), 46–57 (1989)
50. Otsuka, K., Mukawa, N.: Multiview occlusion analysis for tracking densely populated objects based on 2-D visual angles. In: Proceedings of IEEE Conference on Computer Vision and Pattern Recognition, pp. 90–97 (2004)
51. Khan, S.M., Shah, M.: A multiview approach to tracking people in crowded scenes using a planar homography constraint. In: Proceedings of European Conference on Computer Vision, pp. 133–146 (2006)
52. Khan, S.M., Shah, M.: Tracking multiple occluding people by localizing on multiple scene planes. IEEE Trans. Pattern Anal. Mach. Intell. **31**(3), 505–519 (2009)
53. Eshel, R., Moses, Y.: Tracking in a dense crowd using multiple cameras. Int. J. Comput. Vis. **88**(1), 129–143 (2010)
54. Ge, W., Collins, R.T.: Crowd detection with a multiview sampler. In: Proceedings of European Conference on Computer Vision, pp. 324–337 (2010)
55. Utasi, A., Benedek, C.: A Bayesian approach on people localization in multicamera systems. IEEE Trans. Circuits Syst. Video Technol. **23**(1), 105–115 (2013)
56. Ge, W., Collins, R.T.: Marked point processes for crowd counting. In: Proceedings of IEEE Conference on Computer Vision and Pattern Recognition, pp. 2913–2920 (2009)
57. Liu, C.W., Chen, H.T., Lo, K.H., Wang, C.J., Chuang, J.H.: Accelerating vanishing point-based line sampling scheme for real-time people localization. IEEE Trans. Circuits Syst. Video Technol. **27**(3), 409–420 (2017)
58. Fleuret, F., Berclaz, J., Lengagne, R., Fua, R.: Multicamera people tracking with a probabilistic occupancy map. IEEE Trans. Pattern Anal. Mach. Intell. **30**(2), 267–282 (2008)
59. Evans, M., Li, L., Ferryman, J.M.: Suppression of detection ghosts in homography based pedestrian detection. In: Proceedings of IEEE International Conference on Advanced Video and Signal-Based Surveillance, pp. 31–36 (2012)
60. Alahi, A., Jacques, L., Boursier, Y., Vandergheynst, P.: Sparsity driven people localization with a heterogeneous network of cameras. J. Math. Imaging Vis. **41**(1–2), 39–58 (2011)
61. Golbabaee, M., Alahi, A., Vandergheynst, P.: Scoop: a real-time sparsity driven people localization algorithm. J. Math. Imaging Vis. **48**(1), 160–175 (2014)
62. Peng, P., Tian, Y., Wang, Y., Huang, T.: Multi-camera pedestrian detection with multi-view Bayesian network model. In: Proceedings of British Machine Vision Conference, pp. 69.1–69.12 (2012)

63. Peng, P., Tian, Y., Wang, Y., Li, J., Huang, T.: Robust multiple cameras pedestrian detection with multi-view Bayesian network. Pattern Recognit. **48**(5), 1760–1772 (2015)
64. Yan, Y., Xu, M., Smith, J.S., Shen, M., Xi, J.: Multicamera pedestrian detection using logic minimization. Pattern Recognit. **112**, Article no. 107703 (2021)
65. Crow, F.C.: Summed-area table for texture mapping. Comput. Graph. **18**(3), 207–212 (1984)
66. Xu, Y., Liu, X., Liu, Y., Zhu, S.: Multi-view people tracking via hierarchical trajectory composition. In: Proceedings of IEEE Conference on Computer Vision and Pattern Recognition, pp. 4256–4265 (2016)
67. Lopez-Cifuentes, A., Escudero-Vinolo, M., Bescos, J., Carballeira, P.: Semantic driven multi-camera pedestrian detection (2021). arXiv preprint arXiv:1812.10779v2
68. Baque, P., Fleuret, F., Fua, P.: Deep occlusion reasoning for multi-camera multi-people tracking. In: Proceedings of IEEE International Conference on Computer Vision, pp. 271–279 (2017)
69. Chavdarova, T., Baque, P., Bouquet, S., Maksai, A., Jose, C., Bagautdinov, T., Lettry, L., Fua, P., Van Gool, L., Fleuret, F.: WILDTRACK: a multi-camera HD dataset for dense unscripted pedestrian detection. In: Proceedings of IEEE/CVF Conference on Computer Vision and Pattern Recognition, pp. 5030–5039 (2018)
70. Chavdarova, T., Fleuret, F.: Deep multi-camera people detection. In: Proceedings of IEEE International Conference on Machine Learning and Applications, pp. 848–853 (2017)
71. Hou, Y., Zheng, L., Gould, S.: Multiview detection with feature perspective transformation. In: Proceedings of European Conference on Computer Vision, pp. 42–59 (2020)
72. He, K., Zhang, X., Ren, S., Sun, J.: Deep residual learning for image recognition. In: Proceedings of IEEE/CVF Conference on Computer Vision and Pattern Recognition, pp. 770–778 (2016)
73. PETS2009: http://www.cvg.reading.ac.uk/PETS2009
74. Terrace: https://www.epfl.ch/labs/cvlab/data-pom-index-php/
75. Wildtrack: https://www.epfl.ch/labs/cvlab/data/data-wildtrack
76. Kasturi, R., Goldgof, D., Soundararajan, P.: Framework for performance evaluation of face, text, and vehicle detection and tracking in video: data, metrics, and protocol. IEEE Trans. Pattern Anal. Mach. Intell. **31**(2), 319–336 (2009)

Index

W. Pedrycz and S. Chen (eds.), *Recent Advancements in Multi-View Data Analytics*, Studies in Big Data 106, https://doi.org/10.1007/978-3-030-95239-6

N

O

P

S

T

V

W

Printed in the United States
by Baker & Taylor Publisher Services